AF694348

General Relativity and Gravitation

General Relativity and Gravitation

GR17
DUBLIN 2004

Proceedings of the 17th International Conference

RDS Convention Centre, Dublin • 18 – 23 July 2004

Edited by

Petros Florides
Trinity College, Dublin

Brien Nolan
Dublin City University, Dublin

Adrian Ottewill
University College Dublin, Dublin

World Scientific

NEW JERSEY • LONDON • SINGAPORE • BEIJING • SHANGHAI • HONG KONG • TAIPEI • CHENNAI

Published by

World Scientific Publishing Co. Pte. Ltd.

5 Toh Tuck Link, Singapore 596224

USA office: 27 Warren Street, Suite 401-402, Hackensack, NJ 07601

UK office: 57 Shelton Street, Covent Garden, London WC2H 9HE

British Library Cataloguing-in-Publication Data
A catalogue record for this book is available from the British Library.

GENERAL RELATIVITY AND GRAVITATION
Proceedings of the 17th International Conference

Copyright © 2005 by World Scientific Publishing Co. Pte. Ltd.

All rights reserved. This book, or parts thereof, may not be reproduced in any form or by any means, electronic or mechanical, including photocopying, recording or any information storage and retrieval system now known or to be invented, without written permission from the Publisher.

For photocopying of material in this volume, please pay a copying fee through the Copyright Clearance Center, Inc., 222 Rosewood Drive, Danvers, MA 01923, USA. In this case permission to photocopy is not required from the publisher.

ISBN 981-256-424-1

Printed in Singapore by World Scientific Printers (S) Pte Ltd

PREFACE

The 17th International Conference on General Relativity and Gravitation, GR17, was held in Dublin, Ireland, from July 18th-23rd, 2004, in the convention centre of the Royal Dublin Sociey. Like all previous conferences in the GRn series, it was held under the auspices of the International Society on General Relativity and Gravitation, ISGRG. The Local Organising Committee, LOC, would like to thank ISGRG most sincerely for choosing Dublin as the venue of GR17. This choice was not, of course, made without the best of reasons. Dublin is, after all, the city of Sir William Rowan Hamilton, George Francis Fitzgerald and John Lighton Synge, and the adopted city of Erwin Schrödinger and Cornelius Lanczos, to mention but a few, and the home of the Dublin Institute for Advanced Studies.

The conference was opened on the morning of Monday, 19th July, by Her Excellency Mrs Mary McAleese, President of the Republic of Ireland, whose warm, witty and informative address captivated the 680 delegates. To mark the occasion and express our deep appreciation to the President (and "spread the gospel according to Einstein further afield", as the Chair put it) Petros Florides presented her with a copy of the "Book of Abstracts" of GR17 and three best sellers on relativity "A Brief History of Time", "The Road to Reality: A Complete Guide to the Laws of the Universe" and "Black Holes and Time Warps: Einstein's Outrageous Legacy" by Stephen Hawking, Roger Penrose and Kip Thorne, respectively. These she accepted graciously commenting that never in her life had she been given so much homework. Florides then handed the President a crystal paper-weight with the GR17 logo engraved on it, cautioning her that it was quite heavy. The instant she received it she remarked, quite spontaneously and amid great laughter from the delegates, "Gee!, it *is* heavy. I guess it is a matter of gravity". The LOC would like to thank the above three authors for agreeing so readily and enthusiastically to donate and sign their respective books.

The opening ceremony was followed by the presentation by Robert Wald, the (outgoing) President of ISGRG, to Eanna Flanagan of the Xanthopoulos Award, a memorial lecture on JL Synge (1897-1995) by Petros

Florides and a tribute to Peter Bergmann (1915-2002) by Ted Newman.

Going by the kind comments of numerous colleagues, the general consensus is that GR17 was very successful and, in more ways than one, unique. As well as having been honoured by having the meeting opened by the Head of State of the host country, never before has any conference in the series generated so much media interest world-wide. Responsibility for this huge interest lies, of course, with Stephen Hawking who chose GR17 to present his breakthrough on the thirty-year old black hole information paradox. The speed with which the announcement of Hawking's impending lecture spread around the globe prompted Petros Florides, the chair of LOC, to say (jokingly) in introducing the actual lecture, that the basic relativity principle that no information can travel faster than light seems to be wrong! Hawking's lecture, on July 22nd, was attended by 150 correspondents and television crews from 50 countries, in addition to most of the delegates. At the end of the lecture Hawking (but not Thorne) conceded the famous bet between Hawking and Kip Thorne versus John Preskill, giving Preskill a copy of "Total Baseball: The Ultimate Baseball Encyclopaedia". This Preskill waved triumphantly over his head, adding shortly afterwards, that "I always hoped that when Stephen conceded, there would be a witness - this really exceeds my expectations".

The success of GR17 was due to the effort and hard work of many people. To start with, the LOC gratefully acknowledges the excellent work done by the Scientific Organising Committee, SOC, chaired by Curt Cutler (Albert Einstein Institute, Golm) in drawing up a fascinating scientific programme which, as is evident from these proceedings, covered the entire spectrum of the present state of theoretical and experimental/observational work on gravity, and for their excellent choice of the 16 plenary speakers and the chairs of the 19 workshops. We thank these speakers and workshop chairs whose work in the preparation of lectures and the selection of parallel session speakers and posters made for the scientific success of the week. We also thank them for the onerous task of preparing their submissions for this volume. The workshops were divided between afternoon parallel sessions and two evening poster sessions, exhibiting some 200 posters. The high popularity of the poster sessions may not be unrelated to the availability of a glass or two of wine during the two evenings!

The chairs of the workshops had the arduous task of going through nearly 600 submitted abstracts and for drawing up the presentation schedules for their individual sessions: in all, there were over 360 short talks presented during the parallel sessions, as well as the nearly 200 poster presentations. The volume of submissions reflects well on the health General Relativity and we thanks all those who submitted abstracts and presented talks and posters. Once the parallel session programmes were finalised, it was the equally arduous task of the LOC to compile the Book of Abstracts. Here, special thanks are due to Dr Raül Vera of Dublin City University. We also thank Niall O'Leary of DCU for assistance with setting up the conference website (which will be mirrored for the foreseeable future at www.dcu.ie/~nolanb/gr17.htm) and Karen O'Shea of DCU for help with innumerable secretarial details.

The LOC was also responsible for securing the Convention Centre of the RDS as the venue of GR17. This was made possible through the good offices of Miss Olivia O'Reilly of the RDS, who was, also, instrumental in securing for us "academic" rather than "commercial" rates for the premises. To her, and to the staff of the RDS who fine-tuned the premises to meet our requirements to the perfect pitch, we express our deep and sincere thanks.

In addition, the LOC was responsible for organising two highly successful public lectures by Professor Kip Thorne on "Probing the Universe and Black Holes with Gravitational Waves" (on July 19), and Professor Sir Roger Penrose on "Fashion, Faith, and Fantasy in Modern Physical Theories" (on July 23). To our two public lecturers Kip and Roger we say a huge 'Thank You' for firing the imagination of so many people in the city of Dublin.

Perhaps the most arduous and nail-biting, but in the end successful, task of the LOC was raising sufficient sponsorship. Needless to say that without the generous financial support of our sponsors there would have been no GR17; worse still, the members of the LOC would have gone bankrupt! The complete list of our sponsors is given on a separate page in these Proceedings. We express our sincere thanks to each and every one of them. We feel we can be excused for making an exception and thank, in this Preface, Science Foundation Ireland for their huge contribution and support.

Many aspects of the conference, including financial management, audiovisual, travel, accommodation, catering and public relations were taken care of by the Ovation Group. To its personnel, Dr Noel Mitchell, Eimear Burton, Sinead Hope, Ciara McCabe, Jackie McGann, Evelyn O'Sullivan and David Shirley, we say a huge 'thank you'; without their hard work, their professionalism and their charm, the organisation of GR17 would have been almost impossible. Special thanks also go to Miss Ruth Bestgen, for a long time secretary to the Editor of our GRG Journal, for flying all the way from Bern to help, voluntarily, with the running of GR17.

In summary, we would like to express our deep and sincere thanks to the President of Ireland, Mrs Mary McAleese, to all our colleagues in the LOC, the Ovation Group, the SOC and ISGRG, the Plenary speakers, the two public speakers, the Chairs of the Parallel Sessions, the workshop speakers and poster authors, the Sponsors, and, of course the 680 delegates and their companions, for having honoured us with their presence.

Petros Florides (Chair, LOC)
Brien Nolan (Secretary, LOC)
Adrian Ottewill (Member, LOC)

ORGANISING COMMITTEES

Scientific Organising Committee

Curt Cutler (Chair - Albert Einstein Institute, Golm, Germany)

Abhay Ashtekar (Penn State, Pennsylvania, USA)

Sean M. Carroll (University of Chicago, USA)

Matthew W. Choptuik (University of British Columbia, Canada)

Éanna É. Flanagan (Cornell University,USA)

Gary T. Horowitz (University of California at Santa Barbara, USA)

Seiji Kawamura (National Astronomical Observatory,Mitaka, Japan)

Carlos N. Kozameh (University of Cordoba, Argentina)

Malcolm A.H. MacCallum (Queen Mary, University of London, UK)

Niall Ó Murchadha (University College Cork, Ireland)

Adrian C. Ottewill (University College Dublin, Ireland)

E. Sterl Phinney (California Institute of Technology, USA)

Jorge Pullin (University of Louisiana, USA)

Alan D. Rendall (Albert Einstein Institute, Golm, Germany)

Susan M. Scott (Australian National University, Canberra, Australia)

Kip S. Thorne (California Institute of Technology, USA)

Stefano Vitale (University of Trento, Italy)

Robert M. Wald (University of Chicago, USA)

Local Organising Committee

Petros Florides (Chair - Trinity College Dublin)

Brendan Guilfoyle (Institute of Technology, Tralee)

Peter Hogan (University College Dublin)

Brien Nolan (Secretary - Dublin City University)

Niall Ó Murchadha (University College Cork)

Adrian Ottewill (University College Dublin)

SPONSORS

- Science Foundation Ireland
- Royal Dublin Society
- Department of Education and Science, Republic of Ireland
- International Society on General Relativity and Gravitation (Rosenblum Fund)
- International Union of Pure and Applied Physics
- Gravity Research Foundation
- Convention Bureau of Ireland
- Aer Lingus
- Dublin Institute for Advanced Studies
- Classical and Quantum Gravity
- Cambridge University Press
- Oxford University Press
- Living Reviews in Relativity
- Springer Science and Business Media
- University College Dublin
- School of Mathematics, Trinity College Dublin
- School of Mathematical Sciences, Dublin City University
- Hamilton Mathematics Institute, Trinity College Dublin
- Enterprise Ireland
- Institute of Physics
- Irish Mathematical Society
- Institute for Numerical and Computational Analysis, Trinity College Dublin
- The Irish Times
- The Embassy of France in Ireland
- The Embassy of Poland in Ireland
- The Italian Institute of Culture
- Yamanouchi Ireland
- Royal Irish Academy

OPENING ADDRESS

HER EXCELLENCY MRS MARY MCALEESE,
PRESIDENT OF IRELAND
Uachtarán na h-Éireann, Dublin

Introduction by Professor Petros Florides

Your Excellency, Mrs Mary McAleese, President of Ireland, distinguished delegates, ladies and gentlemen,

I am Petros Florides, the Chairman of the Local Organising Committee of the 17th International Conference on General Relativity and Gravitation, GR17. It is my pleasure and my privilege to extend to you all a very warm welcome, a thousand Irish welcomes, to GR17 and to the City of Dublin; the City of Sir William Rowan Hamilton, John Lighton Synge, George Fitzgerald; the adopted City of Erwin Schrödinger and Cornelius Lanczos, to mention but a few, and the Home of the Dublin Institute for Advanced Studies.

We are immensely grateful to you, your Excellency, for giving us the ultimate honour of opening what promises to be one of the most exciting conferences in the series of International Conferences on General Relativity and Gravitation.

Before attaining the Highest Office in the land Mrs McAleese was a fellow academic. She is a distinguished lawyer and was the Reid Professor of Criminal Law, Criminology and Penology in Trinity College Dublin from 1975-1979 and, again, from 1981-1987. It was during those years that I had the privilege of being a colleague of Professor McAleese. She was, at times, a current affairs journalist and presenter with the Irish National Television (1979-1981), and the Director of the Institute of Professional Legal Studies at the Queen's University of Belfast (1987-1997). She served as Pro Vice Chancellor of the Queen's University of Belfast from 1994 to 1997 when she was elected President of Ireland.

Mrs McAleese is one of the most popular Presidents in the history of the Irish Nation. She has proved herself a master "bridge builder" across the two cultures in Northern Ireland, and across the Northern and Southern

Ireland. She has articulated her vision of a peaceful and tolerant society in the whole of Ireland, indeed in the whole world, with clarity and passion.

Mrs McAleese is the incumbent of the High Office once occupied by Mr Eamon de Valera, the father of the modern Irish State. He was an able mathematician and, quite literally, a worshipper of William Rowan Hamilton. He had the vision and the courage to establish, in 1940, under almost impossible conditions, the Dublin Institute for Advanced Studies. The fact that Dublin was chosen as the venue of GR17 is very much connected with this Institute. It is, therefore, perfectly fitting, and a very happy coincidence, that a successor of President de Valera would now formally open GR17.

Ladies and gentlemen, it gives me immense pleasure in calling upon the President of Ireland to open officially the 17th International Conference on General Relativity and Gravition. Your Excellency.

Opening Address by Her Excellency Mrs Mary McAleese

Dia dhíbh a chairde go léir.

Ladies and gentlemen, I am delighted to welcome each one of you to Ireland for the 17th International Conference on General Relativity and Gravitation. You come to us from forty-eight countries across the world and I extend to you the traditional Irish céad míle fáilte - a hundred thousand welcomes. It is a singular honour that Ireland has been chosen as the venue for this prestigious conference, which brings so many great brains and imaginations together from the fields of general relativity, gravitation, cosmology and astrophysics. Here in Dublin I hope you will be inspired, challenged and renewed in your vocation as scientific thinkers and researchers.

I have my dear old friend and former Trinity College colleague Professor Florides and the organising committee to thank for their kind invitation. I have myself to thank for daring to say yes and to present myself as a scientifically illiterate lawyer to such an audience. When I first met Professor Florides thirty years ago I knew nothing about relativity and gravitation. I still don't - my fault, not his!

Back in the early 1970's the then third President of Ireland was a mathematician Eamon de Valera and he would surely have been comfortable in this company for it was he who instigated the setting up, in 1938, when he was Prime Minister, of the Irish institution most particularly associated with developments in the areas of relativity and gravitation - the Dublin Institute for Advanced Studies, and more specifically the School of Theoretical Physics within that Institute. Erwin Schrödinger was the first Senior Professor and Director of the School of Theoretical Physics and he was also the first of many distinguished foreign physicists and mathematicians to work there, among them people of the stature of Cornelius Lanczos, Walter Heitler and Paul Dirac. And of course the School was also home to a number of notable Irish scientists, including Lochlainn O'Raifeartaigh and J L Synge, nephew of our distinguished playwright, J M Synge. By coincidence it was here in the Royal Dublin Society that the scientific Synge was awarded the Boyle medal for excellence in science in 1972 and the School, along with the excellent mathematics and physics departments of our other academic institutions, has had an impressive record of contributions to the developments in theoretical physics which took place in the latter half of the twentieth century.

Apart from his love of mathematics, Eamon de Valera was motivated to establish the Institute for Advanced Studies by his huge admiration for one of Ireland's greatest mathematicians, Sir William Rowan Hamilton. He died of course before the development of relativity and quantum theory but his work underpins much of modern theoretical physics. Among his many contributions to the development of mathematics was his discovery of quaternions, the fundamental formula which came to him as he was walking along the Royal Canal with his wife on his way to a meeting in the Royal Irish Academy, not far from here. For the rest of us mortals who generally grumble about the weather or discuss what we are having for tea on such walks that marital ramble is one of life's great mysteries nearly as big as relativity itself. Even more intriguing was Hamilton's excitement which provoked him to carve the formula with a knife on a stone of Brougham Bridge as he was passing. For classy, upmarket graffiti I think that one is hard to beat. I doubt if any graffiti artist since has achieved the same level of erudition.

Your presence here today is evidence of and vindication of Ireland's long and proud yet often overlooked tradition of contribution to the sciences. So often when the name of Ireland is mentioned we think first of literature, of Joyce, Yeats, Beckett, Shaw, Wilde, Heaney. Yet it is vital that we see our scientific heritage as a proud part of our broad cultural heritage and that each generation is challenged to make its own contribution, to add the fruits of its own genius to that living story, that endless search for fresh insight and understanding.

It was Einstein who said, "Imagination is more important than knowledge." Arguably the greatest scientist of all time, he recognised that imagination is as important to progress and intellectual provocation in the sciences as in the arts.

There is no doubt that it was inspired, imaginative thinking that brought you to Dublin to try and resolve among other things such matters as the inconsistency, if not contradiction, between the theory of relativity and quantum mechanics - after all we in Ireland have some notable experience in reconciling the apparently irreconcilable! You couldn't have come to a better place. I hope the legendary Irish luck will make all the difference to your deliberations and the even more legendary hospitality will not inhibit the clear heads your work obviously requires.

I am sure that this conference will be the great success its organisers have not only hoped for but have worked so hard for. They deserve great credit and thanks but their real reward will come in the enjoyment, the fulfilment and the memories you take home with you from Dublin and in the seeds of powerful scientific discourse which will be the legacy of these days in Ireland. Enjoy it all.

It gives me great pleasure now to declare this conference officially open.

Go raibh míle maith agaibh go léir.

JOHN LIGHTON SYNGE: A MEMORIAL LECTURE

PETROS FLORIDES
School of Mathematics, Trinity College, Dublin 2

One of the main reasons for the choice of Dublin as the venue of GR17 is the immense contribution that Ireland made to the fields of knowledge associated with GR17. To quote Professor Malcolm MacCallum, the Secretary of the International Society on General Relativity and Gravitation, "... Ireland has a long and distinguished history in the areas of relativity and gravitation. The conference will, in part, furnish an opportunity to pay tribute to this history." The chief protagonist in this "distinguished history" is the late Professor John Lighton Synge, a name which is still revered in the world of relativity. It is, therefore, proper, indeed mandatory, to celebrate the man and his work, however briefly, in this short Memorial Lecture. It is a happy coincidence that the 1972 Boyle Medal of the Royal Dublin Society was awarded to Professor Synge in the very premises in which GR17 is now taking place.

Synge was arguably the greatest Irish mathematician and theoretical physicist since Sir William Rowan Hamilton (1805-1865). He was a prolific researcher of great originality and versatility, and a writer of great lucidity and striking 'clarity of expression', as Herman Bondi put it in the first J L Synge Public Lecture which he gave in Trinity College Dublin in 1992. And Bondi continued: "Some of us I may say have at times been daunted by this clarity, because it sets a standard that the rest of us can strive for but it is very hard to attain."

Synge made oustanding contributions to a vast range of subjects and particularly to the theory of relativity. His approach to relativity, and to theoretical physics in general, is characterised by his extraordinary geometrical insight. In addition to bringing clarity and new insights to relativity, his geometrical approach influenced profoundly the development of the subject from the late fifties.

Synge was born in Dublin on March 23, 1897, the youngest of a family of four. At that time the family lived in Kingscourt, county Cavan, where his father, Edward Synge (1859-1939), was land agent of a number of estates.

Synge's mother was Ellen Frances Price (1861-1935), daughter of the distinguished Irish engineer James Price. The Price family can be traced back to the Stuarts of Scotland and, in particular, to Sir William Stuart who settled in Ireland in the early 17th century.

In the male line, Synge's family can be traced back to the 15th century to Thomas Millington, 'corruptly called Singe of Bridgnorth' in Shropshire, England. According to tradition, the changing of the name from Millington to Synge originated with Henry VIII, who commanded a favourite choir boy to 'Singe, Millington, Singe'.

Synge's family were members of the Church of Ireland, and incredibly many of his distant ancestors attained high Office in the Church. Preeminent among them were his direct ancestor Edward Synge (1659-1741), Archbishop of Tuam, and Hugh Hamilton (1729-1805), grandfather of Isabella Hamilton, Synge's great-grandmother. Hugh Hamilton FRS (no relation to the mathematician Sir William Rowan Hamilton) "was the most intellectual Irish Bishop of the eighteenth century", and wrote extensively, and with great distinction, on mathematics, physics and chemistry, and theology. His book on *Conic Sections*, published in 1758, was hailed by the great Euler as a perfect book.

Of J L Synge's more immediate relatives the most distinguished ones are, undoubtedly, his uncle John Millington Synge (1871-1909), the world renowned playwright and poetic dramatist, his distant cousin Richard Lawrence Millington Synge FRS (1914-1994), the 1952 Nobel laureate (with Archer John Martin) for chemistry, and his daughter, Cathleen Synge Morawetz (1923-), an eminent applied mathematician and the first woman to hold the Directorship of the famous Courant Institute of New York.

In stark contrast to so many of his illustrious ancestors Synge was a complete atheist, stating in his characteristic style: "I am a Protestant to the marrow of my bones, holding the essence of Protestantism to consist, not in the recitation of this creed or that, but in the assertion of the right of the individual to hold his own views on all matters and express them as he thinks fit, with the prudential reservation that one does not preach vegetarianism (at least not too violently) in the lion's den."

In 1903 the Synge family moved to Dublin where J L Synge entered Trinity College Dublin (TCD) in 1915; by the end of his *first* year he won a Foundation Scholarship in mathematics, probably an unprecedented achievement. In TCD he met Elizabeth Eleanor Mabel Allen, a history student who shared his religious and political beliefs; they got married in

1918 in a registry office ('certainly not in a church'), with his class mate and friend Charles Rowe as witness. The wedding dinner consisted of scrambled eggs in Synge's college rooms. He graduated in 1919 with a Double Senior Moderatorship in Mathematics *and* Experimental Physics and a Large Gold Medal. He sold the medal shortly afterwards.

In January 1920 he was appointed Lecturer in Mathematics in TCD, and in the late Summer of the same year he left for Canada to join the University of Toronto as Assistant Professor of Mathematics. It was soon after his arrival at Toronto that Synge became interested in Einstein's theory of relativity. His approach to the theory was the elegant geometrical and visual approach which Herman Minkowski initiated in 1908; this geometrical approach was to become the most distinct characteristic of his subsequent work in theoretical physics.

In 1925 Synge returned to Dublin to a Fellowship and the Erasmus Smith Chair of Natural Philosophy in Trinity. In his first year in Trinity, 1925-1926, he felt privileged to have in his class two brilliant students, A J McConnell (the author of *Absolute Differential Calculus*, published in 1931), who was to become Professor of Natural Philosophy and Provost of Trinity College, and E T S Walton who was to become Professor of Physics in Trinity, and shared the Nobel Prize for Physics with J D Cockroft in 1951.

Of his many contributions to differential geometry, dynamics and relativity in the following five years, the most important one was his paper 'On the Geometry of Dynamics' which was published in the *The Philosophical Transactions of the Royal Society of London* in 1926. In it, Synge regarded the configuration space of a dynamical system as a Riemannian manifold and used the method of tensor calculus throughout. A by-product of this work was the derivation of the all important equation of the *geodesic deviation.* A major undertaking of Synge during this time was the editing, with Professor A W Conway FRS, of University College Dublin, of the first volume of the mathematical papers of Sir W R Hamilton; it was published by the Royal Irish Academy in 1931.

In 1930 Synge was invited to head the newly established Department of Applied Mathematics of the University of Toronto. Apart from a number of short visits to Brown University in Maryland and to Princeton University, and a brief appointment as a ballistics mathematician in the US Air Force during the war, he was to remain in Toronto until 1943. In 1933 he applied the theory of elasticity to investigate the problem of 'traumatic occlusion' connected with the periodontal membrane; the result was a number of short papers and a major paper entitled 'On the Tightness of the Teeth, Considered as a Problem Concerning the Equilibrium of a Thin Incompressible Elastic Medium', a supreme example of mathematical modeling. When his friend Charles Rowe, now professor of Pure Mathematics in Trinity, heard of this paper he remarked: "His dentures must be troubling him".

A theorem, known now as Synge's theorem, on even-dimensional Riemannian manifolds, published in 1936 is acclaimed as 'one of the most beautiful results in global differential geometry of the twentieth century'.

Another important and influential paper on 'Relativistic Hydrodynamics', published in 1937, became a classic; it was recently reproduced in the Journal of General Relativity and Gravitation (**34**(12), 2002) as one of the '*golden oldies*' of relativity.

He moved to the Ohio State University in 1943 to head the Mathematics Department for the following three years; the method of the *Hypercircle*, a precursor of today's *finite elements method* in numerical analysis, was developed, with W Prager, during this time. In 1946 Synge accepted an invitation to build up and head the Mathematics Department of the Carnegie Institute of Technology (now the Carnegie-Mellon University) in Pittsburgh. His most brilliant students there were the internationally renowned mathematicians R Bott and J F Nash; the latter received the 1994 Nobel Prize in Economics, and the recent award-winning film *A Beautiful Mind* is based on his life.

Synge left America in 1948 to return permanently to his native Dublin as a Senior Professor in the School of Theoretical Physics of the Dublin Institute for Advanced Studies (DIAS); he officially retired in 1972 but continued his reseach, mainly on relativity now, for another twenty years or so. What must be one of Synge's most remarkable achievements during these years is his paper on the 'Gravitational Field of a Particle' (*Proceedings of the Royal Irish Academy* **53A**(6),1950); in it he was able, for the first time, to penetrate and explore in detail the region inside the Schwarzschild radius (today's black hole). It was, in essence, the first maximal analytic extension of the Schwarzschild solution. May I suggest that this paper be reproduced in the Journal of General Relativity and Gravitation as a *'golden oldie'?*.

Synge's reputation as a relativist attracted research scholars, collaborators and eminent visitors from all over the world, making DIAS one of the great centres in relativity theory. It is reckoned that up to the mid-1960s 12% of the world's relativists passed, physically, through the Dublin Institute for Advanced Studies.

Professor Synge made outstanding contributions to widely varied fields: Classical Mechanics and Geometrical Optics, Gas Dynamics, Hydrodynamics, Elasticity, Electrical Network and Antenna Theory, Mathematical Methods and, above all, Differential Geometry and Einstein's Relativity Theory. He published eleven books, including the extraordinary semi-popular book *Kandleman's Krim*, and over two hundred papers, the last one at the age of 92; it was, appropriately enough, on geometry. Every book and every paper is a remarkable work of art, characterised by his striking 'clarity of expression' and the sheer beauty of his prose and, of course, by Synge's geometric spirit. He was, also, a superb lecturer, perhaps the greatest of his generation.

The almost universal geometrical approach to the theory of relativity in the last forty years or so is due primarily to Synge's influence, especially to his two epoch-making books on the *special* and on the *general theory of relativity* which were published in the late fifties. It is on record, for example, that the outstanding relativist Professor Sir Roger Penrose was drawn into the field of relativity after reading Synge's books. Characteristically, Professor Synge himself had this to say in 1972: *"If you were to ask me what I have contributed to the theory of relativity, I believe that I could claim to have emphasized its geometrical aspect"*.

The two hobbies which Professor Synge was to pursue passionately throughout his life were cycling and sailing. Cycling, especially, formed an integral part of his life, both as a form of recreation and as an object of dynamical investigation. He was, also, an accomplished painter. Well after retirement he took up the mandoline, but without much success.

Synge died on March 30, 1995, exactly one week after his 98th birthday. His mind was lively and vivid to the very end of his life, reading avidly and thinking about mathematical problems. He was a kind and generous man who helped and inspired several generations of students. In old age he suggested that a significant part of his epitaph might read:

'He encouraged younger men'.

Alas, there is no tomb to engrave an epitaph, having bequethed his body to the Medical School of TCD. It is, however, deeply and permanently engraved in the heart and mind of each one of his students, and of those who were fortunate to come in touch with him.

COMMENTS ON THE DEATH OF PETER BERGMANN

E. T. NEWMAN
Department of Physics and Astronomy,
University of Pittsburgh,
Pittsburgh, PA 15260,
USA.

On Oct.19, 2002, Prof. Peter G. Bergmann died, at the age of 87, from complications from Parkinson's disease. He had been one of the outstanding research workers and teachers in General Relativity for half a century. Just a few weeks earlier, he had been honored as the first recipient (with John Wheeler) of the Einstein Prize awarded by the American Physical Society. Though he had been missing from the active research community in recent years, he will be greatly missed by the myriad number of his former graduate students, post-doctoral fellows, colleagues, friends and family.

To me, Prof. Peter G. Bergmann was and always will be simply Peter. Though a summary of his life and work is clearly needed (and has been given elsewhere) I would like to say a few personal words about Peter.

I first met him in 1951 when I was applying for graduate work under his supervision at Syracuse University. I clearly remember him as a cheerful chubby man who quickly put me, a nervous undergrad, at ease. I saw him virtually every year – and often many times a year – since then, up to the time of the onset of his illness. Two years ago, months before his death, I was invited to give a talk at a History of General Relativity Conference in Amsterdam: I chose the topic, "The History of GR at Syracuse University, 1951–61", the years of my closest association with Peter. At one point in my talk the following were my words:

"I believe that he was among the clearest and deepest thinkers in relativity. He played a key role in developing the directions the field took – from Quantum Gravity to Radiation Theory – through his publications, his university courses, his gentle but strong influence on his students during the long talks and walks, his conference reports and lectures. His influence in keeping the field alive was inestimable. In addition, he was one of the kindest, most intellectually honest and honorable scientists I have known –

and he emphasized to his students the importance of these attributes. And he was loved by his students, post-docs and colleagues – and they carried on the traditions and love for physics, and GR in particular, that he had imparted to them."

In the 1940s the subject of general relativity (GR) was virtually dead - or at best dormant. Peter set up the first active research group in GR in the USA. Single-handedly he resurrected the field. He developed a research school with, at first, many students (who then also went on to be leaders and teachers of the next generation of students of GR) and then, with his growing reputation, a research center where post-doctoral fellows and colleagues from around the world came to learn, talk, argue and establish lasting friendships. Several years later other schools of GR developed around J. Wheeler, H. Bondi, L. Infeld and P. Jordan, but it is clear that Peter was the first to understand the importance of resurrecting GR from its dormancy and placing it as an important part of fundamental physics and then actively pursuing this goal.

During this period of intense research activity virtually every major researcher in GR passed through Syracuse to interact with Peter. The atmosphere was quite remarkable: ideas were exchanged freely, the discussions were open, help and suggestions were exchanged with no thought of priorities – I never saw professional jealousies. That was Peter's way. Peter was not interested in public relations and he had no pretensions; he knew he and his students were doing excellent physics but he did not try to oversell what they were doing – as, unfortunately, now occurs too often. Peter loved physics for its own sake, and we do not have many like him today.

In addition to his many contributions to physics, perhaps an equally great legacy he leaves is his inspiration to the many physicists who carry on his values, his integrity and his love for physics.

Let me end with a few words extracted from letters of Josh Goldberg and Alan Held about Peter's role in our GRG Society:

"The International Society on General Relativity and Gravitation was formed in 1971 (in Copenhagen) and is the successor to the International Committee on General Relativity and Gravitation formed in the late 1950s. Peter was a founding member of both groups. A Paris meeting in (1973) of the International Committee approved the Constitution which was written largely by Peter. He was the first President (1977–80).

"The Society owes a debt of gratitude to Peter, not only as one of the fathers of Relativity in America, but also as one of the founders of the Society itself."

CONTENTS

PLENARY LECTURES

THE STATUS OF NUMERICAL RELATIVITY

MIGUEL ALCUBIERRE

Instituto de Ciencias Nucleares, Universidad Nacional Autónoma de México, A.P. 70-543, México D.F. 04510, México

Numerical relativity has come a long way in the last three decades and is now reaching a state of maturity. We are gaining a deeper understanding of the fundamental theoretical issues related to the field, from the well posedness of the Cauchy problem, to better gauge conditions, improved boundary treatment, and more realistic initial data. There has also been important work both in numerical methods and software engineering. All these developments have come together to allow the construction of several advanced fully three-dimensional codes capable of dealing with both matter and black holes. In this manuscript I make a brief review of the current status of the field.

1. Introduction

General relativity is a highly successful theory. It has radically modified our vision of space and time and possesses an enormous predictive power. Its more important achievements include the prediction of exotic objects such as black holes and neutron stars, and the cosmological model of the Big Bang. In addition, it predicts the existence of gravitational waves, which will probably be detected for the first time before the end of this decade. Still, the field equations of general relativity are extremely complex. These equations form a system of ten coupled non-linear partial differential equations in four dimensions. When fully expanded in a general coordinate system they have thousands of terms. Because of this, exact solutions to these equations usually involve high degrees of symmetry, either in space or in time: spherical or axially symmetry, homogeneity, isotropy, staticity, stationarity, etc. When one is interested in studying highly dynamic systems that have little or no symmetry it is simply impossible to solve the equations exactly and one must fall back on numerical solutions.

Numerical relativity started in the 1960's with the pioneering work of Hahn and Lindquist [1], but it wasn't until the mid 1970's that the first successful simulations were performed by Smarr and Eppley in the context of the head-on collision of two black holes [2,3]. Since then numerical relativity has come a long way. The availability of powerful super-computers, together with an improved understanding of the fundamental theoretical issues that underlie the field, have finally allowed simulations of fully three-dimensional (3D) systems with strong and highly dynamical gravitational fields. All this activity has arrived at the right time, as the first generation of advanced gravitational wave detectors are finally coming on line [4–7]. We are living a very exciting stage in the development of this field.

In this manuscript I will discuss the current status of numerical relativity, starting from studies of theoretical issues having to do with formulations of the equations, gauge conditions and initial data. I will also discuss the latest achievements in the simulation of astrophysical systems such as neutron stars and black holes. Of necessity this report will be very brief. I will therefore limit myself to discussing the main ideas and results, and will leave all the details to the list of references at the end.

2. The 3+1 decomposition

The Einstein field equations are usually written in fully covariant form, with no distinction between space and time. This is elegant and mathematically powerful, but it is not very useful when one is interested in studying the evolution in time of a gravitational system starting from some appropriate initial data, the so called "initial value problem".

It is well known from the seminal work of Choquet-Bruhat [8] that general relativity does allow an initial value formulation. Today there are three main procedures in which one can obtain such a formulation: the "Cauchy" or "3+1" formalism, the "conformal" formalism, and the "characteristic" formalism. The 3+1 and conformal approaches share the property of splitting spacetime into a foliation of spacelike hypersurfaces. The main difference lies in the fact that in the 3+1 approach these hypersurfaces reach spatial infinity, while in the conformal approach the hypersurfaces are hyperboloidal and reach future null infinity instead. This allows the hypersurfaces to be conformally compactified, permitting a rigorous treatment of fields at infinity and eliminating the need for the boundary conditions at a finite distance that have to be introduced in the standard 3+1 approach. The characteristic formalism, on the other hand, foliates spacetime using null cones emanating from a central world tube and also allows a compactification that brings null infinity to a finite coordinate distance.

Each approach has its strong points, but to date most work in numerical relativity has been done using the 3+1 formalism, so here I will concentrate on this approach. However, interested readers are referred to the excellent review of Winicour on the characteristic approach [9] and the articles of Friedrich, Fraundiener and Husa on the conformal approach [10–12].

In the 3+1 approach one introduces a global time function t whose levels sets are the hypersurfaces defining the foliation. One then defines three main ingredients: 1) The three-dimensional metric γ_{ij} $(i,j=1,2,3)$ that measures distances within a given hypersurface, 2) the "lapse" function α that measures proper time between adjacent hypersurfaces, and 3) the "shift vector" β^i that measures the relative speed between observers moving along the normal direction to the hypersurfaces, and those keeping constant spatial coordinates. One then writes the four dimensional metric as

$$ds^2 = \left(-\alpha^2 + \beta_i \beta^i\right) dt^2 + 2\,\beta_i\, dt dx^i + \gamma_{ij}\, dx^i dx^j \,, \tag{1}$$

where $\beta_i := \gamma_{ij}\, \beta^j$.

Having split the metric, one introduces the "extrinsic curvature" tensor K_{ij} (also known as the "second fundamental form") that measures how the spatial hypersurfaces are immersed in spacetime. This tensor is given by the Lie derivative of γ_{ij} along the time lines, or in 3+1 language:

$$\partial_t \gamma_{ij} = -2\alpha K_{ij} + D_i \beta_j + D_j \beta_i \,, \tag{2}$$

with D_i the covariant derivative associated with γ_{ij}.

The next step is to decompose the Einstein equations. Doing this one finds that they naturally split in two groups. One group involves no time derivatives and represents constraints that must be satisfied at all times. The "Hamiltonian constraint" is given by (in units such that $G = c = 1$)

$$R + (\mathrm{tr\,K})^2 - K_{ij} K^{ij} = 16\pi\rho \,, \tag{3}$$

with R the scalar curvature of the spatial geometry, $\mathrm{tr\,K} \equiv \gamma^{\mathrm{ij}} \mathrm{K_{ij}}$ the trace of the extrinsic curvature and ρ the energy density of matter measured by the normal observers. The "momentum" constraints take the form

$$D_j \left[K^{ij} - \gamma^{ij} \mathrm{tr\,K}\right] = 8\pi j^i \,, \tag{4}$$

with j^i the momentum flux of matter measured by the normal observers. The existence of the constraints implies, in particular, that one is not free to specify the 12 dynamical quantities $\{\gamma_{ij}, K_{ij}\}$ as arbitrary initial conditions. The remaining 6 Einstein equations contain the dynamics of the system:

$$\begin{aligned} \partial_t K_{ij} &= \beta^a D_a K_{ij} + K_{ia} D_j \beta^a + K_{ja} D_i \beta^a - D_i D_j \alpha \\ &\quad + \alpha \left[R_{ij} - 2K_{ia} K^a_j + K_{ij}\, \mathrm{tr\,K}\right] + 4\pi\alpha \left[\gamma_{ij} \left(\mathrm{tr\,S} - \rho\right) - 2S_{ij}\right] \,, \end{aligned} \tag{5}$$

with S_{ij} the stress tensor of matter.

It is important to mention that the Bianchi identities imply that the evolution equations preserve the constraints, that is, if they are satisfied initially they will remain satisfied at subsequent times. The equations just described are know as the Arnowitt-Deser-Misner equations, or ADM for short. They represent the starting point of practically all of 3+1 numerical relativity. The reader interested in seeing how these equations are derived is referred to the original ADM article [13] or the classic review by York [14].

3. Hyperbolic reductions

The ADM evolution equations introduced in the previous section are in fact highly non-unique. This is because one can easily add arbitrary multiples of the constraints to them, that is multiples of zero, without affecting their physical solutions. They key word here is "physical", since non-physical solutions, *i.e.* those that do not satisfy the constraints, will certainly change. Constraint violating solutions, though not of interest physically, are nevertheless inevitable in numerical simulations since truncation errors imply that the constraints are never satisfied exactly. Adding constraints to the evolution equations can also seriously alter their mathematical properties. This non-uniqueness of the evolution equations is well known. For example, the original equations of ADM [13] differ from those of York [14] precisely by the addition of a multiple of the Hamiltonian constraint. The reformulation of York can be shown to be better behaved mathematically [15] and has become the standard form used in numerical relativity.

A key point that one has to worry about when studying the Cauchy problem is the well-posedness of the system of evolution equations, by which one understands that solutions exist (at least locally) and are stable in the sense that small changes in the initial data produce small changes in the solution. In this respect, one usually looks for either symmetric or strongly hyperbolic systems of equations as such systems are well posed under very general conditions [16]. In the case of general relativity, the problem of well-posedness was studied already by Choquet-Bruhat in the 1950's [8]. By the mid 1980's a number of hyperbolic reductions were known [17–20], though they played a very minor role in numerical relativity (an excellent review of the subject can be found in a recent article by Friedrich and Randall [21]). In the early 1990's, Bona and Masso started studying hyperbolic formulations for numerical relativity [22–24], and this effort was later followed by a number of people [25–28]. However, a generalized interest of the numerical community in hyperbolicity had to wait until 1999, when Baumgarte and Shapiro [29] showed that a reformulation of the ADM equations proposed by Nakamura, Oohara and Kojima [30], and Shibata and Nakamura [31], had far superior numerical stability properties than ADM. Baumgarte and Shapiro attributed this to the fact that the new formulation, known as BSSN, had "more hyperbolic flavor". This was later put on firmer ground in [32–34], and today it is understood that ADM is only weakly hyperbolic (and thus not well posed) [35], whereas BSSN is strongly hyperbolic [33,34].

The search for ever more general hyperbolic reductions of the Einstein evolution equations has continued in the last few years, and today many such formulations exist, several of which have dozens of free parameters [36, 37]. The focus has now changed to finding hyperbolic formulations that allow dynamical gauge choices, both for the lapse and the shift [38–40], and formulations that work well with fewer free parameters, maybe keeping a second order in space form [41, 42], or motivated by covariant approaches like the recently developed "Z4" system of the Majorca group [43]. At the same time, there is a growing realization that well-posedness is not enough, as empirically some hyperbolic formulations have proven to be far more robust than others. Some work has been done on the analytic side trying to understand what makes some hyperbolic formulations better suited for numerical work. In particular one can mention the work of Shinkai and Yoneda [44] and the work of Lindblom and Scheel [45]. Work in this area continues, and today there is no consensus as to which, if any, is the best formulation of the evolution equations for numerical purposes.

4. Constrained evolution

There is another approach, parallel but not inconsistent with the drive for hyperbolic formulations. This approach comes from the observation that numerical instabilities are usually related with the violation of the constraints. It would then seem that imposing the constraints at every time step should improve the stability of the simulations.

Such "constrained evolution" is in fact a old approach, and is commonly used in situations with high symmetry. In fact, it is the standard way to do numerical simulations in spherical symmetry, where for specific gauge choices, *e.g* an areal radial coordinate, one can solve for the metric directly from the Hamiltonian constraint and ignore the evolution equations (in this case the Hamiltonian constraint is an ordinary differential equation and therefore easy to solve numerically). In situations with less symmetry, and in particular in fully 3D simulations, constrained evolution has not been very popular since it involves solving systems of coupled elliptic equations every time step, which makes it a hard problem. The elliptic equations can be solved, but it makes the evolution code extremely slow (though this problem can be considerably ameliorated by using different numerical techniques such as spectral methods).

Another reason why constrained evolution is less popular than "free evolution" is that mathematical results on the well-posedness of mixed elliptic-hyperbolic systems are sparse, and a priori there is no good reason to expect that such constrained systems will be any better behaved than standard approaches. Still, a number of groups are exploring the use of constrained evolution in axi-symmetry and even in full 3D [46–49], and the idea seems to be gaining popularity. It will be very interesting to see if this approach can cure the stability problems faced by free evolution codes.

5. Gauge

The Einstein equations provide us with evolution equations for the spatial metric γ_{ij} and extrinsic curvature K_{ij}. However, they do not say anything about the evolution of the lapse function α and shift vector β^i. This is as it should be, as the lapse and shift represent our freedom in choosing the coordinate system, *i.e.* they are "gauge" functions. Choosing a good gauge is crucial when one solves the Einstein equations numerically, and one must choose carefully if one wants to avoid coordinate (and physical) singularities and to cover the interesting regions of spacetime. One can't just specify the lapse and shift as *a priori* known functions of spacetime for a simple reason: Which functions are a good choice? The lapse and shift must therefore be chosen dynamically as functions of the evolving geometry, that is, we choose the coordinates as we go.

5.1. *Slicing conditions*

Since the seminal work of Smarr and York in 1978 [50, 51], a lot has been learned about slicing conditions. Some classical choices such as maximal slicing are still in use today. Maximal slicing asks for the trace of the extrinsic curvature to remain equal to zero, which implies that the normal volume elements do not change. Through the ADM equations one finds that this implies that the lapse function must satisfy the elliptic equation:

$$D^2\alpha = \alpha K_{ij}K^{ij} \,, \tag{6}$$

with D^2 the spatial Laplace operator. Maximal slicing is an extremely robust slicing condition, producing smooth evolutions whenever in can be applied. It also has the important property of being "singularity avoiding". This means that when a physical singularity is approached, the slicing reacts by making the lapse collapse to zero in that region, thus halting the evolution there while allowing it to continue in regions away from the singularity. Of course, there is a price to pay, and singularity avoiding slicings develop a problem known as "slice stretching", where the radial metric components grow rapidly as the slices are stretched between the frozen inner regions and the evolving outer ones. Maximal slicing continues to be used to this day, and recently there has been some important analytical insight into numerical results for black hole evolutions using maximal slicing [52].

The drive towards hyperbolicity and computational efficiency has made hyperbolic slicing conditions increasingly popular in recent years, and there have been a number of important developments in this respect. In the first place, we have learned that it is much better to prescribe the "densitized lapse" $\tilde{\alpha} := \alpha / \det(\gamma)^{1/2}$ than the lapse itself, as a prescribed densitized lapse allows hyperbolic formulations to be constructed while a prescribed lapse does not. Anderson and York have given important insight as to why this is so [53], and Sperhake *et al.* have recently shown dramatic numerical evidence of this [54]. In the case of dynamical lapse conditions, gauges of the Bona-Masso type [24], $\partial\alpha = -\alpha^2 f(\alpha)\mathrm{tr}K$, have also been shown to allow hyperbolic formulations to be constructed [37, 38, 55]. These type of gauge conditions are becoming very popular, in particular "1+log" slicing ($f = 2/\alpha$), which has singularity avoiding properties similar to maximal slicing but is far easier to implement numerically. Finally, there is now a better understanding of the singularity avoidance properties of Bona-Masso type slicing conditions, as well as the possibility of gauge shock formation [56] (*i.e.* coordinate singularities caused by the crossing of the characteristics associated with the propagation of the gauge).

5.2. *Shift conditions*

Considerably less in known about shift conditions, and the classic conditions of Smarr and York [50] are still in use today in one way or another. In particular, the "minimal distortion" shift condition, which was designed to minimize the distortion in the volume elements over the hypersurface and requires that one solves a system of three coupled elliptic equations, is still a common choice. However, some recent results are important.

The main development has been the introduction of hyperbolic type shift conditions associated with the BSSN formulation. This formulation introduces as auxiliary variables the contracted conformal Christoffel symbols $\tilde{\Gamma}^i := \gamma^{jk}\tilde{\Gamma}^i_{jk}$, and shift conditions have been designed that make the second time derivatives of the shift proportional to the time derivatives of the $\tilde{\Gamma}^i$: $\partial_t^2 \beta^i \sim \partial_t \tilde{\Gamma}^i$. It turns out that the principal part of the operator on the right hand side of this equation is identical with that of the minimal distortion condition, so these "Gamma driver" conditions can be seen as a hyperbolic version of minimal distortion. These type of shift conditions are extremely easy to implement numerically, in contrast with the difficulty involved in solving the minimal distortion elliptic equations. Moreover, they have proven to be extremely robust in practice, controlling the slice stretching associated with singularity avoiding slicings, and allowing very long term evolutions of black hole spacetimes [57]. Because of this, such shift conditions are now being used by a large fraction of existing 3D codes. Recently, shift conditions of this type have been shown to yield well-posed systems when considered together with Einstein's evolution equations [38, 39].

It has also become clear that 3D simulations of rotating and/or orbiting compact objects require non-trivial shift vectors to avoid large shears in the metric and allow long lived simulations. The use of co-rotating coordinates using a shift that approaches a rigid rotation far away has been found to be crucial for achieving long lived binary black hole evolutions [58, 59].

6. Boundaries

Most physical systems one is interested in extend all the way to infinity, but 3+1 simulations have a boundary at a finite distance (however far). One must therefore impose some artificial boundary condition at the edge of the computational domain. This is not the case in the conformal and characteristic approaches that allow one to reach all the way out to null infinity, but it represents a serious problem in standard 3+1 simulations.

For a very long time boundary conditions where not given the attention they deserved by the numerical community. The reason was that the interior evolution was already extremely difficult and was usually plagued with instabilities, so people considered themselves lucky if the boundaries remained stable, however inconsistent they could be with the physics. With the development of more stable and robust hyperbolic formulations for the interior evolution, the boundary problem has come back to the attention of researchers in the field, and today there is a lot of work in trying to ensure that the boundaries are compatible with Einstein's equations (so-called "constraint preserving" boundaries). The use of hyperbolic formulations has also implied that one can separate incoming and outgoing fields at the boundaries, allowing for boundary conditions that are well-posed. This means that there is now a rigorous way to apply the intuitive notion of giving boundary data only for incoming modes while allowing outgoing modes to leave the computational domain undisturbed.

The crucial result in this field has been the first demonstration by Friedrich and Nagy in 1999 of the well posedness of the initial-boundary value problem for Einstein's equations [60]. Since then a number of studies have been made of maximally dissipative boundary conditions for linearized gravity by Szilagyi *et al.* [61] and Calabrese *et al.* [62] and also non-linear gravity with harmonic coordinates [63,64]. More recently other groups have looked at the problem of boundaries, in particular one can mention the work of Gundlach and Martin-Garcia in studying boundary conditions for second order in space systems [42], and the work of Novak and Bonazzola on absorbing boundary conditions for gravitational waves [65].

7. Initial data

Before starting a numerical simulation we need initial data that represents the physical situation we want to evolve. The existence of the constraint equations implies that one needs to solve a system of coupled elliptic equations to find such initial data. Recently, there have been important developments both in the general methods for finding initial data, and in the search for better initial data for black hole spacetimes.

7.1. *The thin sandwich formalism*

The classic method for finding initial data is the York-Lichnerowicz conformal decomposition that finds initial data in the form of the 3-metric and the extrinsic curvature [14,66]. In this approach one gives as free data the conformal metric, the trace of the extrinsic curvature and its transverse part, and solves the constraints for the conformal factor and the longitudinal part of the extrinsic curvature.

A very important recent development has been the introduction of the "conformal thin sandwich" approach by York, where initial data is obtained in the form of the 3-metric and its time derivative [67]. Pfeiffer and York have also shown the equivalence of both conformal approaches [68].

7.2. *Binary black holes*

One of the most active areas of research in numerical relativity is the simulation of the collision of orbiting black holes. In the following sections I will come back to this problem, but at this point I will concentrate on the fact that in order to do those simulations one must have some initial data.

Traditionally, initial data for black hole binaries has been obtained using a topological construction. This idea is based on the fact that even a single Schwarzschild black hole contains a wormhole (the "Einstein-Rosen" bridge) joining two asymptotically flat universes. Initial data for multiple black holes can then be constructed by adding one new throat for each hole. A formal procedure for doing this is well known and comes in two "flavors": Misner type data joins two universes by multiple wormholes, whereas Brill-Lindquist type data joins one universe though a series of wormholes with disjoint multiple other universes. In the case of time-symmetric initial data (corresponding to momentarily static black holes), both types of data have an analytic expression. When the black holes have momentum, as in the case of black holes in orbit around each other, exact solution do not exist, but numerical solutions are not difficult to find. In 1994 Cook found Misner type data for black holes in quasi-circular orbits using an effective potential method [69]. Later, in 1997 Brandt and Bruegmann showed that Brill-Lindquist type data could be obtained much more easily by factoring out analytically the poles representing each black hole [70], and in 2000 this "puncture" method was used by Baumgarte to construct Brill-Lindquist type data for orbiting black holes [71]. There has also been an effort to construct initial data for binary black holes that does not use the topological approach. Matzner, Huq and Shoemaker have introduced the idea of superposing single black hole data using Kerr-Schild type coordinates and the solving the constraints [72]. This Kerr-Schild type data has the advantage of reducing to stationary Kerr for separated holes.

More recently, the thin sandwich approach has been used to construct initial data for black holes that is more "astrophysically motivated". The main idea comes from the fact that black holes in orbit have an approximate helical symmetry as they rotate. The symmetry is not exact since the orbit shrinks as the system emits gravitational waves, but for sufficiently separated holes this happens very slowly. One can then exploit the fact that an approximate helical Killing field exists to obtain black hole initial data in a co-rotating frame. Grandclement, Gourgoulhon and Bonazzola have used this approach to construct binary black hole initial data of the Misner type [73]. This data seems to agree closer with high order post-Newtonian predictions than the data based on the effective potential method. Tichy and Bruegmann have also applied this approach to puncture type data [74], and Yo *et al.* have done the same for Kerr-Schild type data [75].

8. Horizons

When studying the numerical simulations of black hole spacetimes it is important to locate the horizons associated with them. Traditionally, the apparent horizon has been used as an indicator of the presence of a black holes since such horizons are defined locally and can be found on a given hypersurface during the evolution. On the other hand, the existence of an event horizon represents the true definition of a black hole, so locating the event horizon would be much better. However, the global nature of the definition of an event horizon means that one can not locate them during the evolution since one needs to know the full history of the spacetime, and that is precisely what one is trying to solve for. Still, algorithms exist to find event horizons *a posteriori*. The basic idea was developed by Libson *et al.* [76], and consists on evolving a null surface backwards in time once an evolution is finished. As the event horizon is an attractor for light rays moving back in time, such null surfaces rapidly converge to the event horizon and track it all the way to its initial formation (or to the initial data if already present there). An important recent event has been the development of a 3D event horizon finder using these ideas by Diener [77].

Another development has been the introduction of the theoretical framework of isolated and dynamic horizons of Ashtekar *et al.* [78,79] to numerical relativity by Dreyer *et al.* [80]. The formalism permits to calculate both the mass and the angular momentum associated with horizons found during a numerical simulation. These techniques have also recently been applied to the binary coalescence of orbiting black holes [59].

9. Evolving black hole spacetimes

The simulation of black holes spacetimes has been a major theme in numerical relativity since its beginnings. The pioneering work of Hahn and Lindquist in the 1960's [1], as well as the work of Smarr and Eppley in the 1970's [2,3], was precisely on simulations of collisions of black holes.

Black hole spacetimes are in vacuum, so the simulation of these systems has the advantage of not having to deal with complicated hydrodynamics. On the other hand, black holes have a singularity where the gravitational field becomes infinite, and dealing with this singularity is far from trivial. The traditional technique to evolve black holes has been the use of singularity avoiding slicing conditions such as maximal slicing, but this has the weakness of leading to slice stretching which eventually becomes a serious problem on its own. An alternative approach is known as "singularity excision" (previously called "apparent horizon boundary condition"). The original idea has been attributed to Unruh by Thornburg [81,82], and consists on placing a boundary inside the black hole, thus excising its interior from the computational domain, and also using a non-zero shift vector that keeps the horizon roughly in the same coordinate location during the evolution. Since no information can leave the interior of the black hole, excision should have no effect on the physics outside. Black hole excision was first attempted successfully by Seidel and Suen in spherical symmetry [83], and was later studied in more detail by Anninos et.al. [84]. Work in singularity excision for 3D codes continues to this day.

In the early 1990's, a lot of work was done in simulating single static or perturbed black holes [85], as well as more advanced versions of the head-on collision [86, 87]. By the middle of that decade, the simulation of the inspiral collision of orbiting black holes was identified as a key project, motivated by the fact that the large interferometric gravitational wave detectors were finally starting construction. This led to the NSF "Binary Black Hole Grand Challenge" project, that brought together a large number of institutions in the United States. The problem turned out to be considerably more complicated than at first thought, among other reasons because the issue of the well-posedness of the evolution equations was not yet fully understood. Still, a number of important developments did take place, among which one can mention improved singularity excision techniques [88], the long term stable evolution of a single black hole using the characteristic formalism [89], and the development of gravitational wave extraction techniques by perturbative matching [90].

Since the end of the Grand Challenge project, a number of groups have continued independently trying to simulate the inspiral collision of two black holes. In the last few years there has been rapid progress, coming on the one hand from the use of more robust well-posed formulations such as BSSN [29], a better understanding of black hole excision techniques [91–95], and the development of improved gauge conditions, particularly related to the choice of the shift vector [57]. This led to the first simulations of the grazing collision of two black holes done in 2000 by Brandt *et al.* [96] and Alcubierre *et al.* [97].

Research has now moved into the realm of orbiting black holes, with two different approaches taking the lead, both of them based on the BSSN formulation. One approach uses advanced excision techniques to move black holes across the computational domain and has recently achieved a true milestone by having single black holes move in oscillating and even circular trajectories (controlled by gauge choices) in a stable way for essentially unlimited time [98] (See Fig. 1). A second approach uses less advanced excision techniques and attempts to keep the orbiting black holes on a fixed coordinate location by using instead advanced gauge choices and a co-rotating frame. This approach is further along the road, and has finally led to the first ever simulation of a full orbit of two black holes by Bruegmann *et al.* [58] (See Fig. 2). More recently, using the same approach but an independent code, Alcubierre *et al.* have studied sequences of black holes in quasi-circular orbits of the Cook/Baumgarte type and have come to the surprising discovery that the black holes are in fact not orbiting, but are instead in an extended plunge [59] (See Fig. 3) . Other groups are also working on the binary black hole problem using a variety of techniques, among them one can mention the NASA-Goddard group that uses a BSSN code with mesh refinement [99], and the Cornell/Caltech collaboration that uses advanced pseudo-spectral methods together with a multi-parameter hyperbolic formulation of the evolution equations [100].

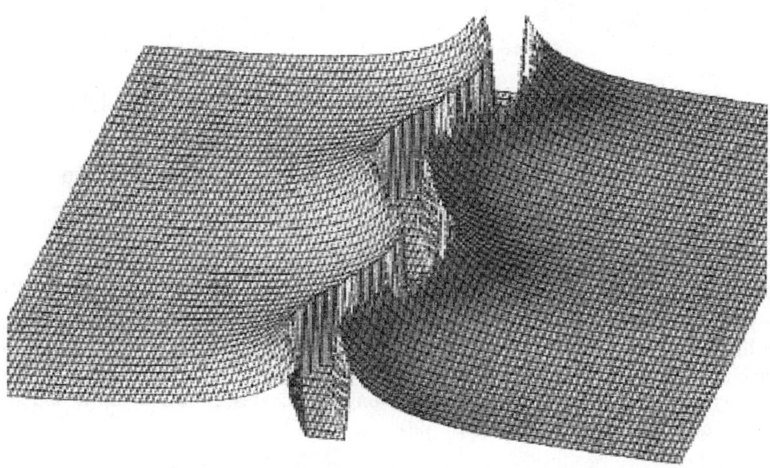

Figure 1. Space-time plot of the BSSN conformal factor ϕ for a single black hole oscillating in coordinate space. (Figure from Ref. [98])

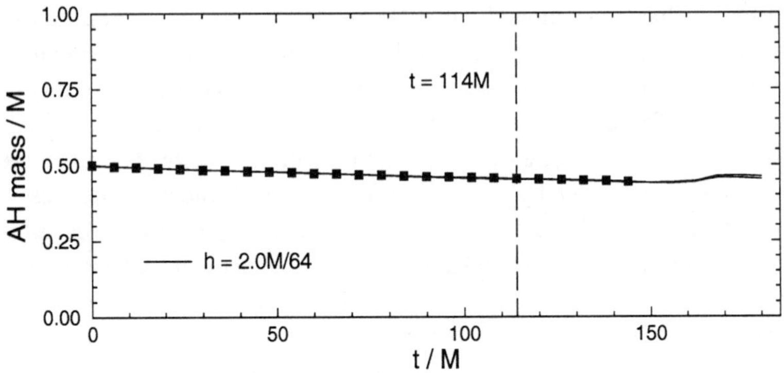

Figure 2. Evolution of the AH mass for one of the individual black holes in an orbiting binary. The evolution lasts longer than one orbital period of $114M$ defined by the initial data, and no common horizon is found (Figure from Ref. [58])

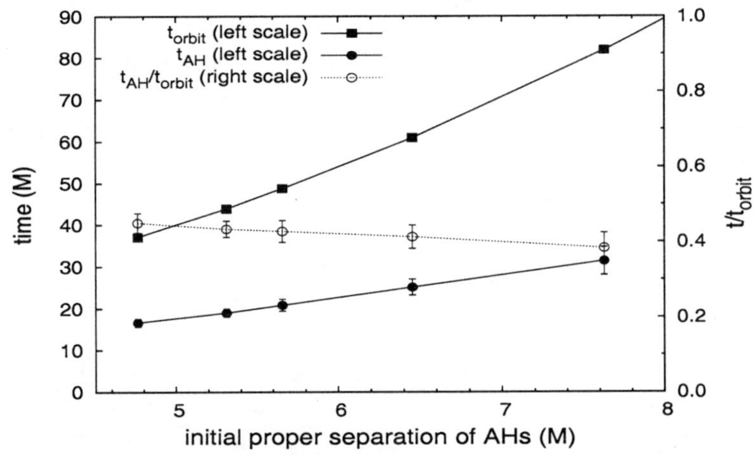

Figure 3. The time to appearance of a common apparent horizon for black hole in quasi-circular orbits for progressive separations, starting from the Cook/Baumgarte innermost stable circular orbit. Filled circles indicate the results of numerical simulations. The upper line indicates the expected orbital period, based on the initial angular velocity. Empty circles indicate the fraction of an orbit before common horizon formation. As the black holes are further separated, they take longer to coalesce, but the do it in a smaller fraction of an orbit (Figure from Ref. [59])

A very important development has been the combination of fully non-linear techniques with perturbation techniques by Baker *et al.* to study the black hole coalescence problem [101–104]. This approach evolves the black holes using a full non-linear code only to a "linearization" time where the two holes are close enough that the system can be treated as a single distorted Kerr black hole, after which a perturbative approach is used to complete the evolution. The approach has been called "Lazarus", making reference to the fact that a "dying" non-linear evolution (about to crash because of instabilities) can be resurrected by continuing the evolution using perturbation methods. The Lazarus approach has already given the first complete gravitational wave-forms to come from the coalescence of two orbiting black holes (See Fig. 4).

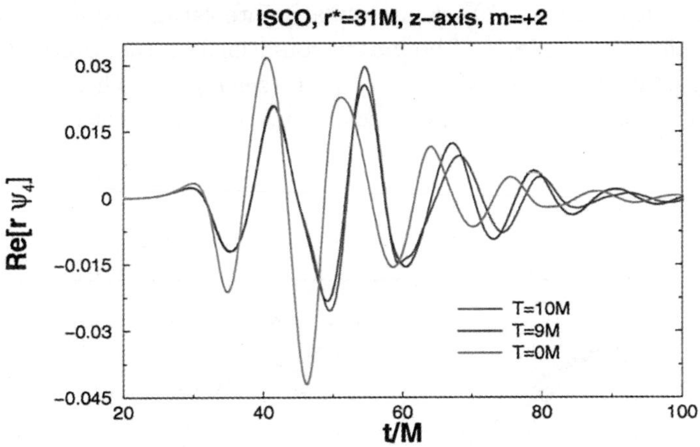

Figure 4. Wave forms for the Cook/Baumgarte innermost stable circular orbit using the Lazarus technique. The different lines show the effects of extracting data for a perturbative evolution at different times. (Figure from Ref. [104])

10. Relativistic hydrodynamics

Apart from black hole spacetimes, there is also a lot of interest in systems with matter. In particular, there has been considerable work on fully 3D general relativistic hydro-dynamical codes. At this time, several codes exist that are capable of simulating core collapse, rotating stars and binary star mergers. Among the groups with advanced 3D hydro codes one can mention the Tokyo group [105–107], the Illinois group [108, 109], the EU-Network group [110] and the JPL-WashU group [111].

It is interesting to notice the convergence of the techniques used in the different codes: All the codes mentioned above use some variety of the BSSN evolution system for the evolution of the geometry; all use either maximal slicing or a hyperbolic slicing condition of the Bona-Masso type (1+log for the most part); all use a non-zero shift vector, with the shift controlled by an evolution equation of the Gamma-driver type; advanced numerical techniques of the type of high-resolution shock-capturing schemes are becoming standard for the hydro-dynamical equations; several codes have already, or are implementing, fixed mesh refinement techniques; several codes have excision techniques for handling black hole formation.

Among the most important results so far one can mention the studies of collapse of super-massive stars [109] and the studies of collapsing rapidly rotating neutron stars [110]. A particularly interesting result has been the study of the strong effects of boundary conditions on the orbital dynamics of binary neutron stars by Miller *et al.* [111] (see Fig. 5). The field is moving rapidly, and many interesting results are sure to follow in the near future.

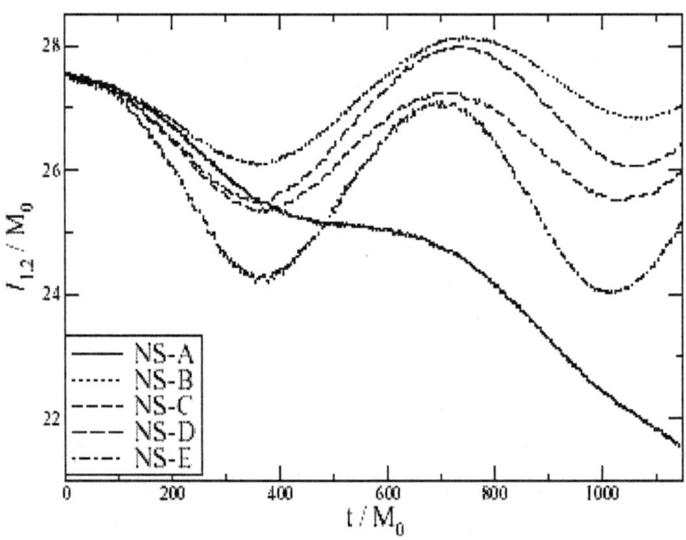

Figure 5. Evolution of geodesic distance between the centers of two orbiting neutron stars as a function of time, for various initial data sets representing the same physical situation but differing in grid resolution and position of outer boundary. Model NS-A shows a qualitatively very different behavior, attributed to the fact that in this case the boundaries of the computational domain were much further out than in the other cases. (Figure from Ref. [111])

11. Conclusions

Numerical relativity has come a long way in the last three decades. We are gaining a deeper understanding of the fundamental theoretical issues that underlie the field, such as well posedness of the evolution equations, gauge conditions, boundary conditions and initial data. At the same time, several advanced 3D codes exist that can handle both black holes and hydrodynamics. These codes are capable of singularity excision and mesh refinement, and have complex analysis tools incorporated: apparent and event horizon finding, horizon studies (mass, angular momentum), gravitational wave extraction, etc. The field is now reaching a state of maturity. In the next few years we hope to be able to simulate interesting astrophysical systems and extract gravitational waveforms that can be compared with the observations coming from the large interferometric detectors. Numerical relativity is truly entering a golden age!

Acknowledgments

It is a great pleasure to thank the organizers of the GR17 conference for their invitation to give a talk and write these notes. I also wish to thank Pablo Laguna, Bernd Bruegmann, Carlos Lousto, Manuela Campanelli, John Baker and Mark Miller for allowing me to show plots from their work in this manuscript. Finally, I want to thank the whole numerical relativity community for their enthusiastic response to my request for information about their current work. This work was supported in part by DGAPA-UNAM through grants IN112401 and IN122002.

References

1. S. G. Hahn and R. W. Lindquist, Ann. Phys. **29**, 304 (1964).
2. K. Eppley, Ph.D. thesis, Princeton University, Princeton, New Jersey, 1975.
3. L. Smarr, A. Čadež, B. DeWitt, and K. Eppley, Phys. Rev. D **14**, 2443 (1976).
4. LIGO - http://www.ligo.caltech.edu/.
5. VIRGO - http://www.virgo.infn.it/.
6. GEO600 - http://www.geo600.uni-hannover.de/.
7. TAMA - http://tamago.mtk.nao.ac.jp/.
8. Y. Bruhat, Acta Mathematica **88**, 141 (1952).
9. J. Winicour, Living Reviews in Relativity **1**, 5 (1998), [Online article].
10. H. Friedrich, Lect. Notes Phys. **604**, 1 (2002).
11. J. Frauendiener, Lect. Notes Phys. **604**, 261 (2002).
12. S. Husa, Lect. Notes Phys. **604**, 239 (2002).

13. R. Arnowitt, S. Deser, and C. W. Misner, in *Gravitation: An Introduction to Current Research*, edited by L. Witten (John Wiley, New York, 1962), pp. 227–265.
14. J. W. York, in *Sources of Gravitational Radiation*, edited by L. L. Smarr (Cambridge University Press, Cambridge, UK, 1979), pp. 83–126.
15. S. Frittelli, Phys. Rev. D **55**, 5992 (1997).
16. H.-O. Kreiss and J. Lorenz, *Initial-Boundary Value Problems and the Navier-Stokes Equations* (Academic Press, New York, 1989).
17. Y. Choquet-Bruhat, in *Gravitation: An Introduction to Current Research*, edited by L. Witten (John Wiley, New York, 1962).
18. Y. Choquet-Bruhat and J. W. York, Jr., in *General Relativity and Gravitation*, edited by A. Held (Plenum, New York, 1980), Vol. 1, pp. 99–172.
19. Y. Choquet-Bruhat and T. Ruggeri, Comm. Math. Phys **89**, 269 (1983).
20. H. Friedrich, Comm. Math. Phys. **100**, 525 (1985).
21. H. Friedrich and A. D. Rendall, Lect. Notes Phys. **540**, 127 (2000).
22. C. Bona and J. Massó, Phys. Rev. D **40**, 1022 (1989).
23. C. Bona and J. Massó, Phys. Rev. Lett. **68**, 1097 (1992).
24. C. Bona, J. Massó, E. Seidel, and J. Stela, Phys. Rev. Lett. **75**, 600 (1995).
25. S. Frittelli and O. Reula, Phys. Rev. Lett. **76**, 4667 (1996).
26. A. Abrahams, A. Anderson, Y. Choquet-Bruhat, and J. York, C.R. Acad. Sci. Paris **t. 323, Serie II**, 835 (1996).
27. H. Friedrich, Class. Quantum Grav. **13**, 1451 (1996).
28. A. Anderson and J. W. York, Phy. Rev. Lett. **82**, 4384 (1999).
29. T. W. Baumgarte and S. L. Shapiro, Phys. Rev. D **59**, 024007 (1999).
30. T. Nakamura, K. Oohara, and Y. Kojima, Progress of Theoretical Physics Supplement **90**, 1 (1987).
31. M. Shibata and T. Nakamura, Phys. Rev. D **52**, 5428 (1995).
32. M. Alcubierre *et al.*, Phys. Rev. D **62**, 124011 (2000).
33. O. Sarbach, G. Calabrese, J. Pullin, and M. Tiglio, Phys. Rev. D **66**, 064002 (2002).
34. H. Beyer and O. Sarbach, Phys. Rev. **D70**, 104004 (2004).
35. S. Frittelli and R. Gómez, J. Math. Phys. **41**, 5535 (2000).
36. L. E. Kidder, M. A. Scheel, and S. A. Teukolsky, Phys. Rev. D **64**, 064017 (2001).
37. O. Sarbach and M. Tiglio, Phys. Rev. D **66**, 064023 (2002).
38. L. Lindblom and M. A. Scheel, Phys. Rev. **D67**, 124005 (2003).
39. C. Bona and C. Palenzuela, Phys. Rev. **D69**, 104003 (2004).
40. M. Alcubierre *et al.*, Class. Quantum Grav. **20**, 3951 (2003).
41. G. Nagy, O. E. Ortiz, and O. A. Reula, Phys. Rev. **D70**, 044012 (2004).
42. C. Gundlach and J. M. Martin-Garcia, Phys. Rev. **D70**, 044032 (2004).
43. C. Bona, T. Ledvinka, C. Palenzuela, and M. Zacek, Phys. Rev. D **67**, 104005 (2003).
44. H. Shinkai and G. Yoneda, Class. Quantum Grav. **19**, 1027 (2002).
45. L. Lindblom and M. A. Scheel, Phys. Rev. D **66**, 084014 (2002).
46. M. W. Choptuik, E. W. Hirschmann, S. L. Liebling, and F. Pretorius, Class. Quant. Grav. **20**, 1857 (2003).

47. S. Bonazzola, E. Gourgoulhon, P. Grandclement, and J. Novak, Phys. Rev. **D70**, 104007 (2004).
48. M. Holst *et al.*, Phys. Rev. **D70**, 084017 (2004).
49. R. A. Matzner, Phys. Rev. D **71**, 024011 (2005).
50. L. Smarr and J. York, Phys. Rev.D **17**, 1945 (1978).
51. L. Smarr and J. W. York, Jr., Phys. Rev. D **17**, 2529 (1978).
52. B. Reimann and B. Brügmann, Phys. Rev. D **69**, 044006 (2004).
53. A. Anderson and J. York, James W., Phys. Rev. Lett. **81**, 1154 (1998).
54. U. Sperhake *et al.*, Phys. Rev. **D69**, 024012 (2004).
55. M. Alcubierre *et al.*, Phys. Rev. D **67**, 104021 (2003).
56. M. Alcubierre, Class. Quantum Grav. **20**, 607 (2003).
57. M. Alcubierre *et al.*, Phys. Rev. D **67**, 084023 (2003).
58. B. Brügmann, W. Tichy, and N. Jansen, Phys. Rev. Lett. **92**, 211101 (2004).
59. M. Alcubierre *et al.*, submitted to Physical Review D (2004).
60. H. Friedrich and G. Nagy, Commun. Math. Phys. **201**, 619 (1999).
61. B. Szilágyi, R. Gómez, N. T. Bishop, and J. Winicour, Phys. Rev. D **62**, 104006 (2000).
62. G. Calabrese *et al.*, Communications in Mathematical Physics **240**, 377 (2003).
63. B. Szilagyi and J. Winicour, Phys. Rev. D **68**, 041501 (2002).
64. B. Szilagyi, B. Schmidt, and J. Winicour, Phys. Rev. D **65**, 064015 (2002).
65. J. Novak and S. Bonazzola, J. Comp. Phys. **197/1**, 186 (2004).
66. A. Lichnerowicz, J. Math Pures et Appl. **23**, 37 (1944).
67. J. W. York, Phys. Rev. Lett. **82**, 1350 (1999).
68. H. P. Pfeiffer and J. York, James W., Phys. Rev. **D67**, 044022 (2003).
69. G. B. Cook, Phys. Rev. D **50**, 5025 (1994).
70. S. Brandt and B. Brügmann, Phys. Rev. Lett. **78**, 3606 (1997).
71. T. W. Baumgarte, Phys. Rev. D **62**, 024018 (2000).
72. R. A. Matzner, M. F. Huq, and D. Shoemaker, Phys. Rev. D **59**, 024015 (1999).
73. P. Grandclément, E. Gourgoulhon, and S. Bonazzola, Phys. Rev. D **65**, 044021 (2002).
74. W. Tichy and B. Brügmann, Phys. Rev. D **69**, 024006 (2004).
75. H.-J. Yo, J. N. Cook, S. L. Shapiro, and T. W. Baumgarte, (2004).
76. J. Libson *et al.*, Phys. Rev. D **53**, 4335 (1996).
77. P. Diener, Class. Quantum Grav. **20**, 4901 (2003).
78. A. Ashtekar, C. Beetle, and S. Fairhurst, Class. Quantum Grav. **16**, L1 (1999).
79. A. Ashtekar and B. Krishnan, Phys. Rev. Lett. **89**, 261101 (2002).
80. O. Dreyer, B. Krishnan, D. Shoemaker, and E. Schnetter, Phys. Rev. D **67**, 024018 (2002).
81. J. Thornburg, Class. Quantum Grav. **4**, 1119 (1987).
82. J. Thornburg, Ph.D. thesis, University of British Columbia, Vancouver, British Columbia, 1993.
83. E. Seidel and W.-M. Suen, Phys. Rev. Lett. **69**, 1845 (1992).
84. P. Anninos *et al.*, Phys. Rev. D **51**, 5562 (1995).

85. P. Anninos *et al.*, in *Computational Astrophysics: Gas Dynamics and Particle Methods*, edited by W. Benz, J. Barnes, E. Muller, and M. Norman (Springer-Verlag, New York, 1997), in press.
86. P. Anninos *et al.*, Phys. Rev. Lett. **71**, 2851 (1993).
87. P. Anninos *et al.*, Phys. Rev. D **52**, 2044 (1995).
88. G. B. Cook *et al.*, Phys. Rev. Lett **80**, 2512 (1998).
89. R. Gómez *et al.*, Phys. Rev. Lett. **80**, 3915 (1998).
90. A. M. Abrahams *et al.*, Phys. Rev. Lett. **80**, 1812 (1998).
91. M. Alcubierre and B. Brügmann, Phys. Rev. D **63**, 104006 (2001).
92. M. Alcubierre *et al.*, Phys. Rev. D **64**, 61501 (R) (2001).
93. G. Calabrese *et al.*, Class. Quantum Grav. **20**, L245 (2003).
94. J. Thornburg, Class. Quantum Grav. **21**, 3665 (2004).
95. M. Alcubierre *et al.*, submitted to Phys. Rev. D (2004).
96. S. Brandt *et al.*, Phys. Rev. Lett. **85**, 5496 (2000).
97. M. Alcubierre *et al.*, Phys. Rev. Lett. **87**, 271103 (2001).
98. D. Shoemaker *et al.*, Class. Quantum Grav. **20**, 3729 (2003).
99. B. Imbiriba *et al.*, Phys. Rev. D **70** 124025 (2004).
100. M. A. Scheel *et al.*, Phys. Rev. D **66**, 124005 (2002).
101. J. Baker, B. Brügmann, M. Campanelli, and C. O. Lousto, Class. Quantum Grav. **17**, L149 (2000).
102. J. Baker *et al.*, Phys. Rev. Lett. **87**, 121103 (2001).
103. J. Baker, M. Campanelli, and C. O. Lousto, Phys. Rev. D **65**, 044001 (2002).
104. J. Baker, M. Campanelli, C. O. Lousto, and R. Takahashi, Phys. Rev. D **65**, 124012 (2002).
105. M. Shibata, Phys. Rev. **D67**, 024033 (2003).
106. M. Shibata, Astrophys. J. **595**, 992 (2003).
107. M. Shibata, K. Taniguchi, and K. Uryu, Phys. Rev. **D68**, 084020 (2003).
108. M. D. Duez, S. L. Shapiro, and H.-J. Yo, Phys. Rev. D **69**, 104016 (2004).
109. M. D. Duez, Y. T. Liu, S. L. Shapiro, and B. C. Stephens, Phys. Rev. D **69**, 104030 (2004).
110. L. Baiotti *et al.*, Phys.Rev. D **71** 024035 (2005).
111. M. Miller, P. Gressman, and W.-M. Suen, Phys. Rev. D **69**, 064026 (2004).

LOOP QUANTUM GRAVITY, QUANTUM GEOMETRY AND SPIN FOAMS

JOHN BAEZ
University of California, Riverside, USA

The last few years have seen the techniques of loop quantum gravity applied to a growing number of problems. They have been used to compute the entropy of various sorts of black holes, including those with non-minimally coupled matter, for which the entropy is not proportional to the area. Perhaps more importantly, loop quantum gravity has given us a clear *picture* of the quantum geometry of the horizon, which accounts for the microscopic degrees of freedom responsible for black hole entropy. There are tantalizing connections to Hod's work on quasinormal modes, but these remain mysterious. Combined with traditional ideas on quantum cosmology, loop quantum gravity has led to new insights on how quantization can eliminate the singular behavior of geometry at the big bang - perhaps with testable consequences. Predictions of Lorentz symmetry violation may also be experimentally testable, but these remain controversial. Finally, the dynamics of the theory is being studied with the help of spin foam models.

TOWARDS DETECTION OF GRAVITATIONAL WAVES

BARRY C. BARISH
*California Institute of Technology,
Pasadena, CA 90405, USA*

The status, progress and plans for the Interferometer Gravitational Wave Detectors are presented. Resonant Bar and Interferometric detectors are complementary approaches to gravitational wave detection. Resonant Bars are generally narrow band cryogenic detectors sensitive at frequencies of about 1 KHz. Interferometers are long baseline broad-band detectors covering a frequencies range of about 10 to 10^4 Hz, having sensitivity to lower frequency gravitational radiation. In this presentation I review the techniques and status of both the resonant bar detectors and gravitational wave interferometers worldwide.

1. Introduction

Gravitational waves are a necessary consequence of Special Relativity with its finite speed for information transfer. Einstein in 1916 and 1918 [1,2] put forward the formulation of gravitational waves in General Relativity. He showed that time dependent gravitational fields come from the acceleration of masses and propagate away from their sources as a space-time warpage at the speed of light. This propagation is called gravitational waves.

The formulation of this concept in general relativity is described by the Minkowski metric, but where the information about space-time curvature is contained in the metric as an added term, $h_{\mu\nu}$. In the weak field limit, the equation can be described with linear equations. If the choice of gauge is the *transverse traceless gauge* the formulation becomes a familiar wave equation

$$\left(\nabla^2 - \frac{1}{c^2}\frac{\partial^2}{\partial t^2}\right) h_{\mu\nu} = 0.$$

The strain $h_{\mu\nu}$ takes the form of a plane wave propagating with the speed of light (c). The speed is the same for electromagnetic and gravitational radiation in Einstein's theory. Since the underlying theory of gravity is spin 2, the waves have two components, like electromagnetic waves, but rotated by 45° instead of 90° from each other. It is an interesting fact that if gravitational waves are observed and the two components are decomposed, this classical experiment will be capable of observing the underlying quantum spin 2 structure of gravity. The solutions for the propagation of gravitational waves can be written as

$$h_{\mu\nu} = h_{+}(t - z/c) + h_{\times}(t - z/c)$$

where z is the direction of the propagation and h_{+} and $h_{\times}$ are the two polarizations.

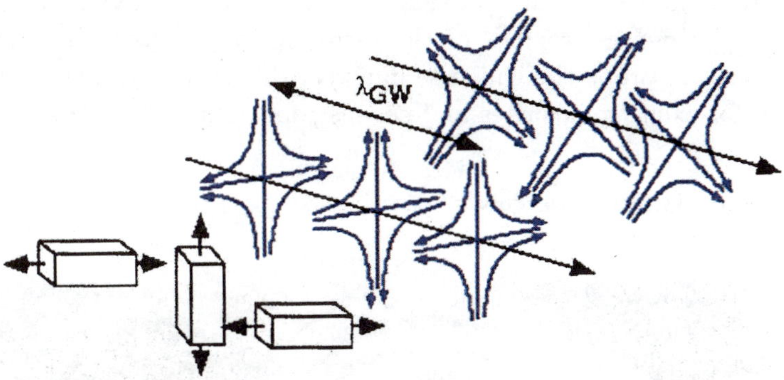

Figure 1. The propagation of gravitational waves illustrating the two polarizations rotated 45° from each other.

Evidence for gravitational waves resulted from the beautiful observations of Russell Hulse and Joseph Taylor in their studies of a neutron star binary system PSR1913+16. Although the experimental results of Hulse and Taylor make a very strong case for the existence of such waves, they result from an indirect measurement — the observation that a binary neutron star system radiates away energy consistent with that expected from general relativity. The next step for gravitational waves is to make direct observations of the waves themselves. For fundamental physics, the direct observation of gravitational waves offers the possibility of studying gravitation in highly relativistic settings, offering tests in the strong field limit, where the effects are not merely a correction to Newtonian gravity but produce fundamentally new physics through the strong curvature of the space-time geometry. Of course, the waves at Earth are not expected to be other than weak perturbations on the local flat space; however they provide information on the conditions at their strong field sources. The detection of the waves will also allow determination of the wave properties such as their propagation velocity and polarization states.

In terms of astrophysics, the observation of gravitational waves will provide a very different view of the Universe. These waves arise from motions of large aggregates of matter, rather than from particulate sources that are the source of electromagnetic waves. For example, the types of known sources from bulk motions that can lead to gravitational radiation include gravitational collapse of stars, radiation from binary systems, and periodic signals from rotating systems. The waves are not scattered in their propagation from the source and provide information of the dynamics in the innermost and densest regions of the astrophysical sources. So, gravitational waves will probe the Universe in a very different way, increasing the likelihood for exciting surprises and new astrophysics.

2. Resonant Bar Detectors

Figure 2. The Nautilus resonant bar gravitational wave detector.

Detection of gravitational waves by a resonant-mass detector is based on detecting the vibrational mode of a resonant body that has a mass quadrupole moment, for example the fundamental longitudinal mode of a cylindrical bar. The mechanical oscillation from a bar made of a high Q material is detected as an electrical signal from the strain sensed by a transducer and then amplified.

The sensitivity of the detector is limited by noise sources, including thermal noise internal to the detector and by amplifier noise in the readout. The bandwidth of such detectors is narrow, but technical advances are broadening the response beyond the very narrow width of the high Q mechanical resonance. The amplifier noise is what limits the bandwidth and reduction in the amplifier noise increase the antenna bandwidth.

Several improvements have been made in recent years, improving both the sensitivity and bandwidth. These include increasing the duty cycle and lowering the background noise, by upgrading the cryogenics leading to more stable operating temperature, and by acting on the vibration attenuation system, including the thermal links between refrigerating stages and the electromechanical sensitive components. They have also improved both the sensitivity and the useful bandwidth, by improving the readout systems.

Bar detectors have been taking data with a typical burst sensitivity $h \sim 4 \times 10^{-19}$ with a bandwidth of about 1 Hz, and duty cycle only limited by cryogenic operations. The recent technical improvements in the readout electronics have increased the detection bandwidth to $\sim$ 20 Hz, and even larger bandwidths are expected in the near future. Resonant-mass detectors of spherical shape are being investigated and give the promise of building a unidirectional detector.

During 2001 EXPLORER and NAUTILUS were the only two operating resonant detectors, with the best ever reached sensitivity. An algorithm based on energy compatibility of the event was applied to reduce the "background." They reported a possible excess of events as shown in Figure 3 in the direction of the galactic disk. More data will be required both from the resonant detectors and the interferometers to check this result.

3. Suspended Mass Interferometers

A new generation of detectors (LIGO, GEO, TAMA and VIRGO) based on suspended mass interferometry promise to attain the sensitivity to observe gravitational waves.

The Laser Interferometer Gravitational-wave Observatory (LIGO) is a joint Caltech-MIT project supported by the National Science Foundation and the Virgo interferometer facility is a joint French-Italian collaboration. To unambiguously make detections of these rare events a time coincidence

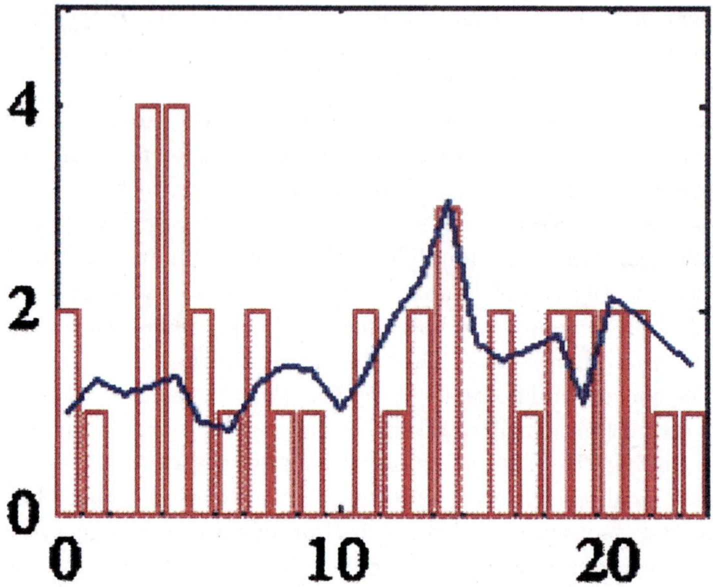

Figure 3. Nautilus-Explorer report of a possible excess in the direction of the galactic disk.

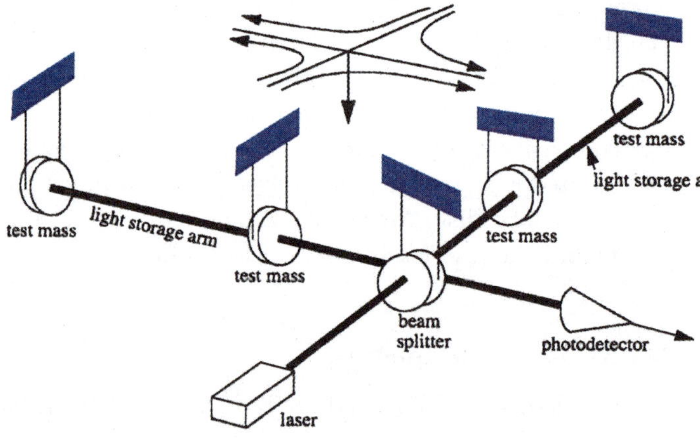

Figure 4. A schematic view of a suspended mass interferometer used for the detection of gravitational waves. A gravitational wave causes one arm to stretch and the other to squash slightly, alternately at the gravitational wave frequency. This difference in length of the two arms is measured through precision interferometry.

between detectors will be sought. The Virgo facilities can be seen in the aerial photograph in Figure 5.

Figure 5. Aerial photograph of the VIRGO 3km long suspended mass interferometers installed at Cascina, Italy.

The facilities developed to support the initial interferometers will allow for the evolution of the detectors to probe the field of gravitational wave astrophysics with more and more sensitivity over the next two decades. Sensitivity improvements and special purpose detectors will be needed either to enable detection if strong enough sources are not found with the initial interferometer, or following detection, to increase the rate in order to enable the detections to begin to become a new tool for astrophysical research. The first stage of incremental improvements using improved test masses, lasers, suspensions, etc should bring the sensitivity to $h \sim 10^{-22}$, which should make detection likely (if they have not been seen with the initial detectors) or significantly improve the rate, which scales as the improved sensitivity cubed.

The network of detectors in the world offer potential simultaneous detection of astrophysical sources and there are plans underway to correlate signals from all operating detectors in a worldwide network. This will enable much greater confidence in detection for early observations and will also both give information on the location of the source by comparing the relative timing and information on the polarization of the gravitational wave signal detected by comparing detailed waveforms from the different detectors, which have different orientations to the incoming gravitational wave.

4. Initial Performance

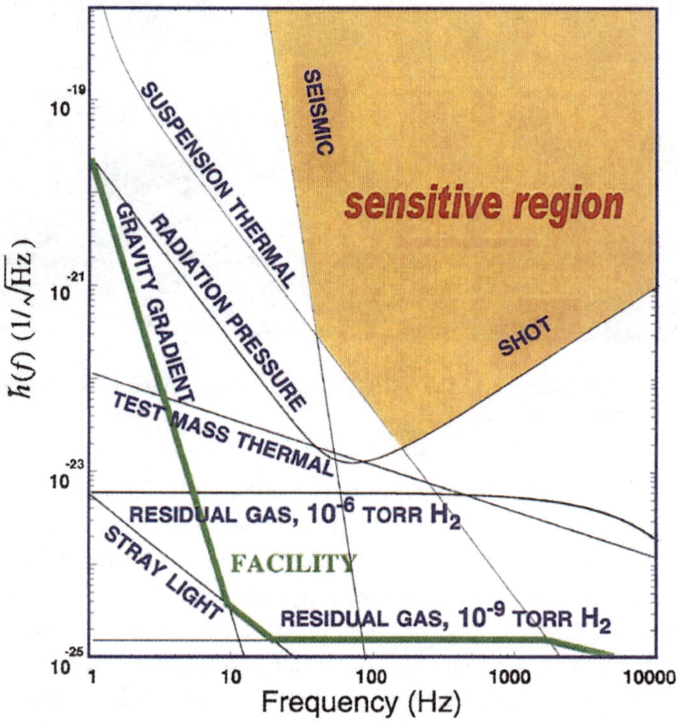

Figure 6. The limiting and other noise sources for LIGO are shown. The shaded area indicates the detector sensitivity.

The success of the detector ultimately will depend on how well we are able to control the noise in the measurement of very small strains. Noise is broadly but also usefully categorized in terms of those phenomena, which limit the ability to sense and register the small motions (sensing noise limits) and those that perturb the masses by causing small motions (random force noise). Eventually one reaches a practical but not fundamental limiting noise, the quantum limit, which combines the sensing noise with a random force limit. The primary noise sources for the initial LIGO detector are shown in Figure 6.

The initial LIGO interferometers are nearing design sensitivity and we now expect to enter our first extended, one to two year, data run before the end of 2005. Much progress has been made in reducing the technical noise as is illustrated in Figure 7.

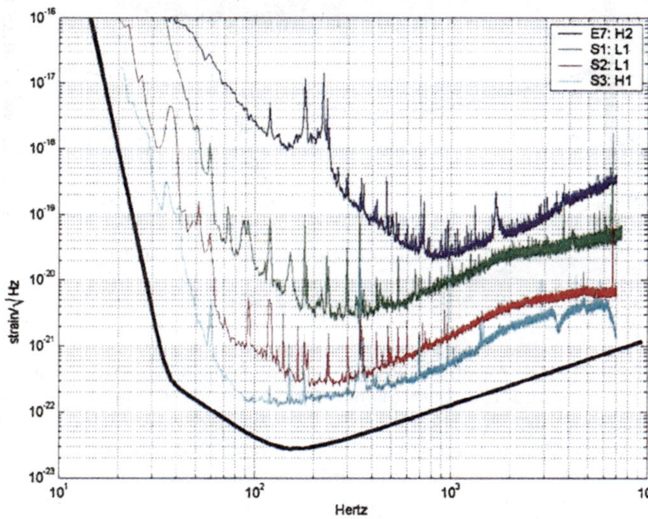

Figure 7. Sensitivity improvements are shown for LIGO since 2001. The data is for our E7, S1, S2 and S3 data as the sensitivity has improved. The analysis discussed in this presentation is from the S2 data set and the solid line represents the design goal we expect to reach by the end of 2005.

5. Astrophysical Signals from Spinning Neutron Stars in our Galaxy

There are a many known astrophysical processes in the Universe that produce gravitational radiation. Terrestrial interferometers, like LIGO, will search for signals from in the 10 Hz – 10 KHz frequency band. Characteristic signals from astrophysical sources will be sought over background noise from continuously recorded time-frequency series of the strain.

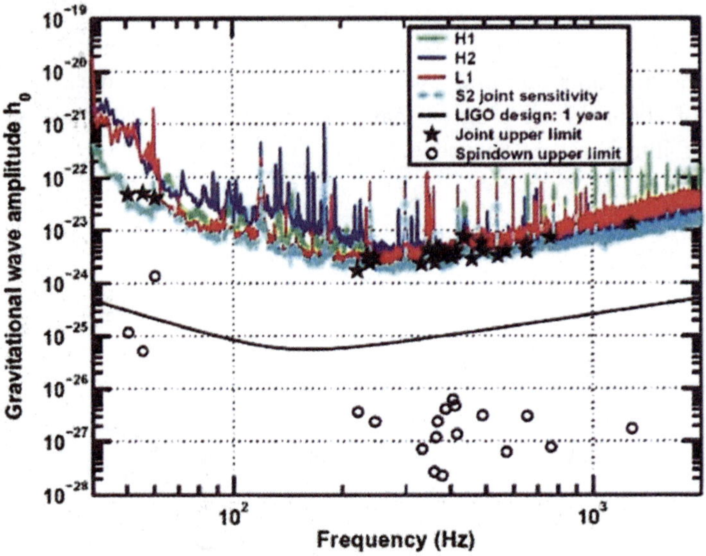

Figure 8. Upper curves: h_0 amplitudes detectable from a known generic source with a 1% false alarm rate and 10% false dismissal rate for single detector analyses and for a joint detector analysis. All the curves use typical S2 sensitivities and observation times. Lower curve: LIGO design sensitivity for 1 yr of data. Stars: upper limits for 28 known pulsars. Circles: spindown upper limits for the pulsars with negative frequency derivative values if all the measured rotational energy loss were due to gravitational waves and assuming a moment of inertia of 10^{45} g cm^2.

Radiation from rotating non-axisymmetric neutron stars will produce periodic signals in the detectors. The emitted gravitational wave frequency is twice the rotation frequency. For many known pulsars, the frequency falls within the LIGO sensitivity band. Searches for signals from spinning neutron stars will involve tracking the system for many cycles, taking into account the Doppler shift for the motion of the Earth around the Sun, and including the effects of spin-down of the pulsar. Both targeted searches for known pulsars and general sky searches are anticipated.

The very first scientific data run for both LIGO and GEO-600 was carried out in September 2002. Data was collected over a period of 17 days. Data collected by the GEO600 and LIGO interferometric gravitational wave detectors during their first observational science run were searched for continuous gravitational waves from the pulsar J1939+2134 at twice its rotation frequency. Two independent analysis methods were used and are demonstrated in this paper [3]: a frequency domain method and a time domain method. Both achieve consistent null results, placing new upper limits on the strength of the pulsar's gravitational wave emission.

In our second data run, S2, we place direct upper limits on the amplitude of gravitational waves from 28 isolated radio pulsars by a coherent multi-detector analysis of the data collected during the second science run of the LIGO interferometric detectors (Figure 8). These are the first direct upper limits for 26 of the 28 pulsars. We use coordinated radio observations for the first time to build radio-guided phase templates for the expected gravitational wave signals. The unprecedented sensitivity of the detectors allows us to set strain upper limits as low as a few times 10^{-24}. These strain limits translate into limits on the equatorial ellipticities of the pulsars, which are smaller than 10^{-5} for the four closest pulsars.

As expected, none of these upper limits improves on those inferred from simple arguments based on the gravitational luminosities achievable from the observed loss of pulsar rotational kinetic energy. However, for pulsars in globular clusters such arguments are complicated by cluster dynamics, which the direct limits presented here avoid.

The result for the Crab pulsar (B0531+21) is within a factor of about 30 of the spindown limit and over an order of magnitude better than the previous direct upper limit of. The equatorial ellipticities of the four closest pulsars (J0030+0451, J2124+3358, J1024-0719, and J1744-1134) are constrained to be less than 10^{-5}.

These limits are about three orders of magnitude higher than the upper limit obtained from the pulsar's measured spindown rate. However, such a large ellipticity could in principle be generated by an interior magnetic field of $\sim 10^{-16}$ gauss or it could probably be sustained in a neutron star with a solid core. Therefore, these early limits are already of interest and demonstrate the ability and techniques for extracting such source signals. Over the next few years the LIGO sensitivity will improve to where it will begin to confront the expectations from spin down rates for some selected pulsars.

References

1. B.C. Barish, and R. Weiss, Physics Today, October 1999 and LIGO Web site http://www.ligo.caltech.edu/
2. K.S. Thorne, *300 Years of Gravitation* Edited by S.W. Hawking and W. Israel

Cambridge University Press, Cambridge, England 1987 Chapter 9, *Gravitational radiation*
3. LIGO Science Collaboration, *Limits on gravitational wave emission from selected pulsars using LIGO data*, Phys. Rev. Lett. **94** (2005) 181103.

ADVANCES IN RELATIVISTIC ASTROPHYSICS

LARS BILDSTEN
KITP, University of California, Santa Barbara, USA

The advent of more sensitive X-ray telescopes in space has allowed for new discoveries, as well as detailed measurements of the properties of compact objects. These breakthroughs include the first gravitational redshift measurement of a neutron star, evidence that accreting neutron stars have a limit to their rate of rotation, and data on neutron star cooling that constrains their interior properties. All of these raise new puzzles for the theoretical astrophysics community, and provide new opportunities for the gravitational wave detectors presently online. The masses of stellar mass black holes are now regularly measured to be in the 7–15 solar mass range, and there are indirect indications of rapid rotation. I will close by showing some of the remarkable data from the recently discovered double radio pulsar, PSR J0737-3039.

RECENT RESULTS IN MATHEMATICAL RELATIVITY

PIOTR T. CHRUŚCIEL

Département de Mathématiques
Faculté des Sciences
Parc de Grandmont
F37200 Tours, France
e-mail: Piotr.Chrusciel@lmpt.univ-tours.fr
URL: www.phys.univ-tours.fr/~piotr

We review selected recent results concerning the global structure of solutions of the vacuum Einstein equations. The topics covered include quasi-local mass, strong cosmic censorship, non-linear stability, new constructions of solutions of the constraint equations, improved understanding of asymptotic properties of the solutions, existence of solutions with low regularity, and construction of initial data with trapped surfaces or apparent horizons.

1. Introduction

Reviewing recent progress in mathematical relativity is a difficult task, in view of the large number of excellent papers appearing in the field. Choices have to be made because of obvious time limits set for a lecture. In order to narrow down the number of topics covered I will concentrate on those results that concern the *global* properties of solutions of the *vacuum* Einstein equations, and which have appeared in preprint or final form within the last three years.

2. Quasi-local mass

The question of localisation of mass in general relativity has a long history, with no unanimously accepted candidate emerging so far, see [116,119] and references therein. There are at least two strategies which one might adopt here: trying to isolate a mathematically interesting object, or trying to find a physically relevant one. In the best of the worlds the same quantity would result, but no such thing has been found yet. From a physical point of view the strongest case can be made, I believe, for definitions obtained by Hamiltonian methods. Recall that the geometric symplectic framework of Kijowski and Tulczyjew [79] has been applied to general relativity by Kijowski and collaborators [37,49,76–78]), and it allows a systematic treatment of boundary terms, together with associated Hamiltonians, at least at a formal level[a]. One of the Hamiltonians that emerges in this way is the following [78]: Consider a three dimensional initial data set (M, g, K) in a four-dimensional space-time $(\mathscr{M}, {}^4g)$. Let Σ be a two dimensional surface within $\mathscr{M}$ and suppose that the mean extrinsic curvature vector κ of Σ is *spacelike.* Let

$$\lambda := \sqrt{{}^4g(\kappa, \kappa)}$$

be the ${}^4g$-length of κ. Assuming that the dominant energy condition holds in $(\mathscr{M}, {}^4g)$, it follows from the embedding equations that the Gauss curvature of the metric induced by g on Σ is positive. One can then invoke the Weyl embedding theorem [99,102] to isometrically embed $(\Sigma, {}^4g|_\Sigma)$ into $\mathbb{R}^3$. We shall denote by λ_0 the associated λ as calculated using the flat metric in $\mathbb{R}^3 \subset \mathbb{R}^{3,1}$. Let $m_{\rm K}$ be the Kijowski mass of Σ,

$$m_{\rm K} = \frac{1}{8\pi} \int_\Sigma (\lambda_0 - \lambda) d^2\mu \, . \tag{1}$$

A surprising theorem of Liu and Yau [91] asserts that

$$m_{\rm K} \geq 0 \, ,$$

with equality if and only if (M, g, K) *is a subset of Euclidean* $\mathbb{R}^3 \subset \mathbb{R}^{3,1}$. The key to the proof is a similar result by Shi and Tam [114], which is the Riemannian analogue of this statement: Shi and Tam prove that for manifolds of positive scalar curvature, the mean curvature H of a convex surface bounding a compact set satisfies

$$m_{\rm BY} = \frac{1}{8\pi} \int_\Sigma (H_0 - H) d^2\mu \geq 0 \, . \tag{2}$$

Here H_0 is the mean curvature of an isometric embedding of ∂M into $\mathbb{R}^3$ (thus H_0 coincides with λ_0). Liu and Yau show that the inequality (1) can be reduced to the Shi-Tam inequality using Jang's equation, in a way somewhat similar to the transition from the "Riemannian" to the "full" Schoen-Yau positive mass theorems [112,113]. The "quasi-local mass" $m_{\rm BY}$

[a]Kijowski's analysis leads to symplectic structures on spaces of fields with prescribed boundary data. To obtain a bona fide Hamiltonian system one should prove that the resulting initial-boundary value problems are well posed, which has not been done so far. It would be of interest to analyse how the Friedrich-Nagy [65] initial-boundary value problems fits into this framework.

appearing in (2) has been introduced, and studied, by Brown and York [26, 27].

In [100] O'Murchadha, Szabados and Tod show that $m_{\mathrm{K}} \neq 0$ for some surfaces in Minkowski space-time, which indicates that the normalisation in (1) is not optimal; this leaves room for future improvements.

The celebrated papers by Bray [24] and Huisken and Ilmanen [70] proving the Riemannian Penrose inequality have appeared recently (see also [25] and references therein). Those papers settle the problem for initial data sets with *positive Ricci scalar* (this is a restrictive condition which is satisfied, e.g., for maximal initial data sets), proving that the ADM mass is not less than the square root of the area of the outermost minimal surface divided by 4π. Suggestions how to approach the *full* Penrose inequality, without the $R \geq 0$ restriction, have recently been made by Frauendiener and by Malec, Mars and Simon in [61, 92]. In [93] Malec and Ó Murchadha show that a direct approach based on Yang's equation cannot succeed. In [121] Weinstein and Yamada point out that for *connected* charged black holes, with charge Q and with $R = \sqrt{A/4\pi}$ – the area radius of the outermost minimal surface, a Penrose-type inequality involving global charge

$$m \geq \frac{1}{2}\left(R + \frac{Q^2}{R}\right) \tag{3}$$

follows from the Huisken-Ilmanen proof. They also prove that (3) fails for some initial data sets with two black holes; the proof is yet another application of the Corvino-Schoen perturbation technique, discussed in Section 5 below.

We finish this section by noting some recent improvements in our understanding of the mass of asymptotically hyperbolic manifolds [44, 50, 52, 120]. Recall, now, that the asymptotically flat positive mass theorem can be used to prove uniqueness of static asymptotically flat black holes [28]; similarly the positive mass theorem for asymptotically hyperbolic manifolds can be used to prove uniqueness results in this context [5, 23] (compare [2, 3, 66, 67] for completely different approaches). A particularly elegant proof of rigidity of hyperbolic space, reducing the problem to the standard asymptotically Euclidean positive energy theorem, has been recently given by Qing [103]. The approach has been exploited in [22] by Bonini, Miao and Qing to considerably extend the rigidity result. Yet another recent approach to rigidity of asymptotically hyperbolic manifolds, based on volume comparison, has been presented by Shi and Tian in [115].

3. Strong cosmic censorship

The *strong cosmic censorship (SCC) problem* concerns predictability: it is a fundamental requirement that solutions of good physical theories should be *uniquely determined by initial data.* This is not the case for Einstein equations — in general relativity there exist examples where uniqueness fails [45, 96]. In this context, a key result is a theorem of Choquet-Bruhat and Geroch [32], which states that *to any initial data one can associate, uniquely up to a diffeomorphism, a maximal globally hyperbolic development of those data.* The resulting space-time is unique in the class of globally hyperbolic space-times, but in some situations it can be extended in more than one way to strictly larger vacuum solutions. In such cases the extension always takes places across a null hypersurface called *Cauchy horizon.*

A mathematical formulation of strong cosmic censorship, essentially due to Moncrief and Eardley [98] (*cf.* also [39, 40, 104]), is the following:

> *Consider the collection of initial data for, say, vacuum or electro–vacuum space–times, with the initial data surface $\mathscr{S}$ being compact, or with the initial data $(\mathscr{S}, {}^3g, K)$ — asymptotically flat. For generic such data the maximal globally hyperbolic development thereof is inextendible.*

The reader is referred to [4] for an excellent recent review of SCC.

Because of the difficulty of the strong cosmic censorship problem, a full understanding of the issues which arise in this context seems to be completely out of reach at this stage. There is therefore some interest in trying to understand that question under various restrictive hypotheses, *e.g.*, symmetry. Such a program has been initiated by Moncrief in [97, 98], and some further results in the spatially compact case have been obtained in [39, 46, 53, 73]. The simplest case, of spatially homogeneous space-times, has turned out to be surprisingly difficult, because of the intricacies of the dynamics of some of the Bianchi models. Spectacular progress in the understanding of this last class of solutions has been made a few years ago by Ringström. His results imply, amongst others, that curvature blows up in the contracting direction in generic Bianchi models. This forbids Cauchy horizons (this was already known, by completely different abstract arguments, from the work in [53]), and also provides further information about the geometry near "the end of space-time".

The next simplest case is that of Gowdy metrics on $\mathbb{T}^3 := S^1 \times S^1 \times S^1$:

in coordinates $t \in (-\infty, 0)$ and $(\theta, x^1, x^2) \in \mathbb{T}^3$ the metrics are of the form

$$g = e^{-\gamma/2} \mid t \mid^{-1/2} (-dt^2 + d\theta^2) + \mid t \mid e^P (dx^1)^2 + 2 \mid t \mid e^P Q \, dx^1 dx^2 \\ + \mid t \mid \left(e^P Q^2 + e^{-P}\right) (dx^2)^2 ,$$

with ∂_{x^1} and ∂_{x^2} being Killing vectors. The essential part of the evolution equations consists of two coupled non-linear equations:

$$\partial_t^2 P - \partial_\theta^2 P = -\frac{\partial_t P}{t} + e^{2P} \left((\partial_t Q)^2 - (\partial_\theta Q)^2\right) ,$$
$$\partial_t^2 Q - \partial_\theta^2 Q = -\frac{\partial_t Q}{t} - 2 \left(\partial_t P \partial_t Q - \partial_\theta P \partial_\theta Q\right) .$$

The question of SCC in this class of metrics has been settled by Ringström [110], who proved that the set of smooth initial data for Gowdy models on $\mathbb{T}^3$ that do *not* lead to the formation of Cauchy horizons contains a set which is open and dense within the set of all smooth initial data. Some key steps in Ringström's analysis are provided by the analysis of Fuchsian PDEs of Kichenassamy and Rendall [75, 105], and the analysis of the action of Geroch transformations by Rendall and Weaver [108] (compare [29]). See also [51] for the related problem of an exhaustive description of Cauchy horizons in those models.

Now, the existence of two Killing vectors is also compatible with S^3, $L(p, q)$ ("lens" spaces), and $S^1 \times S^2$ topologies. Thus, to achieve a complete understanding of the set of spatially compact initial data with precisely two Killing vectors one needs to extend Ringström's analysis to those cases. There is an additional difficulty that arises because of the occurrence of axes of symmetry, where the $(1 + 1)$–reduced equations have the usual singularity associated with polar coordinates. Nevertheless, in view of the analysis by Christodoulou and Tahvildar-Zadeh [35, 36] (see also [38]), the global geometry of *generic* maximal globally hyperbolic solutions with those topologies is reasonably well understood. This leads one to expect that one should be able to achieve a proof of SCC in those models using simple abstract arguments, but this remains to be seen.

Recall, finally, that general models with two Killing vectors X_1 and X_2 on $\mathbb{T}^3$ have non-vanishing *twist constants*:

$$c_a = \epsilon_{\alpha\beta\gamma\delta} X_1^\alpha X_2^\beta \nabla^\gamma X_a^\delta , \qquad a = 1, 2 .$$

The Gowdy metrics are actually "zero measure" in the set of all $U(1) \times U(1)$ symmetric metrics on $\mathbb{T}^3$ because $c_a \equiv 0$ for the Gowdy models. The equations for the resulting metrics are considerably more complicated when the c_a's do not vanish, and only scant rigorous information is available on the global properties of the associated solutions [18, 74]. It seems urgent to study the dynamics of those models, as they are expected to display [19] highly oscillatory behavior as the singularity is approached. Thus, they should provide the simplest non-isotropic model in which to study this behavior.

There has also been progress in the understanding of models with *exactly one* Killing vector. Here one only has *stability* results in the expanding direction, away from the singularity: In [31] Choquet-Bruhat considers $U(1)$ symmetric initial data for the vacuum Einstein equations on a manifold of the form $\Sigma \times S^1$, where Σ is a compact surface of genus $g > 1$. One assumes that the initial data have constant mean curvature and are sufficiently close to (g_0, K_0), where g_0 is a product metric

$$g_0 = h + dx^2 ,$$

with h — a metric of constant Gauss curvature on Σ, and with K_0 — pure trace. The sign of the trace of K_0 determines an expanding time direction and a contracting one. Under those conditions, Choquet-Bruhat proves that the solution exists for an infinite proper time in the expanding direction. The analysis builds upon previous work by Choquet-Bruhat and Moncrief [33], where a supplementary polarisation condition has been imposed. Nothing is known in the contracting direction in those models, where mixmaster behavior is expected [17, 20].

The proof of the above result bears some similarity to the *future stability* theorem of Andersson and Moncrief [8] for a class of hyperbolic models *without any symmetries*. Those authors consider initial data near a negatively curved compact space form satisfying a specific rigidity condition, with the extrinsic curvature being close to a multiple of the metric, obtaining future geodesic completeness in the expanding direction. The control of the solution is obtained by studying the Bel-Robinson tensor, and its higher-derivatives analogues. One of the ingredients of the proof is the use of an elliptic-hyperbolic system of equations to obtain local existence in time [7].

This last reference is interesting in its own right because of the following: the standard local existence theory for the general relativistic Cauchy problem proceeds through two steps. First one solves the equations in some chosen gauge, e.g. in harmonic coordinates. As a second step one patches the resulting solutions together to construct the space-time manifold. When working with harmonic coordinates, this second step leads to an increase by one of the differentiability threshold for existence and uniqueness of the solution. The elliptic-hyperbolic system of [7] is expected to solve that problem, the elliptic character of some of the equations providing improved regularity of the solution. There is, however, some more work needed to settle this issue, related to the fact that Andersson and Moncrief use a constant-mean curvature (CMC) slicing. The current state-of-the art uniqueness theory [12] for CMC slices requires, roughly speaking, $C^{1,1}$ metrics, and this provides a threshold for the analysis involved. It is conceivable that some weak form of the maximum principle such as, e.g., that in [6], could be used to lower the $C^{1,1}$ threshold, but this remains to be seen.

4. Stability of Minkowski space-time

One of the flagship results in mathematical general relativity is nonlinear stability of Minkowski space-time, first proved by Christodoulou and Klainerman [34] on some 500 pages of their celebrated book. An alternative approach to the problem, due to Klainerman and Nicolò [80] has appeared a year ago in print. That proof is based on an analysis of outgoing and ingoing null cones, and takes around 380 pages. A completely new argument of Lindblad and Rodnianski [90] is available now. Their proof takes some 65 Latex pages, though it should be said that the conditions on the initial data there are much stronger than in previous work: those authors require that the initial data coincide with Schwarzschildian ones outside of a compact set. We note that under that asymptotic conditions, together with a smallness condition, the global existence follows already from the stability results of Friedrich [63]. However, it has been announced by the authors (private communication) that the method is flexible enough to allow the inclusion of a scalar field, and to handle the following, rather weak, asymptotic behavior of the initial data:

$$g = (1 + 2M/r)\delta + O(r^{-1-\alpha}) , \qquad K = O(r^{-2-\alpha}) . \tag{4}$$

It should be emphasised that the decay conditions (4) are much weaker than those in [34, 80].

5. Initial data engineering

General relativistic initial data sets have to satisty the constraint equations,

$$\begin{pmatrix} J \\ \\ \rho \end{pmatrix} (K, g) := \begin{pmatrix} 2(-\nabla^j K_{ij} + \nabla_i \, \mathrm{tr} K) \\ \\ R(g) - K^{ij} K_{ij} + (\mathrm{tr} K)^2 \end{pmatrix} = \begin{pmatrix} 0 \\ \\ 0 \end{pmatrix} . \tag{5}$$

This makes it difficult to construct space-times with controlled global properties, such as geodesically complete space-times with conformally smooth asymptotics (Penrose's "asymptotic simplicity"), or solutions containing many black holes, or wormholes, or trapped surfaces, or apparent horizons, and so on. A new technique for deforming initial data sets has been invented by Corvino and Schoen [54, 55]. One of the applications of that technique is the unexpected statement [55] that

> *asymptotically flat initial data with well defined mass, momentum, center of mass, and angular momentum can be deformed in the asymptotic region to a Kerr metric, with arbitrarily small change in global Poincaré charges*

(compare [42]). While this theorem is very interesting in its own, what is even more important is the new technique introduced, which has already led to a few noteworthy applications. One of them is *existence of asymptotically simple space-times*: recall that a space-time is *asymptotically simple* [101] if it has smooth conformal structure at null infinity $\mathscr{I}$, and if all maximally extended null geodesics have initial and final end points on $\mathscr{I}$. Examples of such space-times include Minkowski space-time, and static asymptotically flat stars, and the static solutions of the Einstein – Yang-Mills equations of Bartnik and McKinnon [15]. More generally, spaces-times which are stationary and vacuum outside of a world tube have a smooth conformal structure at null infinity [57, 59], but the Kruszkal-Szekeres-Schwarzschild space-time is not asymptotically simple because of null geodesics terminating in the singularities under the event horizons. In any case, asymptotically simple space-times possess a smooth global $\mathscr{I}$, and various previous attempts to construct asymptotically simple *vacuum* solutions have been unsuccessful. For instance, the C-metrics have both singularities inside space-time and at $\mathscr{I}$ [9, 21]; the Robinson-Trautman metrics possess a smooth $\mathscr{I}^+$ which is complete to the future, but which is expected to be complete to the past only for the Schwarzschild metric; the space-times constructed by Christodoulou-Klainerman are not known to possess smooth asymptotic structure. In fact, the only previous dynamical example satisfying reasonable field equations has been given by Cutler and Wald, in electro-vacuum [56].

The paper [41] gives the first proof of *existence of non-trivial asymptotically simple vacuum space-times.* The construction uses the Corvino-Schoen gluing techniques [42] and Friedrich's stability theorem [63], and the simplest version of the result is the following: One considers a non-trivial initial data set $(\mathbb{R}^3, \mathring{g}, \mathring{K} = 0)$, with ADM mass $\mathring{m}$, satisfying a *parity condition*

$$\mathring{g}_{ij}(x) = \mathring{g}_{ij}(-x) . \tag{6}$$

One also assume that one is given on $\mathbb{R}^3 \setminus B(R)$ a parity-symmetric *reference family* $(\mathbb{R}^3, g^m, K^m = 0)$: this is, by definition, a family of metrics labeled by their ADM masses m, such that the m's cover continuously a neighborhood of $\mathring{m}$. In [41] it is proved that

> *the initial data set* $(\mathbb{R}^3, \mathring{g}, 0)$ *can be deformed to* some *member of the reference family on* $B(2R) \setminus B(R)$ *if the metrics are close enough near* $S(R)$.

If $\mathring{m}$ is sufficiently small one can invoke Friedrich's stability theorem to obtain existence of a global smooth $\mathscr{I}$, as well as asymptotic simplicity.

We note that large families of parity-symmetric, scalar flat metrics $\mathring{g}$, with mass $\mathring{m}$ as small as desired, can be constructed using the conformal method [42]. Similarly, the collection of reference families that can be used here is rather large: one can use Schwarzschild metrics, or Weyl metrics, or other static metrics [85, 109]. In fact, any metric on $\mathbb{R}^3$ with zero scalar curvature and with ADM mass m equal to $\mathring{m}$ is part of a reference family obtained by scaling the metric and the coordinates,

$$x^i \to \lambda x^i , \quad g \to \lambda^{-2} g \qquad \Longrightarrow \qquad m \to \lambda^{-1} \mathring{m} .$$

The stationary character of the reference metrics is needed to be able to assert smoothness of $\mathscr{I}$, but is not needed in the gluing construction.

The time-symmetry condition $K = 0$ has only been made for simplicity of presentation of the result, and the construction also provides families of asymptotically simple space-times which are not time-symmetric.

The parity condition (6) does look rather unnatural at first sight. It can be replaced by the requirement that $\mathring{g}$ is sufficiently close to a metric which satisfies (6). It is related to the fact that in the gluing process one needs to adjust the centers of mass of the metrics which are being glued. Parity guarantees that the centers of mass of all metrics involved are automatically zero, and therefore there is nothing to adjust. There is a simple Newtonian analogue which illustrates the problem: Let ρ be the Newtonian energy density, m the Newtonian mass (integral over the "space manifold" $M = \mathbb{R}^3$ of ρ), and let $\vec{c}$ be the Newtonian center of mass of the gravitating system,

$$\vec{c} = \int_M \vec{x} \rho .$$

If $\vec{x}$ is changed to $\vec{x}+\vec{a}$, then the new center of mass $\vec{c}\,'$ equals

$$\vec{c}\,' = \int_M (\vec{x}+\vec{a})\rho = \vec{c}+m\vec{a}\,.$$

In order to obtain $\vec{c}\,'=0$, one needs to translate the system by $-\frac{\vec{c}}{m}$, which can be very large even if $\vec{c}$ is small when m is small. Because the gluing techniques are based on the implicit function theorem, they break down when the translations cease to be very small.

So far we have been gluing metrics which were close to each other, with the final metric not being drastically different from the original one. It turns out that gluings can lead to objects with global properties rather distinct from the ones we started with. The idea here is to make a small gluing to add a "neck" to a first initial data set (M_1, g_1, K_1), and then another small gluing to add a second initial data (M_2, g_2, K_2) set on the other end of the neck. In this way one obtains a *new initial data set on the connected sum manifold* $M_1\#M_2$. A similar construction adds wormholes within a single initial data set. This has been successfully carried out by Isenberg, Mazzeo and Pollack [71, 72] a few years ago, using the conformal method. Their approach works for CMC initial data (and, with an additional hypothesis, for data which is only CMC near the neck region) in either the compact, asymptotically flat or asymptotically hyperbolic setting. The conformal factor can be made as close to one as desired away from the necks by making the necks small enough, but those techniques lead to deformations of the original data sets which are never localised. The question then arises, whether this can be done by deformations which are supported in a small neighborhood of the neck.

It is easily seen that there is an obstruction to doing that, related to the positive energy theorem: if one could glue, via local deformations, a non-trivial initial data set to $(\mathbb{R}^n, \delta, 0)$, where δ denotes the Euclidean metric, one would end up with a non-trivial initial data set with zero ADM mass, which is impossible.

Nevertheless, in [48] it is shown that

> *generic* initial data sets can be glued together, with *no deformations away from arbitrarily small sets* where the gluing takes place.

Here a data set is *generic* if there are no *local Killing vectors* in the associated space-time. (A local Killing vector field is a Killing vector field defined only on a subset of the space-time, not necessarily globally. The intuitively obvious fact, that non-existence of local Killing vectors is a generic condition, is rigorously justified in [16] in several important cases.) The proofs use a version with boundary of the results in [71, 72], the results by Bartnik [12] on existence of maximal surfaces, together with Corvino-Schoen type techniques [42, 55].

One interesting application of this gluing technique is the construction of spatially compact, *vacuum*, maximal globally hyperbolic space-times without any CMC surfaces [47,48]. The argument follows an idea of Bartnik [13], who obtains such space-times which contain dust.

The Corvino-Schoen gluing technique is based on an analysis of the operator which arises by linearising the map $(g, K) \mapsto (J, \rho)$ of (5), and it is well known that the properties of this map determine the manifold structure of the set of solutions of the constraint equations [60]. The analysis in that last reference allows one to introduce a Fréchet manifold structure on solutions of the constraint equations away from initial data sets with symmetries, but fails to provide a Banach manifold structure. In a recent beautiful paper [14] Bartnik manages to overcome the technical difficulties that arise, obtaining a Hilbert manifold. Various different Banach manifold structures have also been recently constructed in [43], again as an application of the Corvino-Schoen techniques; further applications of the technique can be found in [42, 121].

6. Asymptotic structure

The following questions have been keeping mathematical relativists busy for years:

> *What are the asymptotic conditions on the initial data at $\{t = 0\}$ that guarantee that the resulting space-time has*
>
> *(1) "a piece of smooth $\mathscr{I}^+$"?*
>
> *(2) a global smooth $\mathscr{I}^+$?*

As already pointed out, Corvino-Schoen gluings lead to initial data with a maximal globally hyperbolic development containing a smooth piece of $\mathscr{I}$ when the resulting initial data are stationary at large distances (and even global smooth $\mathscr{I}$'s for small initial data). They provide thus the first example of non-trivial initial data which are relevant in this context. The Christodoulou-Klainerman theorem [34] gives a $\mathscr{I}$ with very low differentiability, not enough for peeling theorems. It is conceivable that one shouldn't care about that, the properties of $\mathscr{I}$ in Christodoulou-Klainerman space-times being enough for e.g. for the Trautman-Bondi mass-loss formula.

However, for the sake of completeness of our understanding of the asymptotics of the gravitational field, one would like to have an answer to questions *1* and *2*.

Some progress on this has been recently made by Klainerman and Nicolò [81], who show that *peeling holds for initial data with asymptotics*

$$g = (1 + \frac{M}{2r})^4 \delta + O(r^{-3-\alpha}) \,, \qquad K = O(r^{-4-\alpha}) \,, \quad \alpha > 0 \,. \tag{7}$$

Large classes of initial data as above, not necessarily stationary outside of a compact set, can be constructed by a variant of the Corvino-Schoen gluing [41]. We emphasise that (7) provides *peeling*, but does not guarantee smoothness of $\mathscr{I}$. In fact, the *first known explicit obstructions to smoothness of $\mathscr{I}$ in terms of data at $t = 0$* have been found by Valiente-Kroon [86–89]. This leads to the striking statement, that

> *Bowen-York* and *Brill-Lindquist* initial data do not lead to a smooth $\mathscr{I}$ (if any).

In Friedrich's Cargèse lecture notes [62, 64] and in [88] the following conjecture is made: There exists $N(k) \in \mathbb{N}$, with $N(k) \to \infty$ if $k \to \infty$, such that

> *time-symmetric* initial data lead to a $\mathscr{I}$ of differentiability C^k if and only if the data are *static* up to terms $O(r^{-N(k)})$.

One expects an analogous statement without the time-symmetry condition, but this requires more care. As discussed in [89], smoothness of $\mathscr{I}^+$ imposes conditions on the initial data that are not identical to those for smoothness of $\mathscr{I}^-$. The requirement that *both* $\mathscr{I}^+$ and $\mathscr{I}^-$ are C^k is expected then to be compatible only with those initial data which are *stationary* up to terms $O(r^{-N(k)})$, for some $N(k) \to_{k\to\infty} \infty$.

So far we have been assuming that the cosmological constant vanishes. Some new results by Rendall on existence and properties of conformal completions for vacuum space-times with $\Lambda \geq 0$ can be found in [106, 107]. Recall that Friedrich's analysis [63] applies only to dimension four, while Rendall's approach works in higher dimensions as well. A new method of approaching the question of existence of conformal completions in *all even space-time dimensions* has been presented by Anderson in [1].

7. Low regularity solutions

Standard theory of hyperbolic equations [69, 118] shows that local existence of solutions of the general relativistic Cauchy problem holds for initial data $(g, K) \in H_s \times H_{s-1}$ with $s > 5/2$. There are several reasons why it is of interest to try to lower this threshold. An improved existence theorem:

- renders certain singularities innocuous
- can potentially be used to simplify and/or extend global existence results
- gives insight into the mathematical structure of the equations.

A first step in this direction has been taken by Klainerman and Rodnianski [82,83], who show that the existence time of the solution can be bounded from below by a function that depends only upon

$$\|(g, K)\|_{H_{2+\epsilon} \times H_{1+\epsilon}}$$

where ϵ is as small as desired. (See [10, 117] and references therein for related results by Bahouri and Chemin, and by Tataru, and others, for large classes of nonlinear wave equations.)

The existence of initial data in the $H_2 \times H_1$ class is proved by Choquet-Bruhat in [30] and by Maxwell in [95]. (In fact, Maxwell's paper goes further and produces initial data in $H_s \times H_{s-1}$ for $s > 3/2$, and shows that one can approximate (in that topology) this data by smooth data.) The already mentioned construction of manifold structure by Bartnik [11] allows $H_2 \times H_1$ differentiability as well. Klainerman conjectured that

L^2 regularity of the Riemann tensor should suffice for existence.

Some partial results towards the proof of this conjecture can be found in the paper [84] by Klainerman and Rodnianski.

8. Vacuum initial data set with apparent horizons and trapped surfaces

A classical statement in Hawking and Ellis [68] asserts that *the boundary of a trapped region is an apparent horizon.* The argument given there does not quite prove that, as it seems to make the implicit hypothesis that the boundary of the trapped region is a C^2 surface, and it is a still an open question whether or not this is correct. Now, the main point is the existence of an apparent horizon enclosing the trapped region, and this has been shown to be correct very recently by Schoen [111]:

Suppose that an asymptotically flat initial data set $(\mathscr{S}, g, K)$ contains a compact trapped surface $\mathscr{T}$, then there exists in $\mathscr{S}$ a smooth, outermost, compact apparent horizon enclosing $\mathscr{T}$.

The proof proceeds by constructing singular solutions of the Jang equation, with delicate control of their blow-up set. This generalises a previous result by Huisken and Ilmanen [70], who prove the corresponding statement in the time-symmetric case.

We finish this section, and this review, by noting that Dain [58] and Maxwell [94] have provided direct constructions of initial data with smooth, trapped or marginally trapped, boundaries.

Acknowledgements

Useful comments by D. Pollack on a previous version of this manuscript are gratefully acknowledged.

This work was partially supported by a Polish Research Committee grant 2 P03B 073 24, and by the Institute of Physics, London.

References

1. M.T. Anderson, *Existence and stability of even dimensional asymptotically de Sitter spaces*, (2004), gr-qc/0408072.
2. M.T. Anderson, P.T. Chruściel, and E. Delay, *Non-trivial, static, geodesically complete vacuum space-times with a negative cosmological constant*, JHEP **10** (2002), 063, 22 pp., gr-qc/0211006.
3. ———, *Non-trivial, static, geodesically complete vacuum space-times with a negative cosmological constant II: $n \geq 5$*, Proceedings of the Strasbourg Meeting on AdS-CFT correspondence (Berlin, New York) (O. Biquard and V. Turaev, eds.), IRMA Lectures in Mathematics and Theoretical Physics, de Gruyter, in press, gr-qc/0401081.
4. L. Andersson, *The global existence problem in general relativity*, 50 years of the Cauchy problem in general relativity (P.T. Chruściel and H. Friedrich, eds.), Birkhäuser, Basel, 2004, pp. 71–120.
5. L. Andersson and M. Dahl, *Scalar curvature rigidity for asymptotically locally hyperbolic manifolds*, Annals of Global Anal. and Geom. **16** (1998), 1–27, dg-ga/9707017.
6. L. Andersson, G.J. Galloway, and R. Howard, *A strong maximum principle for weak solutions of quasi-linear elliptic equations with applications to Lorentzian and Riemannian geometry*, Comm. Pure Appl. Math. **51** (1998), 581–624.
7. L. Andersson and V. Moncrief, *Elliptic-hyperbolic systems and the Einstein equations*, Annales Henri Poincaré **4** (2003), 1–34, gr-qc/0110111.

8. ______, *Future complete vacuum spacetimes*, 50 years of the Cauchy problem in general relativity (P.T. Chruściel and H. Friedrich, eds.), Birkhäuser, Basel, 2004, pp. 299–330, gr–qc/0303045.
9. A. Ashtekar and T. Dray, *On the existence of solutions to Einstein's equation with nonzero Bondi news*, Commun. Math. Phys. **79** (1981), 581–589.
10. H. Bahouri and J.-Y. Chemin, *Quasilinear wave equations and microlocal analysis*, Proceedings of the International Congress of Mathematicians, Vol. III (Beijing, 2002) (Beijing), Higher Ed. Press, 2002, pp. 141–153.
11. R. Bartnik, *The mass of an asymptotically flat manifold*, Comm. Pure and Appl. Math. **39** (1986), 661–693.
12. ______, *Regularity of variational maximal surfaces*, Acta Math. **161** (1988), 145–181.
13. ______, *Remarks on cosmological spacetimes and constant mean curvature surfaces*, Comm. Math. Phys. **117** (1988), 615–624.
14. ______, *Phase space for the Einstein equations*, (2004), gr-qc/0402070.
15. R. Bartnik and J. McKinnon, *Particle-like solutions of the Einstein Yang-Mills equations*, Phys. Rev. Lett. **61** (1988), 41–44.
16. R. Beig, P.T. Chruściel, and R. Schoen, *KIDs are non-generic*, Annales Henri Poincaré, in press, (2004), gr-qc/0403042.
17. V.A. Belinskii, E.M. Lifshitz, and I.M. Khalatnikov, *Oscillatory approach to the singular point in relativistic cosmology*, Sov. Phys. Usp. **13** (1971), 745–765.
18. B. Berger, P.T. Chruściel, J. Isenberg, and V. Moncrief, *Global foliations of vacuum spacetimes with T^2 isometry*, Ann. Phys. (NY) **260** (1997), 117–148.
19. B. Berger, J. Isenberg, and M. Weaver, *Oscillatory approach to the singularity in vacuum spacetimes with T^2 isometry*, Phys. Rev. **D64** (2001), 084006, gr-qc/0104048, erratum-ibid. D67, 129901 (2003).
20. B.K. Berger, D. Garfinkle, J. Isenberg, V. Moncrief, and M. Weaver, *The singularity in generic gravitational collapse is spacelike, local, and oscillatory*, Mod. Phys. Lett. **A13** (1998), 1565–1574, gr-qc/9805063.
21. J. Bičák and B. Schmidt, *Asymptotically flat radiative space–times with boost–rotation symmetry: The general structure*, Phys. Rev. **D40** (1989), 1827–1853.
22. V. Bonini, P. Miao, and J. Qing, *Ricci curvature rigidity for weakly asymptotically hyperbolic manifolds*, (2004), math.DG/0310378.
23. W. Boucher, G.W. Gibbons, and G.T. Horowitz, *Uniqueness theorem for anti–de Sitter spacetime*, Phys. Rev. **D30** (1984), 2447–2451.
24. H.L. Bray, *Proof of the Riemannian Penrose conjecture using the positive mass theorem*, Jour. Diff. Geom. **59** (2001), 177–267, math.DG/9911173.
25. H.L. Bray and P.T. Chruściel, *The Penrose inequality*, 50 years of the Cauchy problem in general relativity (P.T. Chruściel and H. Friedrich, eds.), Birkhäuser, Basel, 2004, pp. 39–70, gr–qc/0312047.
26. J.D. Brown, S.R. Lau, and J.W. York, Jr., *Energy of isolated systems at retarded times as the null limit of quasilocal energy*, Phys. Rev. **D55** (1997), 1977–1984, gr-qc/9609057.

27. J.D. Brown and J.W. York, Jr., *Quasilocal energy and conserved charges derived from the gravitational action*, Phys. Rev. **D47** (1993), 1407–1419.
28. G. Bunting and A.K.M. Masood–ul–Alam, *Nonexistence of multiple black holes in asymptotically euclidean static vacuum space-time*, Gen. Rel. Grav. **19** (1987), 147–154.
29. M. Chae and P.T. Chruściel, *On the dynamics of Gowdy space times*, Commun. Pure Appi. Math. **57** (2004), 1015–1074, gr-qc/0305029.
30. Y. Choquet-Bruhat, *Einstein constraints on n dimensional compact manifolds*, Class. Quantum Grav. **21** (2004), S127–S152, gr-qc/0311029.
31. ———, *Future complete $U(1)$ symmetric Einsteinian spacetimes, the unpolarized case*, 50 years of the Cauchy problem in general relativity (P.T. Chruściel and H. Friedrich, eds.), Birkhäuser, Basel, 2004, pp. 251–298.
32. Y. Choquet-Bruhat and R. Geroch, *Global aspects of the Cauchy problem in general relativity*, Commun. Math. Phys. **14** (1969), 329–335.
33. Y. Choquet-Bruhat and V. Moncrief, *Future global in time Einsteinian spacetimes with $U(1)$ isometry group.*, Ann. Henri Poincaré **2** (2001), 1007–1064.
34. D. Christodoulou and S. Klainermann, *On the global nonlinear stability of Minkowski space*, Princeton University Press, Princeton, 1995.
35. D. Christodoulou and A. Shadi Tahvildar-Zadeh, *On the asymptotic behavior of spherically symmetric wave maps*, Duke Math. Jour. **71** (1993), 31–69.
36. ———, *On the regularity of spherically symmetric wave maps*, Commun. Pure Appl. Math **46** (1993), 1041–1091.
37. P.T. Chruściel, *On the relation between the Einstein and the Komar expressions for the energy of the gravitational field*, Ann. Inst. H. Poincare (1985), 267–282.
38. P.T. Chruściel, *On space–times with $U(1)\times U(1)$ symmetric compact Cauchy surfaces*, Ann. of Phys. (NY) **202** (1990), 100–150.
39. P.T. Chruściel, *On uniqueness in the large of solutions of Einstein equations ("Strong Cosmic Censorship")*, Australian National University Press, Canberra, 1991.
40. ———, *On uniqueness in the large of solutions of Einstein equations ("Strong Cosmic Censorship")*, Cont. Math. **132** (1992), 235–273.
41. P.T. Chruściel and E. Delay, *Existence of non-trivial asymptotically simple vacuum space-times*, Class. Quantum Grav. **19** (2002), L71–L79, gr-qc/0203053, erratum-ibid, 3389.
42. ———, *On mapping properties of the general relativistic constraints operator in weighted function spaces, with applications*, Mém. Soc. Math. de France. **94** (2003), 1–103, gr-qc/0301073v2.
43. ———, *Manifold structures for sets of solutions of the general relativistic constraint equations*, Jour. Geom Phys. (2004), 442–472, gr-qc/0309001v2.
44. P.T. Chruściel and M. Herzlich, *The mass of asymptotically hyperbolic Riemannian manifolds*, Pacific Jour. Math. **212** (2003), 231–264, dg-ga/0110035.
45. P.T. Chruściel and J. Isenberg, *Non–isometric vacuum extensions of vacuum*

maximal globally hyperbolic space–times, Phys. Rev. **D48** (1993), 1616–1628.
46. P.T. Chruściel, J. Isenberg, and V. Moncrief, *Strong cosmic censorship in polarized Gowdy space–times*, Class. Quantum Grav. **7** (1990), 1671–1680.
47. P.T. Chruściel, J. Isenberg, and D. Pollack, *Gluing initial data sets for general relativity*, Phys. Rev. Lett. **93** (2004), 081101, gr-qc/0409047.
48. ______, *Initial data engineering*, Comm. Math. Phys. **257** (2005), 29-42, gr-qc/0403066.
49. P.T. Chruściel, J. Jezierski, and J. Kijowski, *Hamiltonian field theory in the radiating regime*, Lect. Notes in Physics, vol. m70, Springer, Berlin, Heidelberg, New York, 2001, URL `http://www.phys.univ-tours.fr/~piotr/papers/hamiltonian_structure`.
50. P.T. Chruściel, J. Jezierski, and S. Łęski, *The Trautman-Bondi mass of hyperboloidal initial data sets*, Adv. Theor. Math. Phys. **8** (2004), 83–139, gr-qc/0307109.
51. P.T. Chruściel and K. Lake, *Cauchy horizons in Gowdy space times*, Class. Quantum Grav. **21** (2004), S153–S170, gr-qc/0307088.
52. P.T. Chruściel and G. Nagy, *The mass of spacelike hypersurfaces in asymptotically anti-de Sitter space-times*, Adv. Theor. Math. Phys. **5** (2002), 697–754, gr-qc/0110014.
53. P.T. Chruściel and A. Rendall, *Strong cosmic censorship in vacuum space–times with compact, locally homogeneous Cauchy surfaces*, Annals of Phys. **242** (1995), 349–385.
54. J. Corvino, *Scalar curvature deformation and a gluing construction for the Einstein constraint equations*, Commun. Math. Phys. **214** (2000), 137–189.
55. J. Corvino and R. Schoen, *On the asymptotics for the vacuum Einstein constraint equations*, 2003, gr-qc/0301071.
56. C. Cutler and R.M. Wald, *Existence of radiating Einstein-Maxwell solutions which are C^∞ on all of $\mathscr{I}^+$ and $\mathscr{I}^-$*, Class. Quantum Grav. **6** (1989), 453–466.
57. S. Dain, *Initial data for stationary space-times near space-like infinity*, Class. Quantum Grav. **18** (2001), 4329–4338, gr-qc/0107018.
58. ______, *Trapped surfaces as boundaries for the constraint equations*, Class. Quantum Grav. **21** (2004), 555–574, gr-qc/0308009.
59. T. Damour and B. Schmidt, *Reliability of perturbation theory in general relativity*, Jour. Math. Phys. **31** (1990), 2441–2453.
60. A.E. Fischer and J.E. Marsden, *Topics in the dynamics of general relativity*, Structure of Isolated Gravitating Systems (J. Ehlers, ed.), 1979, pp. 322–395.
61. J. Frauendiener, *On the Penrose inequality*, Phys. Rev. Lett. **87** (2001), 101101, gr-qc/0105093.
62. H. Friedrich, lectures given at the Cargèse Summer School, August 2002, online at `http://fanfreluche.math.univ-tours.fr`.
63. ______, *On the existence of n–geodesically complete or future complete solutions of Einstein's field equations with smooth asymptotic structure*, Commun. Math. Phys. **107** (1986), 587–609.
64. ______, *Smoothness at null infinity and the structure of initial data*, 50 years

of the Cauchy problem in general relativity (P.T. Chruściel and H. Friedrich, eds.), Birkhäuser, Basel, 2004, pp. 121–203, gr–qc/0304003.

65. H. Friedrich and G. Nagy, *The initial boundary value problem for Einstein's vacuum field equation*, Commun. Math. Phys. **201** (1998), 619–655.
66. G.J. Galloway, S. Surya, and E. Woolgar, *Non-existence of black holes in certain $\Lambda < 0$ spacetimes*, Class. Quantum Grav. **20** (2003), 1635–1648, gr-qc/0212079.
67. ———, *On the geometry and mass of static, asymptotically AdS spacetimes, and the uniqueness of the AdS soliton*, Commun. Math. Phys. **241** (2003), 1–25, hep-th/0204081.
68. S.W. Hawking and G.F.R. Ellis, *The large scale structure of space-time*, Cambridge University Press, Cambridge, 1973.
69. T.J.R. Hughes, T. Kato, and J.E. Marsden, *Well-posed quasi-linear second-order hyperbolic systems with applications to nonlinear elastodynamics and general relativity*, Arch. Rat. Mech. Anal. **63** (1977), 273–294.
70. G. Huisken and T. Ilmanen, *The inverse mean curvature flow and the Riemannian Penrose inequality*, Jour. Diff. Geom. **59** (2001), 353–437, URL `http://www.math.nwu.edu/~ilmanen`.
71. J. Isenberg, R. Mazzeo, and D. Pollack, *Gluing and wormholes for the Einstein constraint equations*, Commun. Math. Phys. **231** (2002), 529–568, gr-qc/0109045.
72. ———, *On the topology of vacuum spacetimes*, Annales Henri Poincaré **4** (2003), 369–383, gr-qc/0206034.
73. J. Isenberg and V. Moncrief, *Asymptotic behavior of the gravitational field and the nature of singularities in Gowdy spacetimes*, Ann. Phys. (N.Y.) **199** (1990), 84–122.
74. J. Isenberg and M. Weaver, *On the area of the symmetry orbits in T^2 symmetric spacetimes*, Class. Quantum Grav. **20** (2003), 3783–3796, gr-qc/0304019.
75. S. Kichenassamy and A. Rendall, *Analytic description of singularities in Gowdy spacetimes*, Class. Quantum Grav. **15** (1998), 1339–1355.
76. J. Kijowski, *On a new variational principle in general relativity and the energy of the gravitational field*, Gen. Rel. Grav. **9** (1978), 857–877.
77. ———, *Unconstrained degrees of freedom of gravitational field and the positivity of gravitational energy*, Gravitation, geometry and relativistic physics (Aussois, 1984), Springer, Berlin, 1984, pp. 40–50.
78. ———, *A simple derivation of canonical structure and quasi-local Hamiltonians in general relativity*, Gen. Rel. Grav. **29** (1997), 307–343.
79. J. Kijowski and W.M. Tulczyjew, *A symplectic framework for field theories*, Lecture Notes in Physics, vol. 107, Springer, New York, Heidelberg, Berlin, 1979.
80. S. Klainerman and F. Nicolò, *The evolution problem in general relativity*, Progress in Mathematical Physics, vol. 25, Birkhäuser, Boston, MA, 2003.
81. ———, *Peeling properties of asymptotically flat solutions to the Einstein vacuum equations*, Classical Quantum Gravity **20** (2003), 3215–3257.
82. S. Klainerman and I. Rodnianski, *The causal structure of microlocalized*

Einstein metrics, (2001), math.ap/0109174.
83. ———, *Rough solution for the Einstein vacuum equations*, (2001), math.ap/0109173.
84. ———, *Causal geometry of Einstein-vacuum spacetimes with finite curvature flux*, (2003), math.ap/0308123.
85. J. Klenk, *Existence of stationary vacuum solutions of Einstein's equations in an exterior domain*, Jour. Aust. Math. Soc., Ser. B **41** (1999), 231–247.
86. J.A. Valiente Kroon, *Does asymptotic simplicity allow for radiation near spatial infinity?*, Commun. Math. Phys. **251** (2004) 211-234, gr-qc/0309016.
87. ———, *On the nonexistence of conformally flat slices in the kerr and other stationary spacetimes*, (2003), gr-qc/0310048.
88. ———, *A new class of obstructions to the smoothness of null infinity*, Commun. Math. Phys. **244** (2004), 133–156, gr-qc/0211024.
89. ———, *Time asymmetric spacetimes near null and spatial infinity. I. Expansions of developments of conformally flat data*, Class. Quantum Grav. **21** (2004) 5457-5492, gr-qc/0408062.
90. H. Lindblad and I. Rodnianski, *Global existence for the Einstein vacuum equations in wave coordinates*, Commun. Math. Phys. **256** (2005) 43-110, math.ap/0312479.
91. C.M. Liu and S.-T. Yau, *New definition of quasilocal mass and its positivity*, Phys. Rev. Lett. **90** (2003) 231102, gr-qc/0303019.
92. E. Malec, M. Mars, and W. Simon, *On the Penrose inequality for general horizons*, Phys. Rev. Lett. **88** (2002), 121102, gr-qc/0201024.
93. E. Malec and N. Ó Murchadha, *The Jang equation, apparent horizons, and the Penrose inequality*, Class. Quantum Grav. **21** (2004) 5777-5788, gr-qc/0408044.
94. D. Maxwell, *Solutions of the Einstein constraint equations with apparent horizon boundary*, Commun. Math. Phys. **253** (2004) 561-583, gr-qc/0307117.
95. ———, *Rough solutions of the Einstein constraint equations*, (2004), gr-qc/0405088.
96. C.W. Misner, *Taub–NUT space as a counterexample to almost anything*, Relativity Theory and Astrophysics, AMS, Providence, Rhode Island, 1967, Lectures in Appl. Math., vol. 8, pp. 160–169.
97. V. Moncrief, *Global properties of Gowdy spacetimes with $T^3 \times \mathbb{R}$ topology*, Ann. Phys. **132** (1981), 87–107.
98. V. Moncrief and D. Eardley, *The global existence problem and cosmic censorship in general relativity*, Gen. Rel. Grav. **13** (1981), 887–892.
99. L. Nirenberg, *The Weyl and Minkowski problems in differential geometry in the large*, Commun. Pure Appl. Math. **6** (1953), 337–394.
100. N. O'Murchadha, L.B. Szabados, and K.P. Tod, *A comment on Liu and Yau's positive quasi-local mass*, (2003), gr-qc/0311006.
101. R. Penrose, *Zero rest-mass fields including gravitation*, Proc. Roy. Soc. London **A284** (1965), 159–203.
102. A.V. Pogorelov, *Izgibanie vypuklyh poverhnosteĭ*, Gosudarstv. Izdat. Tehn.-Teor. Lit., Moscow-Leningrad, 1951.

103. J. Qing, *On the uniqueness of AdS spacetime in higher dimensions*, Annales Henri Poincaré **5** (2004), 245–260, math.DG/0310281.
104. A.D. Rendall, *Local and global existence theorems for the Einstein equations*, Living Reviews in Relativity **1** (1998), URL `http://www.livingreviews.org`.
105. ———, *Fuchsian analysis of singularities in Gowdy spacetimes beyond analyticity*, Class. Quantum Grav. **17** (2000), 3305–3316, gr-qc/0004044.
106. ———, *Asymptotics of solutions of the Einstein equations with positive cosmological constant*, Ann. Henri Poincaré **5** (2004) 1041-1064, gr-qc/0312020.
107. ———, *Mathematical properties of cosmological models with accelerated expansion*, (2004), gr-qc/0408053.
108. A.D. Rendall and M. Weaver, *Manufacture of Gowdy spacetimes with spikes*, Class. Quantum Grav. **18** (2001), 2959–2975, gr-qc/0103102.
109. O. Reula, *On existence and behaviour of asymptotically flat solutions to the stationary Einstein equations*, Commun. Math. Phys. **122** (1989), 615–624.
110. H. Ringström, *Lecture at the Miami Waves conference*, January 2004.
111. R. Schoen, *Lectures at the Miami Waves conference*, January 2004.
112. R. Schoen and S.-T. Yau, *On the proof of the positive mass conjecture in general relativity*, Commun. Math. Phys. **65** (1979), 45–76.
113. ———, *Proof of the positive mass theorem II*, Commun. Math. Phys. **79** (1981), 231–260.
114. Y. Shi and L.-F. Tam, *Positive mass theorem and the boundary behaviors of compact manifolds with nonnegative scalar curvature*, Jour. Diff. Geom. **62** (2002), 79–125, math.DG/0301047.
115. Y. Shi and G. Tian, *Rigidity of asymptotically hyperboblic manifolds*, (2002), math.DG/0402358.
116. L. Szabados, *Quasi-local energy-momentum and angular momentum in GR: A review article*, Living Rev. **4** (2004), URL `http://relativity.livingreviews.org/Articles/lrr-2004-4`.
117. D. Tataru, *Nonlinear wave equations*, Proceedings of the International Congress of Mathematicians, Vol. III (Beijing, 2002) (Beijing), Higher Ed. Press, 2002, pp. 209–220.
118. M.E. Taylor, *Partial differential equations III*, Applied Mathematical Sciences, vol. 117, Springer, 1996.
119. A. Trautman, *Conservation laws in general relativity*, Gravitation. An introduction to current research (Witten, L., ed.), John Wiley and Sons, New York and London, 1962.
120. X. Wang, *Mass for asymptotically hyperbolic manifolds*, Jour. Diff. Geom. **57** (2001), 273–299.
121. G. Weinstein and S. Yamada, *On a Penrose inequality with charge*, Commun. Math. Phys. **257** (2005) 703-723, math.dg/0405602.

BLACK HOLES AND THE INFORMATION PARADOX

STEPHEN HAWKING
DAMTP, Cambridge University, UK

The Euclidean path integral over all topologically trivial metrics can be done by time slicing and so is unitary when analytically continued to the Lorentzian. On the other hand, the path integral over all topologically non-trivial metrics is asymptotically independent of the initial state. Thus the total path integral is unitary and information is not lost in the formation and evaporation of black holes. The way the information gets out seems to be that a true event horizon never forms, just an apparent horizon.

I want to report that I think I have solved a major problem in theoretical physics, a problem that has been around since I discovered that black holes radiate thermally, thirty years ago. The question is: "is information lost in black hole evaporation?" If it is, the evolution is not unitary and pure quantum states decay into mixed states.

I am grateful to my graduate student Christophe Galfard for help in preparing this talk.

The black hole information paradox started in 1967 when Werner Israel showed that the Schwarzschild metric was the only static vacuum black hole solution. This was then generalized to the no-hair theorem: the only stationary rotating black hole solutions of the Einstein Maxwell equations are the Kerr-Newman metrics. The no-hair theorem implied that all information about the collapsing body was lost from the outside region apart from three conserved quantities: the mass, the angular momentum, and the electric charge.

This loss of information wasn't a problem in the classical theory. A classical black hole would last forever and the information could be thought of as preserved inside it but just not very accessible. However, the situation changed when I discovered that quantum effects would cause a black hole to radiate at a steady rate. At least in the approximation I was using, the radiation from the black hole would be completely thermal and would carry no information. So what would happen to all that information locked inside a black hole that evaporated away and disappeared completely? It seemed the only way the information could come out would be if the radiation was not exactly thermal but had subtle correlations. No one has found a mechanism to produce correlations but most physicists believe one must exist. If information were lost in black holes, pure quantum states would decay into mixed states and quantum gravity wouldn't be unitary.

I first raised the question of information loss in 1975 and the argument continued for years without any resolution either way. Finally, it was claimed that the issue was settled in favour of conservation of information by AdS/CFT. AdS/CFT is a conjectured duality between supergravity in anti-de Sitter space and a conformal field theory on the boundary of anti-de Sitter space, at infinity. Since the conformal field theory is manifestly unitary, the argument is that supergravity must be information preserving. Any information that falls in a black hole in anti-de Sitter space must come out again. But it still wasn't clear how information could get out of a black hole. It is this question I will address.

Black hole formation and evaporation can be thought of as a scattering process. One sends in particles and radiation from infinity and measures what comes back out to infinity. All measurements are made at infinity, where fields are weak, and one never probes the strong field region in the middle. So one can't be sure a black hole forms, no matter how certain it might be in classical theory. I shall show that this possibility allows information to be preserved and to be returned to infinity.

I adopt the Euclidean approach, the only sane way to do quantum gravity non-perturbatively. In this, the time evolution of an initial state is given by a path integral over all positive definite metrics that go between two surfaces that are a distance T apart at infinity. One then Wick rotates the time interval, T, to the Lorentzian.

The path integral is taken over metrics of all possible topologies that fit in between the surfaces. There is the trivial topology: the initial surface cross the time interval. Then there are the non-trivial topologies, all the other possible topologies. The trivial topology can be foliated by a family of surfaces of constant time. The path integral over all metrics with trivial topology can be treated canonically by time slicing. In other words, the time evolution (including gravity) will be generated by a Hamiltonian. This will give a unitary mapping from the initial surface to the final.

The non-trivial topologies can not be foliated by a family of surfaces of constant time. There will be a fixed point in any time evolution vector field on a non-trivial topology. A fixed point in the Euclidean regime corresponds to a horizon in the Lorentzian. A small change in the state on the initial surface would propagate as a linear wave on the background of each metric in the path integral. If the background contained a horizon, the wave would fall through it, and would decay exponentially at late time outside the horizon. For example, correlation functions decay exponentially in black hole metrics. This means the path integral over all topologically non-trivial metrics will be independent of the state on the initial surface. It will not add to the amplitude to go from initial state to final that comes from the path integral over all topologically trivial metrics. So the mapping from initial to final states, given by the path integral over all metrics, will be unitary.

One might question the use, in this argument, of the concept of a quantum state for the gravitational field on an initial or final space-like surface. This would be a functional of the geometries of space-like surfaces, which is not something that can be measured in weak fields near infinity. One can measure the weak gravitational fields on a time-like tube around the system, but the caps at top and bottom go through the interior of the system where the fields may be strong.

One way of getting rid of the difficulties of caps would be to join the final surface back to the initial surface and integrate over all spatial geometries of the join. If this was an identification under a Lorentzian time interval, T, at infinity, it would introduce closed time like curves. But if the interval at infinity is the Euclidean distance, β, the path integral gives the partition function for gravity at temperature $1/\beta$.

The partition function of a system is the trace over all states weighted with $\exp(-\beta H)$, H being the Hamiltonian of the system. One can then integrate β along a contour parallel to the imaginary axis, with the factor $\exp(\beta E_0)$. This projects out the states with energy E_0. In a gravitational collapse and evaporation, one is interested in states of definite energy rather than states of definite temperature.

There is an infrared problem with this idea for asymptotically flat space. The Euclidean path integral with period β is the partition function for space at temperature $1/\beta$. The partition function is infinite because the volume of space is infinite. This infrared problem can be solved by a small negative cosmological constant. It will not affect the evaporation of a small black hole but it will change infinity to anti de Sitter space and make the thermal partition function finite.

The boundary at infinity is then a torus, $S^1 \times S^2$. The trivial topology, periodically identified anti de Sitter space, fills in the torus, but so also do non-trivial topologies, the best known of which is Schwarzschild anti-de Sitter. Providing that the temperature is small compared to the Hawking-Page temperature, the path integral over all topologically trivial metrics represents self-gravitating radiation in asymptotically anti-de Sitter space. The path integral over all metrics of Schwarzschild AdS topology represents a black hole and thermal radiation in asymptotically anti de Sitter.

The boundary at infinity has topology $S^1 \times S^2$. The simplest topology that fits inside that boundary is the trivial topology $S^1 \times D_3$, the three-disk. The next simplest topology, and the first non-trivial topology, is $S^2 \times D_2$. This is the topology of the Schwarzschild anti-de Sitter metric. There are other possible topologies that fit inside the boundary but these two are the important cases: topologically trivial metrics and the black hole. The black hole is eternal. It can not become topologically trivial at late times.

In view of this, one can understand why information is preserved in topologically trivial metrics but exponentially decays in topologically non-trivial metrics. A final state of empty space without a black hole would be topologically trivial and be foliated by surfaces of constant time. These would form a three cycle modulo the boundary at infinity. Any global symmetry would lead to conserved global charges on that three cycle. These would prevent correlation functions from decaying exponentially in topologically trivial metrics. Indeed, one can regard the unitary Hamiltonian evolution of a topologically trivial metric as the conservation of information through a three cycle.

On the other hand, a non-trivial topology, like a black hole, will not have a final three cycle. It will not therefore have any conserved quantity that will prevent correlation functions from exponentially decaying. One is thus led to the remarkable result that late time amplitudes of the path integral over a topologically non-trivial metric are independent of the initial state. This was noticed by Maldacena [1] in the case of asymptotically AdS_3 and interpreted as implying that information is lost in the BTZ black hole metric. Maldacena was able to show that topologically trivial metrics have correlation functions that do not decay and have amplitudes of the right order to be compatible with a unitary evolution. Maldacena did not realize, however, that it follows from a canonical treatment that the evolution of a topologically trivial metric will be unitary.

So in the end everyone was right, in a way. Information is lost in topologically non-trivial metrics, like the eternal black hole. On the other hand, information is preserved in topologically trivial metrics. The confusion and paradox arose because people thought classically, in terms of a single topology for spacetime. It was either $\mathbb{R}^4$ or a black hole. But the Feynman sum over histories allows it to be both at once. One can not tell which topology contributed the observation any more than one can tell which slit the electron went through in the two slits experiment. All that observation at infinity can determine is that there is a unitary mapping from initial states to final, and that information is not lost.

My work with Jim Hartle showed the radiation could be thought of as tunnelling out from inside the black hole. It was therefore not unreasonable to suppose that it could carry information out of the black hole. This explains how a black hole can form, and then give out the information about what is inside it, while remaining topologically trivial. There is no baby universe branching off as I once thought. The information remains firmly in our universe. I'm sorry to disappoint science fiction fans, but if information is preserved, there is no possibility of using black holes to travel to other universes. If you jump into a black hole, your mass-energy will be returned to our universe, but in a mangled form which contains the information about what you were like, but in an unrecognisable state.

There is a problem describing what happens because strictly speaking, the only observables in quantum gravity are the values of the field at infinity. One can not define the field at some point in the middle because there is quantum uncertainty in where the measurement is done. However, in cases in which there are a large number, N, of light matter fields coupled to gravity, one can neglect the gravitational fluctuations because they are only one among N quantum loops. One can then do the path integral over all matter fields in a given metric to obtain the effective action, which will be a functional of the metric.

One can add the classical Einstein-Hilbert action of the metric to this quantum effective action of the matter fields. If one integrated this combined action over all metrics, one would obtain the full quantum theory. However, the semi-classical approximation is to represent the integral over metrics by its saddle point. This will obey the Einstein equations, where the source is the expectation value of the energy momentum tensor of the matter fields in their vacuum state.

The only way to calculate the effective action of the matter fields used to be perturbation theory. This is not likely to work in the case of gravitational collapse. However, fortunately, we now have a non-perturbative method in AdS/CFT. The Maldacena conjecture says that the effective action of a CFT on a background metric is equal to the supergravity effective action of anti- de Sitter space with that background metric at infinity. In the large N

limit, the supergravity effective action is just the classical action. Thus the calculation of the quantum effective action of the matter fields is equivalent to solving the classical Einstein equations.

The action of an anti-de Sitter-like space, with a boundary at infinity, would be infinite, so one has to regularize. One introduces subtractions that depend only on the metric of the boundary. The first counter term is proportional to the volume of the boundary. The second counter term is proportional to the Einstein-Hilbert action of the boundary. There is a third counter term, but it is not covariantly defined.

One now adds the Einstein-Hilbert action of the boundary and looks for a saddle point of the total action. This will involve solving the coupled four and five dimensional Einstein equations. It will probably have to be done numerically.

In this talk, I have argued that quantum gravity is unitary and information is preserved in black hole formation and evaporation. I assume the evolution is given by a Euclidean path integral over metrics of all topologies. The integral over topologically trivial metrics can be done by dividing the time interval into thin slices and using a linear interpolation to the metric in each slice. The integral over each slice will be unitary, and so the whole path integral will be unitary.

On the other hand, the path integral over topologically non-trivial metrics, will lose information and will be asymptotically independent of its initial conditions. Thus the total path integral will be unitary and quantum mechanics is safe.

It is great to solve a problem that has been troubling me for nearly thirty years, even though the answer is less exciting than the alternative I suggested. This result is not all negative however, because it indicates that a black hole evaporates while remaining topologically trivial. However, the large N solution is likely to be a black hole that shrinks to zero. This is what I suggested in 1975.

In 1997, Kip Thorne and I bet John Preskill that information was lost in black holes. The loser or losers of the bet are to provide the winner or winners with an encyclopaedia of their own choice from which information can be recovered with ease. I'm now ready to concede the bet but Kip Thorne isn't convinced just yet. I will give John Preskill the encyclopaedia he has requested. John is all American, so naturally he wants an encyclopaedia of baseball. I had great difficulty in finding one over here, so I offered him an encyclopaedia of cricket, as an alternative, but John wouldn't be persuaded of the superiority of cricket. Fortunately, my assistant, Andrew

Dunn, persuaded the publishers Sportclassic Books, to fly a copy of Total Baseball, The Ultimate Baseball Encyclopedia to Dublin. I will give John the encyclopaedia now. If Kip agrees to concede the bet later, he can pay me back.

References

1. Juan M. Maldacena, JHEP **0109** (2001) 024, hep-th/0106112

GRAVITATION AND COSMOLOGY IN BRANE-WORLDS

DAVID LANGLOIS
GRECO, Institut d'Astrophysique de Paris,
98bis Boulevard Arago, 75014 Paris, France
e-mail:langlois@iap.fr

In brane-worlds, our universe is assumed to be a submanifold, or brane, embedded in a higher-dimensional bulk spacetime. Focusing on scenarios with a curved five-dimensional bulk spacetime, I discuss their gravitational and cosmological properties.

1. Introduction

In this contribution, I review the basic ingredients of the so-called brane-world models. These models are based on the assumption that our universe is a *brane*: a sub-space embedded in a higher dimensional *bulk* spacetime. In contrast with the traditional Kaluza-Klein treatment of extra dimensions, ordinary matter fields are *confined* on the brane. As a consequence, the size of the extra-dimensions in brane-world models is not limited by the usual bound, typically $R \lesssim (1\mathrm{TeV})^{-1}$, based on the non-detection of Kaluza-Klein modes in collider experiments up to energies close to 1 TeV. In brane-worlds, only gravity propagates in the higher-dimensional spacetime and, therefore, the size of the extra-dimensions can be much larger, up to a fraction of a millimeter, corresponding to the current limit of gravity experiments that look for deviations of Newton's law [1].

Although there were some precursor works on brane-worlds, the huge interest they have encountered in the last few years is due to recent developments in string/M-theory. This has in turn opened up new avenues to tackle some fundamental problems.

For example, the fact that the four-dimensional Planck mass M_p is in this context only a "projection" of the higher-dimensional (fundamental) Planck mass M_*, which can thus be lower than M_p, offers a new perspective on the hierarchy problem and suggests the possibility that quantum gravity might be closer than previously thought. This has been emphasized by Arkani-Hamed, Dimopoulos and Dvali [2], who suggested a very simple model based on a brane embedded in a flat geometry with $4+n$ dimensions, where n dimensions are compactified on a torus of size R. Using Gauss' theorem, one can easily show that, on scales $r \gg R$, standard 4D gravity is recovered with $M_P^2 = M_*^{2+n} R^n$. On scales $r \ll R$, gravity becomes $(4+n)$-dimensional.

Another important progress was made by Randall and Sundrum [3, 4], who considered *curved*, or *warped*, bulk geometries. They showed, in particular, that *compact* extra-dimensions are not necessary to obtain a four-dimensional behaviour. The bulk curvature can indeed lead to an *effective compactification.*

In this contribution, I will focus my attention on brane-worlds characterized by a single extra dimension where the bulk space-time is *curved* instead of flat and where the self-gravity of the brane is taken into account. This includes the configurations discussed by Randall and Sundrum.

The outline is the following. In the next section, I will present in some detail the Randall-Sundrum models and discuss the corresponding effective gravity on the brane. Then, in the subsequent sections, I will turn my attention to cosmological brane models, reviewing a few important topics in brane cosmology. In section 3, the bulk and brane geometries will be presented. Section 4 will be devoted to the so-called dark radiation, or Weyl radiation. Inflation in the brane will be discussed in section 5. Section 6 will discuss some aspects of brane cosmology when the assumption of isotropy or homogeneity is relaxed. Finally, section 7 will provide some conclusions and perspectives.

Only some aspects of brane-worlds are discussed in this contribution. Much more can be learnt on this subject from several detailed reviews [5].

2. Gravity in brane-worlds

With extra-dimensions, gravity is intrinsically different from 4D gravity and the models must be able to mimic standard gravity in the regimes tested by experiments. The standard approach is to compactify the extra-dimensions on a size smaller than the scales probed by gravity experiments. However, as shown by Randall and Sundrum a few years ago, the extra-dimension can remain infinite, the "effective compactification" being the consequence of the bulk spacetime curvature.

2.1. *The Randall-Sundrum model*

The (second) Randall-Sundrum [4] model is based on the following ingredients

- a five-dimensional bulk spacetime, empty, but endowed with a negative cosmological constant
$$\Lambda = -\frac{6}{\ell^2}, \tag{1}$$
- a self-gravitating brane, which represents our world, endowed with a tension σ, and assumed to be Z_2-symmetric.

The five-dimensional Einstein equations are given by

$$G_{AB} + \Lambda g_{AB} = \kappa^2 T_{AB}, \tag{2}$$

where κ^2 is the gravitational coupling, and the corresponding five-dimensional Planck mass, M_5, is defined by

$$\kappa^2 = M_5^{-3}. \tag{3}$$

The bulk being empty, only the brane contributes to the energy-momentum tensor T_{AB}. There are two equivalent ways of solving Einstein's equation. Either one solves it directly by taking into account the presence of the brane, assumed to be infinitely thin along the extra-dimension, in the form of a *distributional* energy-momentum tensor. Or, one solves first the *vacuum* Einstein equations, i.e. setting the right hand side to zero, and *afterwards*, one takes into account the brane by imposing appropriate *junction conditions* at the spacetime boundary where the brane is located. These boundary conditions are the generalization, to five dimensions, of the so-called Israel (-Darmois) junction conditions and read

$$[K_{AB}] = -\kappa^2 \left(T_{AB} - \frac{T}{3} h_{AB} \right). \tag{4}$$

They relate the jump, between the two sides of the brane, of the extrinsic curvature tensor, defined by $K_{AB} \equiv h_A^C D_C n_B$ (where n^A is the unit vector normal to the brane and $h_{AB} = g_{AB} - n_A n_B$ is the induced metric on the brane), to the brane energy-momentum tensor. For a Z_2 symmetric brane, the jump of the extrinsic curvature is simply twice the value of the extrinsic curvature on one side of the brane.

Provided the tension satisfies the *constraint*

$$\Lambda + \frac{\kappa^4}{6}\sigma^2 = 0, \tag{5}$$

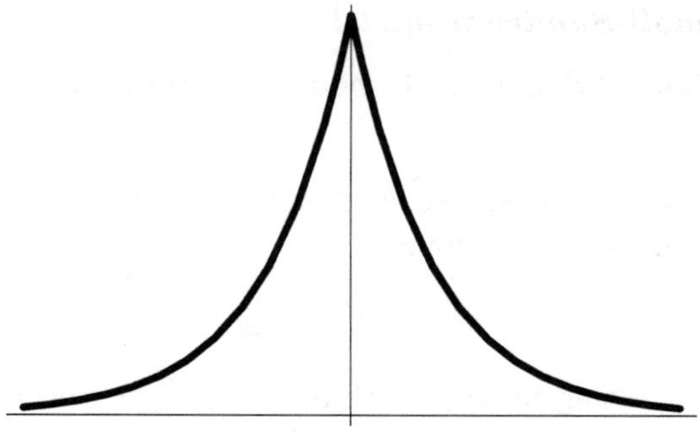

Figure 1. Warping factor $a(y)$.

which implies in particular that $\sigma = 6M^3/\ell$, it can be shown that the five-dimensional Einstein equations admit the following *static* solution

$$ds^2 = a^2(y)\eta_{\mu\nu}dx^\mu dx^\nu + dy^2, \tag{6}$$

where $\eta_{\mu\nu}$ is the usual Minkowski metric and $a(y)$ is a *warping* scale factor, whose explicit dependence on y is given by

$$a(y) = e^{-|y|/\ell}, \tag{7}$$

as shown on Fig. 1. Here, the brane is located at $y = 0$ and the Z_2 symmetry means that y with $-y$ are identified. This bulk solution (6-7) can also be interpreted as two identical portions of AdS (Anti-de Sitter) spacetime glued together at the brane location.

2.2. *Gravity in the Randall-Sundrum model*

Let us now investigate the effective gravity for this model, as measured by an observer located on the brane. A first, and rather simple, step is to compute the effective four-dimensional Planck mass. This can be done by substituting in the five-dimensional Einstein-Hilbert action,

$$S_{\rm grav} = \frac{M^3}{2}\int d^4x\, dy\, \sqrt{-g}\, R, \tag{8}$$

the metric (6) and by integrating over the extra-dimension. The factor in front of the resulting four-dimensional Einstein-Hilbert action (for $\eta_{\mu\nu}$) gives

$$M_{Pl}^2 = M^3 \int_{-\infty}^{+\infty} dy\; a^2(y) = M^3\ell. \tag{9}$$

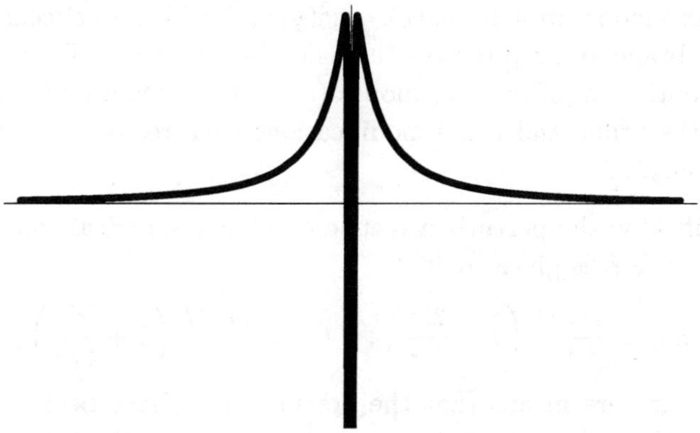

Figure 2. Effective potential of the Schrödinger-like equation that governs the dependence on the fifth dimension.

It is important to emphasize that the extra-dimension extends here to infinity. In the absence of the warping factor $a(y)$ this would lead to an infinite four-dimensional Planck mass. The warping of the extra-dimension, governed by the AdS lengthscale ℓ, thus leads to an *effective compactification*, even if the extra-dimension is infinite.

To explore further the gravitational behaviour and derive for example the effective potential of a point mass located on the brane, one must study the perturbations about the background metric (6). Perturbing the metric, $g_{AB} = \bar{g}_{AB} + h_{AB}$, and working in the gauge $h_{yy} = 0$, $h_{y\mu} = 0$, $h^\mu_\mu = 0$, $\partial_\mu h^\mu_\nu = 0$, one finds that the linearized Einstein equations reduce to

$$\left(a^{-2}\partial^2_{(4)} + \partial^2_y - \frac{4}{\ell^2} + \frac{4}{\ell}\delta(y)\right) h_{\mu\nu} = 0. \tag{10}$$

This equation is separable and the solutions can be written as the superposition of eigenmodes $h(x^\mu, y) = u_m(y)e^{ip_\mu x^\mu}$, with $p_\mu p^\mu = -m^2$. This implies that the dependence on the fifth dimension of the massive modes is governed by the *Schrödinger-like equation*:

$$\frac{d^2\psi_m}{dz^2} - V(z)\psi_m = -m^2\psi_m, \qquad V(z) = \frac{15}{4(|z|+\ell)^2} - \frac{3}{\ell}\delta(z), \tag{11}$$

using the function $\psi_m = a^{-1/2}u_m$ and the variable $z = \int dy/a(y)$. The potential $V(z)$, plotted in Fig. 2, is "volcano"-shaped and goes to zero at infinity.

One can divide the solutions of this Schrödinger-like equation into:

- a zero mode ($m = 0$), $u_0(y) = a^2(y)/\sqrt{\ell}$, which is concentrated near the brane and reproduces the usual behaviour of 4D gravity;
- a continuum of massive modes ($m > 0$), which are weakly coupled to the brane and bring modifications with respect to standard 4D gravity.

More specifically, the perturbed metric outside a spherical source of mass M, and for $r \gg \ell$, is given by [6]

$$\bar{h}_{00} \simeq \frac{2GM}{r}\left(1 + \frac{2\ell^2}{3r^2}\right), \quad \bar{h}_{ij} \simeq \frac{2GM}{r}\left(1 + \frac{\ell^2}{3r^2}\right), \tag{12}$$

where the bar here means that the perturbations have been rewritten in a Gaussian Normal gauge (i.e. $h_{yy} = h_{y\mu} = 0$ and the brane is located at $y = 0$) and thus correspond directly to the quantities measured on the brane. Standard gravity is thus recovered on scales $r \gg \ell$!

On scales of the order of ℓ, and below, one expects deviations from the usual Newton's law. Since gravity experiments [1] have confirmed the standard Newton's law down to scales of the order 0.1 mm, this implies

$$\ell \lesssim 0.1\text{mm}, \tag{13}$$

and thus $M_{(5)} \gtrsim 10^8$ GeV.

Although the above results apply to linearized gravity, other works, based on second order calculations or numerical gravity, have confirmed the recovery of standard gravity on scales larger than ℓ. However, the behaviour of black holes in the Randall-Sundrum model might significantly deviate from the standard picture. Indeed, inspired by the AdS/CFT correspondence, it has been conjectured that Randall-Sundrum black holes should *evaporate classically*, or, in other words, be classically unstable. The underlying argument is that the five-dimensional classical solutions should correspond to *quantum-corrected* four-dimensional black hole solutions, of a conformal field theory (CFT) coupled to gravity [7, 8]. Since there are many CFT degrees of freedom into which the black hole can radiate, its life time is shorter than for a standard black hole:

$$\tau \simeq 10^2 \left(M/M_\odot\right)^3 \left(\ell/1\text{mm}\right)^{-2} \text{years}. \tag{14}$$

2.3. *Two-brane models*

Although I have been considering so far a *single* brane (this will also be the case for most of the discussion on cosmology), it is also worth mentioning models with *two* Z_2 symmetric branes. Because of the Z_2 symmetry, the two branes can be seen as "end-of-the-world" branes and therefore the extra-dimension is explicitly compactified.

In their first model [3], Randall and Sundrum have introduced two such branes, with opposite tensions $\sigma_\pm = \pm 6M^3/\ell$, separated by a distance d. Because both branes satisfy the constraint (5), one can still get a static configuration. However, as far as gravity is concerned, there is a crucial difference between the two branes. The effective gravity for a brane observer, ignoring the corrections due the massive modes, is described by a *scalar-tensor* theory with the Brans-Dicke parameter [6]

$$\omega_{\rm BD}^{(\pm)} = \frac{3}{2}\left(e^{\pm 2d/\ell} - 1\right), \tag{15}$$

where the sign depends on which brane the observer is located (positive or negative tension respectively). If we live on the positive tension brane, this is compatible with the present constraint, $\omega_{\rm BD} > 4 \times 10^5$, provided the distance d between the two branes is large enough. The corresponding effective action is given by [9]

$$S_{\rm eff} = \int d^4x\sqrt{-g}\left[\frac{1}{2}\Psi R - \frac{\omega(\Psi)}{2\Psi}\partial^\mu\Psi\partial_\mu\Psi - \sigma_+ - \sigma_-\left(1-\Psi\right)^2\right], \tag{16}$$

with

$$\omega(\Psi) = \frac{3}{2}\frac{\Psi}{1-\Psi}. \tag{17}$$

However, if we live on the negative tension brane, as was originally assumed by Randall and Sundrum in order to solve the hierarchy problem, then the effective gravity is incompatible with reality. To rescue this scenario, one must invoke a mechanism that stabilizes the distance between the two branes, e.g. by introducing a bulk scalar field coupled to the branes [10].

3. Homogeneous brane cosmology

I now discuss the cosmology of a brane embedded in a five-dimensional bulk spacetime.

3.1. *The model*

As in standard cosmology, *homogeneity* and *isotropy* are assumed along the three ordinary spatial dimensions. One thus requires the bulk spacetime to satisfy the *cosmological symmetry*, which means that one can foliate the bulk with maximally symmetric three-dimensional surfaces. This is in complete analogy with *the spherical symmetry*, associated with (positively curved) maximally symmetric two-dimensional surfaces in a 4D spacetime.

In addition to the three ordinary spatial dimensions, spanning the homogeneous and isotropic surfaces, one introduces a time coordinate t and a spatial coordinate y for the extra dimension. The cosmological symmetry implies that the metric components depend only on t and y. It is convenient to work in a Gaussian Normal (GN) coordinate system, in which the brane is always located at $y = 0$ and the five-dimensional metric takes the form

$$ds^2 = -n^2(t,y)dt^2 + a^2(t,y)d\Sigma_k^2 + dy^2, \tag{18}$$

where $d\Sigma_k^2$ is the metric for the maximally symmetric three-surface ($k = 0, \pm 1$). Note that, in closer analogy with the spherical symmetry mentioned above, another possibility would be to choose a coordinate system where the metric reads

$$ds^2 = -n^2(t,r)dt^2 + b^2(t,r)dr^2 + r^2 d\Sigma_k^2. \tag{19}$$

To obtain the equations governing the cosmological evolution, one substitutes the ansatz (18) into the five-dimensional Einstein equations

$$G_{AB} + \Lambda g_{AB} = \kappa^2 T_{AB} \tag{20}$$

where the energy-momentum tensor, assuming a bulk otherwise empty, is due to the brane matter and thus given by

$$T_A^B = Diag(-\rho_b(t), P_b(t), P_b(t), P_b(t), 0)\delta(y), \tag{21}$$

where ρ_b and P_b are respectively the total energy density and pressure in the brane. The five-dimensional Einstein's equations can be solved explicitly [11] and one gets a solution for the metric components $n(t,y)$ and $a(t,y)$, in terms of $\rho_b(t)$ and $P_b(t)$, defined up to an integration constant.

3.2. *The cosmological evolution in the brane*

On the brane, the metric is given by

$$ds_b^2 = -n_b(t)^2 dt^2 + a_b(t)^2 d\Sigma_k^2, \qquad n_b(t) \equiv n(t,0), \quad a_b(t) \equiv a(t,0). \tag{22}$$

It can be shown that the scale factor $a_b(t)$ satisfies the *modified Friedmann equation* [11, 12]:

$$H_b^2 \equiv \frac{\dot{a}_b^2}{a_b^2} = \frac{\Lambda}{6} + \frac{\kappa^4}{36}\rho_b^2 + \frac{C}{a_b^4} - \frac{k}{a_b^2}, \tag{23}$$

where C is an integration constant. It can also be shown that, for an empty bulk, the usual conservation equation holds, which implies

$$\dot{\rho}_b + 3H_b(\rho_b + P_b) = 0. \tag{24}$$

For $\Lambda = 0$ and $C = 0$, the bulk is 5-D Minkowski and the cosmology is highly unconventional since the Hubble parameter is proportional to the brane energy density [12]. This has the unfortunate consequence to ruin the standard nucleosynthesis scenario, which is based on the evolution of the expansion rate with respect to the relevant microphysical interaction rates.

To obtain a viable brane cosmology scenario, the simplest way is to generalize the Randall-Sundrum model to cosmology [13]. In the static version of the previous section, the energy density of the "Minkowski" brane is $\rho_b = \sigma_{RS} \equiv 6M^3/\ell$. This can be generalized to a "FLRW" brane by adding to the intrinsic tension σ_{RS} the usual cosmological energy density $\rho(t)$ so that the total energy density is given by

$$\rho_b(t) = \sigma_{RS} + \rho(t). \tag{25}$$

Moreover, the bulk is assumed to be endowed with a negative cosmological constant $\Lambda < 0$, satisfying the constraint (5).

Substituting the decomposition (25) into the Friedmann equation (23), one finds

$$H_b^2 = \frac{8\pi G}{3}\rho + \frac{\kappa^4}{36}\rho^2 + \frac{C}{a_b^4} - \frac{k}{a_b^2}. \tag{26}$$

In the expansion in ρ, the constant term vanishes because of the constraint (5), whereas the coefficient of the linear term is the standard one because $8\pi G \equiv \kappa^4\sigma/6$, as implied by (5) and (9). However, the Friedmann equation (26) is characterized by two new features:

- a ρ^2 term, which dominates at high energy;
- a radiation-like term, C/a_b^4, usually called *dark radiation.*

The cosmological evolution undergoes a transition from a high energy regime, $\rho \gg \sigma$, characterized by an unconventional behaviour of the scale factor, into a low energy regime which reproduces our standard cosmology. For $C = 0$, $k = 0$ and an equation of state $w = P/\rho = const$, one can solve analytically the evolution equations and one finds

$$a(t) \propto t^{1/q}\left(1 + \frac{q\,t}{2\ell}\right)^{1/q}, \qquad q = 3(1+w). \tag{27}$$

One clearly sees the transition, at the epoch $t \sim \ell$, between the early, unconventional, evolution $a \sim t^{1/q}$ and the standard evolution $a \sim t^{2/q}$.

In order to be compatible with the nucleosynthesis scenario, the high energy regime, where the cosmological evolution is unconventional, must take place before nucleosynthesis. This requires $\sigma^{1/4} \gtrsim 1$ MeV, and since $\sigma = 6/(\kappa^2\ell) = 6M_5^6/M_P^2$, this gives the constraint $M_5 \gtrsim 10^4$ GeV. One notes that this is much less stringent than the constraint from small-scale gravity experiments, which presently require $\ell \lesssim 0.1$ mm and $M_5 \gtrsim 10^8$ GeV. As will be detailed in the next section, another observational constraint applies to the dark radiation constant C.

3.3. *Another point of view*

If, instead of the GN ansatz (18) for the metric, one starts from the metric (19), in analogy with the spherical symmetry, one recognizes the generalization of the Birkhoff theorem, which states that a vacuum spherical symmetric solution of Einstein's equation is necessarily static and gives the Schwarschild metric: the 5D vacuum cosmologically symmetric solution of 5D Einstein's equations with a (negative) cosmological constant is necessarily static and corresponds to the AdS-Schwarzschild metric in five dimensions:

$$ds^2 = -f(R)dT^2 + \frac{dR^2}{f(R)} + R^2 d\Sigma_k^2, \quad f(R) = k + \frac{R^2}{\ell^2} - \frac{C}{R^2}, \quad k = 0, \pm 1. \tag{28}$$

In this coordinate system, the brane is *moving* and the so-called junction conditions $[K_{\mu\nu}] = -\kappa^2 \left(S_{\mu\nu} - (S/3)g_{\mu\nu}\right)$ give the modified Friedmann equation obtained above [14, 15].

4. Dark radiation

So far, the bulk has been assumed to be *strictly empty*, apart from the presence of the brane. However, the fluctuations of brane matter generate bulk gravitational waves. Equivalently, the scattering of brane particles produce bulk gravitons ($\psi + \bar{\psi} \to G$). Therefore, a realistic model must take into account the presence of these bulk gravitons, which are emitted by the brane and then propagate in the bulk.

4.1. *Emission and propagation of bulk gravitons*

The rate of emission of these gravitons by the brane can be computed explicitly when the brane matter is in thermal equilibrium (with a temperature

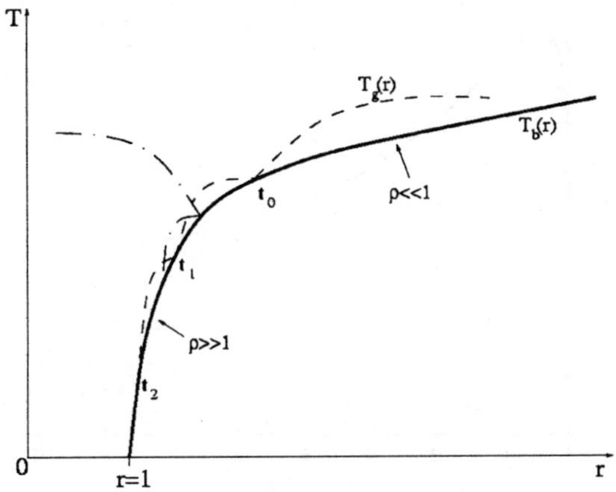

Figure 3. Examples of graviton trajectories in the bulk (dashed line). The brane trajectory, with relativistic matter, is also shown (continuous line).

T). The corresponding energy loss rate is given by [16,17]

$$\dot{\rho} + 4H\rho = -\frac{315}{512\,\pi^3}\,\hat{g}\,\kappa^2\,T^8, \tag{29}$$

with the effective number of degrees of freedom

$$\hat{g} = (2/3)g_s + 4g_v + g_f, \tag{30}$$

which is weighted sum of the scalar, vector and fermionic degrees of freedom.

After their emission, the gravitons propagate freely in the bulk where they follow geodesic trajectories. As illustrated in Fig. 3, some of these gravitons (in fact many) tend to come back onto the brane and bounce off it. All these gravitons contribute to an effective bulk energy-momentum tensor, which can be written as

$$T_{AB}^{(\mathrm{bulk})} = \int d^5p\,\delta\left(p_M p^M\right)\sqrt{-g}\,f\,p_A p_B,, \tag{31}$$

where f is the phase space distribution function.

From the 5D Einstein equations, one can derive effective 4D Einstein equations [18], which in the homogeneous case yield

- the Friedmann equation

$$H^2 = \frac{8\pi G}{3}\left[\left(1+\frac{\rho}{2\sigma}\right)\rho + \rho_{\mathrm{D}}\right], \tag{32}$$

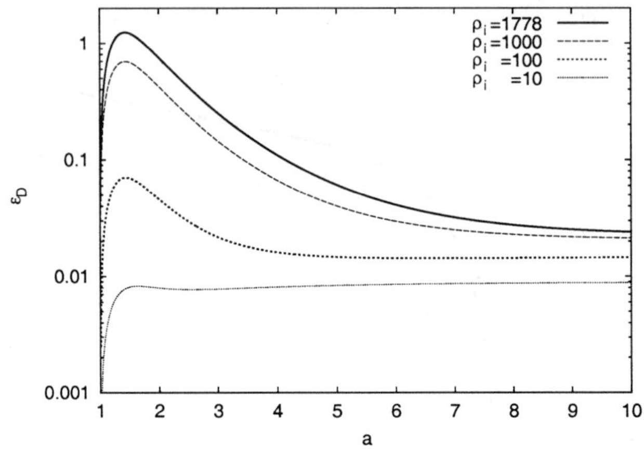

Figure 4. Evolution of the ratio $\epsilon_D = \rho_D/\rho$ for different values of the initial energy density on the brane ρ_i (in units of σ^4).

- the non-conservation equation for brane matter, which must be identified with (29),

$$\dot{\rho} + 3\,H\;(\rho + p) = 2\,T^{(\text{bulk})}{}_{RS}\,n^R\,u^S\,, \tag{33}$$

where n^A is the unit vector normal to the brane and u^A its velocity in the bulk;

- the non-conservation equation for the "dark" component ρ_D (which includes all effective contributions from the bulk):

$$\dot{\rho}_\text{D} + 4H\rho_\text{D} = -2\left(1 + \frac{\rho}{\sigma}\right) T^{(\text{bulk})}_{AB} u^A n^B - 2H\ell T^{(\text{bulk})}_{AB} n^A n^B\,. \tag{34}$$

On the right hand side of this last equation, we find two terms involving the bulk energy-momentum tensor: the first term, due to the energy flux from the brane into the bulk, contributes positively and thus increases the amount of dark radiation whereas the second term, due to the pressure along the fifth dimension, decreases the amount of dark radiation. These terms can be estimated numerically [19]. A striking property is that the gravitons coming back onto the brane and bouncing off it give a significant contribution to the transverse pressure effect, which almost, although not quite, compensates the flux effect. The evolution of the dark radiation, or rather its ratio with respect to the brane radiation density $\epsilon_D \equiv \rho_D/\rho$, is plotted on Fig. 4. One observes that at late times, i.e. far in the low energy regime, the ratio reaches a plateau, because the right hand side of (34) becomes negligible. The dark component then scales exactly like radiation.

4.2. *Observational constraints*

The computed amount of dark radiation can be confronted with observations. Indeed, since dark radiation behaves as radiation, it must satisfy the nucleosynthesis constraint on the number of *additional relativistic degrees of freedom*, usually expressed in terms of the extra number of light neutrinos ΔN_ν. The relation between ΔN_ν and ϵ_D is given by

$$\epsilon_D = \frac{7}{43}\left(\frac{g_*}{g_*^{\rm nucl}}\right)^{1/3}\Delta N_\nu, \tag{35}$$

where $g_*^{\rm nucl} = 10.75$ is the number of degrees of freedom at nucleosynthesis (in fact before the electron-positron annihilation). Assuming $g_* = 106.75$ (standard model), this gives $\epsilon_D \simeq 0.35\Delta N_\nu$. The typical constraint from nucleosynthesis $\Delta N_\nu \lesssim 0.2$ thus implies

$$\epsilon_D \equiv \frac{\rho_D}{\rho_r} \lesssim 0.03\left(\frac{g_*}{g_*^{\rm nucl}}\right)^{1/3}, \tag{36}$$

which gives $\epsilon_D \lesssim 0.09$ with the degrees of freedom of the standard model.

5. Brane inflation

In brane cosmology, the famous horizon problem is much less severe than in standard cosmology, because the gravitational horizon, associated with the signal propagation in the bulk, can be much bigger than the standard photon horizon, associated with the signal propagation on the brane [20]. However, it is still alive because the energy density on the brane is limited by the Planck limit $\rho \sim M^4$. Thus, one must still invoke inflation, although alternative ideas based on the collision of branes [21] have been actively explored (however, the generation of a quasi-scale-invariant fluctuation spectrum, as required by observations, remains problematic).

The simplest way to get inflation in the brane is to detune the brane tension from its Randall-Sundrum value (6) in order to obtain a net effective four-dimensional cosmological constant that is positive. This leads to exponential expansion on the brane. In the GN coordinate system, the metric is separable and can be written as

$$ds^2 = A(y)^2\left(-dt^2 + e^{2Ht}d\vec{x}^2\right) + dy^2, \tag{37}$$

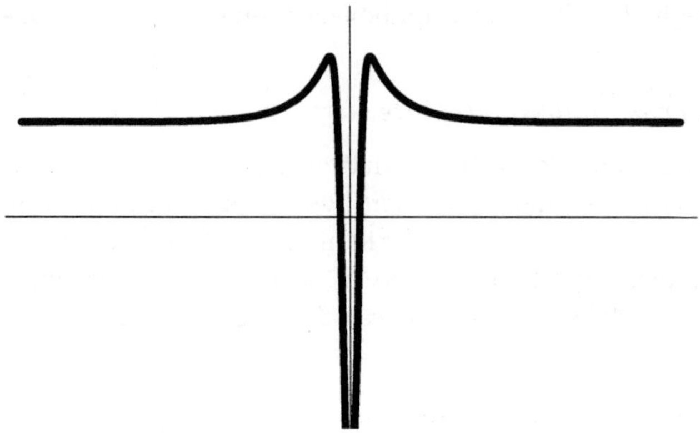

Figure 5. Potential for the graviton modes in a de Sitter brane

with

$$A(y) = \cosh \mu y - \left(1 + \frac{\rho}{\sigma}\right) \sinh \mu |y|. \tag{38}$$

As in the Randall-Sundrum case, the linearized Einstein equations for the tensor modes lead to a *separable* wave equation. The shape along the fifth dimension of the corresponding massive modes is governed by the Schrödinger-like equation

$$\frac{d^2\Psi_m}{dz^2} - V(z)\Psi_m = -m^2\Psi_m\,, \tag{39}$$

after introducing the new variable $z - z_b = \int_0^y d\tilde{y}/A(\tilde{y})$ (with $z_b = H^{-1}\sinh^{-1}(H\ell)$) and the new function $\Psi_m = A^{-1/2}u_m(y)$. The potential is given by

$$V(z) = \frac{15H^2}{4\sinh^2(Hz)} + \frac{9}{4}H^2 - \frac{3}{\ell}\left(1 + \frac{\rho}{\sigma}\right)\delta(z - z_\mathrm{b}) \tag{40}$$

and plotted in Fig. 5. In constrast with the Randall-Sundrum potential, the potential goes asympotically to the non-zero value $9H^2/4$. This indicates the presence of a gap between the zero mode ($m = 0$) and the continuum of Kaluza-Klein modes ($m > 3H/2$).

In practice, inflation is not strictly de Sitter but the de Sitter case discussed above is a good approximation when $\dot{H} \ll H^2$. To get "realistic" inflation in the brane, two main approaches have been considered: either to assume a five-dimensional scalar field which induces inflation in the brane [22], or to suppose a four-dimensional scalar confined on the brane [23].

In the latter case, the cosmological evolution during inflation is obtained by substituting the energy density $\rho_\phi = \dot{\phi}^2/2 + V(\phi)$ in the modified Friedmann equation (26). For slow-roll inflation, this can be approximated by

$$H^2 \simeq \frac{8\pi G}{3}\left(1+\frac{V}{2\sigma}\right)V. \tag{41}$$

Interestingly, because of the modified Friedmann equation, new features appear at high energy ($V > \sigma$): the slow-roll conditions are changed and, because the Hubble parameter is bigger than the standard value, yielding a higher friction on the scalar field, inflation can occur with potentials usually too steep to sustain it [24].

The scalar and tensor spectra generated during inflation driven by a brane scalar field have also been computed [23,25]. They are modified with respect to the standard results:

$$P_S = P_S^{(4D)}\left(1+\frac{V}{2\sigma}\right)^3, \qquad P_T = P_T^{(4D)} F^2(H\ell), \tag{42}$$

with

$$F(x) = \left\{\sqrt{1+x^2} - x^2 \ln\left[\frac{1}{x} + \sqrt{1+\frac{1}{x^2}}\right]\right\}^{-1/2}. \tag{43}$$

$F \simeq 1$ at low energies, i.e. for $H\ell \ll 1$, whereas $F \simeq (3/2)H\ell \sim V/\sigma$ at very high energies, i.e. for $H\ell \gg 1$. At low energies, $F \simeq 1$ and one recovers exactly the usual four-dimensional result but at higher energies the multiplicative factor F provides an *enhancement* of the gravitational wave spectrum amplitude with respect to the four-dimensional result. However, comparing this with the amplitude for the scalar spectrum, one finds that, at high energies ($\rho \gg \sigma$), the *tensor over scalar ratio is in fact suppressed* with respect to the four-dimensional ratio. However, a difficult question is how the perturbations will evolve during the subsequent cosmological phases, the radiation and matter eras.

6. Beyond homogeneous brane cosmology

The homogeneous and isotropic brane cosmology is in fact very simple because of the generalized Birkhoff's theorem mentioned earlier. But, when the cosmological symmetry is relaxed, things become rather difficult because the bulk geometry is no longer Schwarzschild-AdS. This section presents some aspects of this complexity, starting with anisotropic, but still homogeneous, models and then discussing the evolution of the linear cosmological perturbations.

6.1. *Anisotropic brane cosmology*

Let us now discuss configurations where the cosmology in the brane is homogeneous but anisotropic, e.g. of the Bianchi I type with a metric of the form

$$ds_b^2 = -d\tau^2 + \sum_{i=1}^{3} a_i^2(\tau)(dx^i)^2. \tag{44}$$

Although many works in the literature have been devoted to this subject, most of them use the effective four-dimensional equations projected on the brane. It is a more challenging task to solve the 5D Einstein equations for the bulk as well, starting e.g. from an ansatz of the form

$$ds_{\text{bulk}}^2 = -n^2(t,y)dt^2 + \sum_{i=1}^{3} a_i^2(t,y)(dx^i)^2 + dy^2. \tag{45}$$

Assuming that the metric is separable, it turns out that analytical solutions can be obtained [26]. The five-dimensional metric is given by

$$\begin{aligned} ds^2 = \sinh^{1/2}(4y/\ell)\,\Big[& -\tanh{(2y/\ell)}^{2q_0}\,dt^2 \\ & + \sum_i \tanh{(2y/\ell)}^{2q_i}\,t^{2p_i}\left(dx^i\right)^2\Big] + dy^2, \end{aligned} \tag{46}$$

where the seven coefficients q_μ and p_i must satisfy the constraints

$$\sum_{\mu=0}^{3} q_\mu = 0, \quad \sum_\mu q_\mu^2 = \frac{3}{4}, \quad \sum_{i=1}^{3} p_i = 1, \quad \sum_i p_i^2 = 1, \quad \sum_i q_i\,(p_i + 1) = 0. \tag{47}$$

In general, a brane embedded in an anisotropic bulk spacetime must contain matter with *anisotropic stress*, because of the junction conditions:

- isotropic part:

$$\frac{n^{-1}\dot{y}_b\,\dot{A}\Big|_b + \sqrt{1+\dot{y}_b^2}\;A'|_b =}{\kappa^2 6\rho_b,} \tag{48}$$

- anisotropic part:

$$\frac{n^{-1}\dot{y}_b\,\dot{B}_i\Big|_b + \sqrt{1+\dot{y}_b^2}\;B'{}_i|_b =}{\kappa^2 2\pi_i,} \tag{49}$$

where π_i is the anisotropic pressure in the brane,

with the notation $3A \equiv \ln(a_1a_2a_3)$ and $B_i \equiv \ln a_i - A$. Note that the brane position y_b is not assumed to be fixed here: in this sense the coordinate system is not Gaussian Normal. Interestingly, the above solutions include a particular bulk geometry, for $q_0 = \pm\sqrt{3}/4$, in which one can embed a moving brane with perfect fluid as matter. The effective cosmological equation of state $P_{\rm eff}/\rho_{\rm eff}$ is negative but goes to zero at late times.

6.2. *Scalar cosmological perturbations*

A crucial step for brane cosmology is to be confronted with cosmological observations, in particular the CMB fluctuations. Although the *primordial* power spectra for scalar and tensor perturbations have been computed, the subsequent evolution of the cosmological perturbations is non trivial and has not been fully solved yet. The reason for this is that, in contrast with standard cosmology where the evolution of cosmological perturbations can be reduced to ordinary differential equations for the Fourier modes, the evolution equations in brane cosmology are partial differential equations with two variables: time and the fifth coordinate. Another delicate point is to specify the boundary conditions, both in time and space.

An instructive, although limited, approach for the brane cosmological perturbations is the brane point of view, based on the 4D effective Einstein equations on the brane, usually written in the form [18]

$$G_{\mu\nu} + \Lambda_4 g_{\mu\nu} = 8\pi G \tau_{\mu\nu} + \kappa^2 \Pi_{\mu\nu} - E_{\mu\nu}, \tag{50}$$

where $\tau_{\mu\nu}$ is the brane energy-momentum tensor, $\Pi_{\mu\nu}$ is a tensor depending quadratically on $\tau_{\mu\nu}$, giving in particular the ρ^2 term in the Friedmann equation, and $E_{\mu\nu}$, which corresponds to the dark radiation in the homogeneous case, is the projection on the brane of the bulk Weyl tensor.

It is then a straightforward exercise, starting from (50), to write down explicitly the perturbed effective Einstein equations on the brane, which will look exactly as the four-dimensional ones for the geometrical part but with extra terms due to $\Pi_{\mu\nu}$ and $T^{Weyl}_{\mu\nu}$. One thus gets equations relating the perturbations of the metric to the matter perturbations *and* the perturbations of the projected Weyl tensor, which formally can be assimilated to a virtual fluid, with corresponding (perturbed) energy density $\rho_E + \delta\rho_E$, pressure $P_E + \delta P_E = \frac{1}{3}(\rho_E + \delta\rho_E)$ and anisotropic pressure. The contracted Bianchi identities ($\nabla_\mu G^\mu_\nu = 0$) and energy-momentum conservation for matter on the brane ($\nabla_\mu \tau^\mu_\nu = 0$) ensure, using Eq. (50), that

$$\nabla_\mu E^\mu_\nu = \kappa^4\, \nabla_\mu \Pi^\mu_\nu\,. \tag{51}$$

In the background, this tells us that ρ_E behaves like radiation, as we knew already, and for the first-order perturbations, one finds that the effective energy of the projected Weyl tensor is conserved independently of the quadratic energy-momentum tensor. The only interaction is a momentum transfer.

It is also possible to construct [27] gauge-invariant variables corresponding to the curvature perturbation on hypersurfaces of uniform density, both for the brane matter energy density and for the total effective energy density (including the quadratic terms and the Weyl component). These quantities are extremely useful because their evolution on scales larger than the Hubble radius can be solved easily. However, their connection to the large-angle CMB anisotropies involves the knowledge of anisotropic stresses due to the bulk metric perturbations. This means that *for a quantitative prediction of the CMB anisotropies, even at large scales, one needs to determine the evolution of the bulk perturbations.*

In summary, one can obtain a set of equations for the brane linear perturbations, where one recognizes the ordinary cosmological equations but modified by two types of corrections:

- modification of the homogeneous background coefficients due to the additional ρ^2 terms in the Friedmann equation. These corrections are negligible in the low energy regime $\rho \ll \sigma$;
- presence of source terms in the equations. These terms come from the bulk perturbations and cannot be determined solely from the evolution inside the brane. To determine them, one must solve the full problem in the bulk (which also means to specify some initial conditions in the bulk).

One should also mention a very recent work [28] on the post-inflation evolution of gravitational waves, which indicates that, on very small scales, the spectral density of gravitational waves is *reduced* with respect to the standard 4D result.

7. Conclusions

In this contribution, I have presented some aspects of brane-world models, covering both the (static) Randall-Sundrum model and its cosmological extensions. Due to lack of time/space, I have not discussed many other interesting topics in the field. Examples are the brane cosmology of models involving Gauss-Bonnet corrections; the induced gravity models, where one includes a 4D Einstein-Hilbert action for the brane and which can lead to late-time cosmological effects mimicking dark energy.

There are still many open questions in brane cosmology. Even in the simplest set-up, discussed here, based on a cosmological extension of the Randall-Sundrum model, the evolution of cosmological perturbations has not yet been solved, although some significant progress has been made. The situation is still more complicated in more sophisticated models, involving a bulk scalar field and/or collision of branes. It must be emphasized that the predictions for the cosmological perturbations, as observed in the CMB experiments, and their adequation with the present data, is a crucial test for brane-world models for which the early universe is modified. More direct tests of brane-world models involve gravity experiments or collider experiments. However, if the fundamental Planck mass is too high, such direct experiments cannot see extra-dimensional effects and one must turn to cosmology to try to see indirect signatures from the early universe.

Another direction of research is to make contact between the brane-worlds, which are still only phenomenological models, and a fundamental theory like string theory.

References

1. C. D. Hoyle, D. J. Kapner, B. R. Heckel, E. G. Adelberger, J. H. Gundlach, U. Schmidt and H. E. Swanson, Phys. Rev. D **70**, 042004 (2004) [arXiv:hep-ph/0405262].
2. N. Arkani-Hamed, S. Dimopoulos and G. R. Dvali, Phys. Lett. B **429**, 263 (1998) [arXiv:hep-ph/9803315].
3. L. Randall and R. Sundrum, Phys. Rev. Lett. **83** (1999) 3370 [arXiv:hep-ph/9905221].
4. L. Randall and R. Sundrum, Phys. Rev. Lett. **83** (1999) 4690 [arXiv:hep-th/9906064].
5. V. A. Rubakov, Phys. Usp. **44**, 871 (2001) [Usp. Fiz. Nauk **171**, 913 (2001)] [arXiv:hep-ph/0104152]; D. Langlois, Prog. Theor. Phys. Suppl. **148**, 181 (2003) [arXiv:hep-th/0209261]; F. Quevedo, Class. Quant. Grav. **19**, 5721 (2002) [arXiv:hep-th/0210292]. P. Brax and C. van de Bruck, Class. Quant. Grav. **20**, R201 (2003) [arXiv:hep-th/0303095]; R. Maartens, Living Rev. Rel. **7**, 1 (2004) [arXiv:gr-qc/0312059]; P. Brax, C. van de Bruck and A. C. Davis, Rept. Prog. Phys. **67**, 2183-2232 (2004) [arXiv:hep-th/0404011].
6. J. Garriga and T. Tanaka, Phys. Rev. Lett. **84** (2000) 2778 [arXiv:hep-th/9911055].
7. T. Tanaka, Prog. Theor. Phys. Suppl. **148**, 307 (2003) [arXiv:gr-qc/0203082].
8. R. Emparan, A. Fabbri and N. Kaloper, JHEP **0208**, 043 (2002) [arXiv:hep-th/0206155].

9. S. Kanno and J. Soda, Phys. Rev. D **66**, 083506 (2002) [arXiv:hep-th/0207029].
10. W. D. Goldberger and M. B. Wise, Phys. Rev. Lett. **83** (1999) 4922 [arXiv:hep-ph/9907447].
11. P. Binétruy, C. Deffayet, U. Ellwanger and D. Langlois, Phys. Lett. B **477** (2000) 285 [arXiv:hep-th/9910219].
12. P. Binétruy, C. Deffayet and D. Langlois, Nucl. Phys. B **565** (2000) 269 [arXiv:hep-th/9905012].
13. C. Csaki, M. Graesser, C. F. Kolda and J. Terning, Phys. Lett. B **462**, 34 (1999) [arXiv:hep-ph/9906513]; J. M. Cline, C. Grojean and G. Servant, Phys. Rev. Lett. **83**, 4245 (1999) [arXiv:hep-ph/9906523].
14. P. Kraus, JHEP **9912** (1999) 011 [arXiv:hep-th/9910149].
15. D. Ida, JHEP **0009**, 014 (2000) [arXiv:gr-qc/9912002]
16. D. Langlois, L. Sorbo and M. Rodriguez-Martinez, Phys. Rev. Lett. **89**, 171301 (2002) [arXiv:hep-th/0206146].
17. A. Hebecker and J. March-Russell, Nucl. Phys. B **608**, 375 (2001) [arXiv:hep-ph/0103214].
18. T. Shiromizu, K. i. Maeda and M. Sasaki, Phys. Rev. D **62**, 024012 (2000) [arXiv:gr-qc/9910076].
19. D. Langlois and L. Sorbo, Phys. Rev. D **68**, 084006 (2003) [arXiv:hep-th/0306281].
20. R. R. Caldwell and D. Langlois, Phys. Lett. B **511**, 129 (2001) [arXiv:gr-qc/0103070].
21. J. Khoury, B. A. Ovrut, P. J. Steinhardt and N. Turok, Phys. Rev. D **64** (2001) 123522 [arXiv:hep-th/0103239].
22. Y. Himemoto and M. Sasaki, Phys. Rev. D **63**, 044015 (2001) [arXiv:gr-qc/0010035].
23. R. Maartens, D. Wands, B. A. Bassett and I. Heard, Phys. Rev. D **62**, 041301 (2000) [arXiv:hep-ph/9912464].
24. E. J. Copeland, A. R. Liddle and J. E. Lidsey, Phys. Rev. D **64**, 023509 (2001) [arXiv:astro-ph/0006421].
25. D. Langlois, R. Maartens and D. Wands, Phys. Lett. B **489**, 259 (2000) [arXiv:hep-th/0006007].
26. A. Fabbri, D. Langlois, D. A. Steer and R. Zegers, JHEP **0409**, 025 (2004) [arXiv:hep-th/0407262].
27. D. Langlois, R. Maartens, M. Sasaki and D. Wands, Phys. Rev. D **63**, 084009 (2001) [arXiv:hep-th/0012044].
28. K. Ichiki and K. Nakamura, arXiv:astro-ph/0406606.

A FEW WORDS ON ENTROPY, THERMODYNAMICS, AND HORIZONS

DONALD MAROLF

Physics Department
University of California, Santa Barbara USA 93106
e-mail: marolf@physics.ucsb.edu

We review recent progress in understanding certain aspects of the thermodynamics of black holes and other horizons. Our discussion centers on various "entropy bounds" which have been proposed in the literature and on the current understanding of how such bounds are *not* required for the semi-classical consistency of black hole thermodynamics. Instead, consistency under certain extreme circumstances is provided by two effects. The first is simply the exponential enhancement of the rate at which a macrostate with large entropy is emitted in any thermal process. The second is a new sense in which the entropy of an "object" depends on the observer making the measurement, so that observers crossing the horizon measure a different entropy flux across the horizon than do observers remaining outside. In addition to the review, some recent criticisms are addressed. In particular, additional arguments and detailed numerical calculations showing the observer dependence of entropy are presented in a simple model. This observer-dependence may have further interesting implications for the thermodynamics of black holes.

1. Introduction

Most researchers agree that black holes, Hawking radiation, and black hole entropy are some of our strongest clues in the quest to discover and understand ever deeper layers of fundamental physics. However, somewhat less agreement has been reached as to the nature of the fundamental physics to which they will lead. Below, we address only a narrow part of this discussion, reviewing recent work [1–3] which exposes loopholes in certain arguments for so-called "entropy bounds"– conjectures [4–9] such as Bekenstein's proposed bound [4],

$$S < \alpha RE/\hbar c, \tag{1}$$

and the so-called holographic bound [5, 6],

$$S < Ac^3/4\hbar G, \tag{2}$$

that would require the entropy of any thermodynamic system to respect a bound set by the size and, perhaps, the energy of the system[a]. We also review an unexpected piece of new physics that forms the basis of one such loophole: a new dependence [3] of entropy on the observer, and in particular a dependence on the observer of the flux of entropy across a null surface. This aspect of the review is supplemented below with a detailed new analysis of a simple example, using both analytic and numerical methods.

To place this review in its proper context, recall that the above conjectured entropy bounds were originally motivated by claims that they are required for the consistency of black hole entropy with the second law of thermodynamics. These claims have been controversial [11, 12] since they were first introduced. However, last year a new class of very general loopholes was pointed out in Refs. [2, 3]. We review and comment on these loopholes below.

While these works still have not quieted all controversy [13, 14], they have gone some distance toward doing so. In particular, we will see that the criticisms of Ref. [13] (which were stated in the context of a certain simple model) can be refuted in complete detail. This is done twice in section 3.1, once using analytic arguments and once numerically. The numerical results are an excellent match to the analytic predictions. On the other hand, as discussed in section 2.1. the criticisms of Ref. [14] apply only to a special case. Furthermore, even in that context, Ref. [14] makes an assumption about the nature of Hawking radiation which appears to conflict with well-established results [15–17]. Thus, the original loopholes [2, 3] stand firm.

While the discovery of loopholes does not prove the conjectured entropy bounds of Refs. [4–8] to be false[b] it does remove the original motivation for such bounds. Proponents may still appeal to the AdS/CFT correspondence [18] and in particular to the Susskind-Witten calculation [19] for

---

[a]Here we have displayed the fundamental constants explicitly, but below we use geometric units with $k_B = \hbar = c = G = 1$. The original version [4] of Bekenstein's bound has $\alpha = 2\pi$, while some subsequent discussions (e.g. Ref. [10]) weaken the bound somewhat, enlarging α by a factor of order ten.

[b]In particular, there are certain observational bounds (see, e.g. Ref. [2]) which limit the existence of highly entropic objects in our universe. While the form of such observational bounds does not obviously match that of (1) or (2) it would be interesting to explore such observational bounds in more detail. Indeed, Bekenstein stresses that his interest in the proposal (1) is in the context of our particular universe. In contrast, we are more interested here in possible deep ties between the bounds and quantum gravity; i.e., in the question of whether (1) and (2) could in principle be violated in some consistent theory of quantum gravity, which may or may not describe our particular universe. We will therefore discuss (1) and (2) in this strong sense below.

support, as well as the difficulty (see e.g. Refs. [20–22]) of constructing counter-examples to the Covariant Entropy Bound conjecture [8] within the particular regime in which it has been formulated. However, the lessons of these observations may in the end turn out to be more subtle, and it is clear that no compelling evidence for the bounds is available at this time.

The plan of this review is as follows. We first outline the loopholes in the arguments of Refs. [4–6, 10, 23]. Our treatment (section 2) roughly follows that of Ref. [2], and more details can be found in that reference. Section 3 then follows Ref. [3] and moves on to consider a related thought experiment, which one might at first think would require similar restrictions on matter for the consistency of the second law. However, a careful examination of the physics in fact leads to a different conclusion, in which the issue is resolved by the observer-dependence of entropy foreshadowed above.

2. Thermodynamics can hold its own

Below, we turn our attention first to bounds of the form (1), and then to bounds of the form (2). In each case, the essential ingredient is the well established [12, 15, 16, 24–28] point that the radiation surrounding a black hole of temperature T_{BH} is thermal in the sense that, *in equilibrium*, it is described by an ensemble of the form $e^{-\beta H}$, where $\beta = 1/T_{BH}$. As a result, the probability to find a particular *macrostate* (such as a bound-violating box) in a thermal ensemble is not $e^{-\beta E}$ but $e^{-\beta F}$, with $F = E - T_{BH}S$ is the free energy of the macrostate at the black hole temperature T_{BH} and where the phase space factor of e^S arises from associating e^S microstates with a single macrostate.

2.1. *General Comments*

Of interest here will be the somewhat different setting in which a black hole radiates into empty space. It is more difficult in this context to arrive at detailed results for interacting fields, though for free fields one can readily show [15, 17] that the radiation produced by the black hole is just the outgoing component of the above thermal ensemble. This has been established rigorously [17] even for the complicated case where the black hole forms dynamically from the collapse of some object.

Consider now the case of interacting fields. Suppose in particular that the thermal ensemble can be described as a collection of weakly interacting particles or quasi-particles. We would again like to claim that the outward flux of both particles and quasi-particles from a black hole radiating into free space is well-modeled by the outgoing component of a thermal ensemble. Let us therefore consider placing our black hole inside a spherical box whose walls are held at a fixed temperature. In this context, the above results tell us that the black hole will eventually reach thermal equilibrium. Since the particles and quasi-particles interact weakly, to a good approximation we may regard this ensemble as a combination of its ingoing and outgoing parts. Since the ingoing parts incident on the black hole are completely absorbed, equilibrium can only be maintained if the black hole also radiates a flux of both particles and quasi-particles at a rate given by the outgoing part of the thermal ensemble. As discussed in the conclusions of Ref. [17], even in the presence of strong interactions one expects that the outgoing Hawking radiation is determined by properties of the thermal ensemble, though the details in that case will be quite complicated. Thus, any proposal (such as that of Ref. [14]) that Hawking radiation of quasi-particles be treated differently to that of "fundamental" particles, and which results in interesting cases in an outward flux of quasi-particles significantly lower than which arises in equilibrium, is in conflict with standard results [12, 15, 16, 24–28].

Because just such a reduced quasi-particle model (in the context of weak interactions) formed the basis[c] of a recent criticism [14] by Bekenstein of the work [2] to be described below, it is worthwhile to examine this point once more in great detail. The particular proposal of Ref. [14] was that quasi-particles cannot be *directly* Hawking radiated by a black hole, but instead arise in thermal equilibrium only via additional processes in which they are created by the interaction of other (so-called "fundamental") particles. The case of interest [14] is one where kinetics would dictate that such a process happens only extremely slowly, on some timescale which we will call t_{quasi}. But now consider the case of a black hole in Anti-de Sitter (AdS) space. Here we know [16, 25–28] that there is a state (effectively, the Hartle-Hawking vacuum) in which the black hole is at thermal equilibrium. Yet if the interactions are weak, any object with zero angular momentum will fall into the black hole on a timescale t_{AdS} set by the size of the AdS space. Thus, under the proposal of Ref. [14], one may choose $t_{quasi} \gg t_{AdS}$ so that the quasi-particles drain out of the ensemble and into the black hole far faster than they are created. Thus there could be no thermal

[c]We also note below that quasi-particles are not essential for the loopholes we discuss. Thus, it is not strictly necessary to refute Bekenstein's recent comments [14]. However, we do so here in order to show that the loopholes are quite general.

equilibrium, contradicting in particular the theorems of Ref. [16]. One can only conclude that the reduced-quasi-particle proposal [14] is untenable, and that quasi-particles are directly Hawking radiated by the black hole at a rate set by the outgoing flux of such objects in the thermal ensemble determined by the black hole's Hawking temperature.

2.2. *The Bekenstein bound*

Let us now recall the setting for the argument of Ref. [10] in favor of (1). One considers an object of size R, energy E, and entropy S which falls into a Schwarzschild black hole of size R_{BH} from a distance $d \gg R_{BH}$. Following Ref. [10], we parametrize the problem by $\zeta = \frac{R_{BH}}{2R}$. The parameter ζ is taken to be large enough that the object readily falls into the black hole without being torn apart. In other words, we engineer the situation so that the black hole is, at least classically, a perfect absorber of such objects. It is also assumed that the Hawking radiation emitted during the infall of our object is dominated by the familiar massless fields. Assuming that the number of such fields is not overly large, the effects of Hawking radiation are negligible during the infall. Under these assumptions, Ref. [10] shows that the second law is satisfied only if

$$S < 8\pi\nu\zeta RE, \tag{3}$$

where ν is a numerical factor in the range $\nu = 1.35 - 1.64$. Here the energy E has been assumed to be much smaller than the mass M_{BH} of the black hole and, up to the factor ν, the above bound is obtained by considering the increase in entropy of the black hole, $dS_{BH} = dE_{BH}/T_{BH}$. Notice here that the black hole was in particular assumed not to readily emit copies of our object as part of its Hawking radiation.

However, let us now suppose that a "highly entropic object" can exist with $S > 8\pi\nu\zeta RE$. Then, according to Ref. [10], this will lead to a contradiction with the second law in a process involving the particular black hole on which we have chosen to focus. We wish to consider the possibility that, because this entropy is very large, such objects *will* be readily emitted by the black hole so that the Hawking radiation will *not* be dominated by massless fields. In particular, if they are emitted sufficiently rapidly, the black hole will emit such objects faster than we have proposed to drop them in and, as a result, it is clear that the second law will be satisfied. It is natural to consider such highly entropic objects as being built of many parts and thus, in some sense, "composite" so that they may best be described as long-lived quasi-particle excitations of some fundamental theory. However, for clarity we point out that such complications are unnecessary, and the reader is encouraged to keep in mind the concrete example of a system with e^S species of fundamental massive scalar field, all having of mass E. In this example, the entropy arises if one takes the fields to have sufficiently similar interactions such that it is reasonable to coarse-grain over the species label.

To study the production rate, we follow the path set out in section 2.1 above and compute the free energy of our object at the black hole temperature $T_{BH} = (4\pi R_{BH})^{-1} = (8\pi\zeta R)^{-1}$. We have

$$F = E - \frac{S}{4\pi R_{BH}} < E - \nu E < 0. \tag{4}$$

Thus, we expect an exponentially *large* rate of Hawking emission of these objects from this black hole.

In fact, in a naive model the resulting thermal ensemble contains a sum over states with arbitrary numbers n of such highly entropic objects, weighted by factors of $e^{-nF} > 1$. Such an ensemble clearly diverges and is physically inadmissible. Thus, some effect must intervene to cut off the divergence and to stabilize the system. If no other effect intervenes first, this cut-off can be supplied by the quantum statistics of the highly entropic objects. To see this, we note that the above naive model treats the objects as *distinguishable.* In contrast, the partition function for N species of either boson or fermion converges at any temperature no matter how large N may be. Thus, if our "object" were a collection of $N = e^S$ free Bose or Fermi fields, the associated partition function would converge. The divergence of the naive model simply indicates that the true equilibrium ensemble contains so many boxes that they cannot be treated as *distinguishable* and *non-interacting* objects! So long as there is no strong barrier that prevents the Hawking radiation from escaping, the outgoing radiation will also be correspondingly dense, at least in the region near the black hole.

On the other hand, suppose that the boxes interact in such a way that the thermal atmosphere of the black hole largely blocks the passage of outward-moving boxes. Let us assume that it also blocks the passage of the CPT conjugate objects, since these will carry equal entropy. This might happen because the thermal atmosphere already contains many densely packed copies of our object, or alternatively because our objects are blocked by other components of the atmosphere. However, (due to CPT invariance), such an atmosphere will also provide a barrier hindering our ability to drop in a new object from far away. If we attempt to drop in such an object, then with correspondingly large probability it must bounce off the thermal atmosphere or be otherwise prevented from entering the black hole! Again, one expects that the black hole radiates at least one object for every highly entropic object which is successfully sent inward through the atmosphere from outside.

The argument above was stated in terms of the particular scenario suggested in Ref. [10]. However, one might expect that it can be formulated much more generally. This is indeed possible, as may be seen by considering any process in which a given object ("box" above) with entropy S is hidden inside the black hole, with the black hole receiving its energy E and its entropy S otherwise disappearing from the universe. Note that this includes both processes of the original form [4] where the object is slowly lowered into the black hole, as well as the more recent version [10] discussed above where the object falls freely. As before, we suppose that this represents a small change, with E being small in comparison to the total energy of the black hole. From the ordinary second law we have

$$\Delta S_{total} \geq \Delta S_{BH} - S, \tag{5}$$

but the *first* law of black hole thermodynamics tells us that this is just

$$\Delta S_{total} \geq \frac{E}{T_{BH}} - S = \frac{F}{T_{BH}}, \tag{6}$$

where F is the free energy of the object at the Hawking temperature T_{BH}. In particular, since $T_{BH} > 0$, the sign of ΔS_{total} must match that of F. Thus absorption of an object by a black hole violates the second law only if $F < 0$, in which case any of the above mechanisms may either prevent the process from occurring or sufficiently alter the final state so that the second law is satisfied.

2.3. *The holographic bound*

We now turn to the "holographic bound" (2) and consider the analogous arguments of Refs. [5,6]. These works suggest that one compare a (spherical, uncharged) object with $S \geq A/4$ with a Schwarzschild black hole of equal area $A = 4\pi R_{BH}^2$. Since the highly entropic object is not itself a black hole, its energy E must be less than the mass M_{BH} of the black hole. One is then asked to drop the bound-violating object into a black hole of mass $M_{BH} - E$ or otherwise transform it into a black hole of mass M_{BH}. For arguments which drop the object into a pre-existing black hole, one typically[d] (though not always) assumes $E \ll M_{BH}$.

[d]See, e.g., the weakly gravitating case described in Ref. [23].

Let us suppose for the moment that $E \ll M_{BH}$. Then we may estimate the emission rate of such 'highly entropic objects' from a black hole of mass M_{BH} by considering the free energy $F = E - TS$ of our object at the Hawking temperature $T_{BH} = (4\pi R_{BH})^{-1}$. We have

$$F = E - T_{BH}S < E - \frac{A/4}{4\pi R_{BH}} = E - M_{BH}/2 < 0. \tag{7}$$

Thus, we again see that our object is likely to be emitted readily in Hawking radiation. As a result, one may also expect significant radiation during the formation of the black hole so that no violation of the 2nd law need occur.

And what if E and M_{BH} are comparable? Then back-reaction will be significant during the emission of one of our objects from the black hole and reliable information cannot be obtained by considering an equilibrium thermal ensemble at fixed temperature. In fact, we are now outside the domain of validity of Hawking's calculation [15]. However, Refs. [29,30] suggest how an emission rate may be estimated in this regime. These works find that the emission rate for a microstate is proportional to $\exp(\Delta S_{BH})$, where ΔS_{BH} is the change in black hole entropy when the particle is emitted. Note that this change is caused by the loss of energy by the black hole and is typically negative. Following this conclusion allows us to proceed in much the same way as in section 2.2: Forming a black hole from our object would violate the second law only if $S > S_{BH}$, but since $\Delta S_{BH} > -S_{BH}$ we have $\Delta S_{total} = \Delta S_{BH} + S > 0$ for the corresponding emission process. Thus, Refs. [29,30] predict an exponentially large emission rate for our object. We conclude that the combined process of collapse and emission would actually result in a net *increase* in the total entropy.

3. The Observer Dependence of Entropy

In section 2 above we described loopholes in certain arguments [4–6,10,23] which appeared to suggest that (1) and (2) might be required if the 2nd law is to be satisfied in general processes involving black holes. In particular, we argued that during the time it takes to insert an object violating (1) into a black hole of the appropriate size, the black hole will in fact radiate many similar objects. The result with then be a net production of entropy.

It turns out to be instructive to carry this analysis one step further. Following Ref. [3], let us therefore consider a related thought experiment in which we first place our black hole in a reflecting cavity (or, perhaps, in a small anti-de Sitter space) and allow it to reach thermal equilibrium. Note that in equilibrium the rate at which highly entropic objects are absorbed by the black hole exactly balances the emission rate. But what happens if we now alter the state of the system by adding yet one more highly entropic object, say, violating (1) and travelling inward toward the black hole?

To simplify the problem, we take the limit of a large black hole. We may then describe the region near the horizon by Rindler space. Recall [32] that an object crossing a Rindler horizon is associated with a corresponding increase in the Bekenstein-Hawking entropy of this horizon by at least $\Delta S \geq E_{acc}/T$ where E_{acc} (the energy of the object as measured by the accelerated observer) is the Killing energy associated with the boost symmetry ξ (i.e., the Rindler time translation) and the associated temperature is given by $\kappa/2\pi$ where κ is the surface gravity of ξ. As usual, the normalization of ξ cancels so that E_{acc}/T is independent of this choice. Thus, in this setting the interesting highly entropic objects are those with entropy $S > E_{acc}/T$. Might these lead to a failure of the 2nd law for the associated observers when such an object falls across the horizon? Note that because uniformly accelerated observers perceive the Minkowski vacuum as a state of thermal equilibrium at temperature $T \neq 0$, this setting is analogous to the query from the previous paragraph involving black holes.

We now proceed to analyze this situation following Ref. [3]. Let us first review the inertial description of this process, for which the state with no objects present is the Minkowski vacuum $|0_M\rangle$. The presence of an object is then described as an excitation of this vacuum. If there are n possible microstates $|1_M; i\rangle$ ($i \in \{1, ..., n\}$) for one such object, then an object in an *undetermined* microstate is described by the density matrix

$$\rho_M = \frac{1}{n} \sum_{i=1}^{n} |1_M; i\rangle\langle 1_M; i|. \tag{8}$$

The entropy assigned to the object by the inertial observer is then as usual

$$S_M = -\mathrm{Tr}\rho_M \ln \rho_M = \ln n. \tag{9}$$

Of course, the inertial observer can still access the object (and its entropy) after the object crosses the horizon so that there is no possibility of a 2nd law violation from the inertial point of view. Thus, the interesting question is what entropy a Rindler observer assigns to this object. In particular, we wish to compute the change in the amount of entropy accessible to the Rindler observer when the object falls across the horizon. This is just the difference between the entropies of the appropriate two states given by the Rindler descriptions of ρ_M and $|0_M\rangle$. An important point is that, since $|0_M\rangle$ is seen as a thermal state, it already carries a non-zero entropy that must be subtracted. Similarly $E_{acc} = \delta E$ is the corresponding difference in Killing energies. We will work in the approximation where gravitational back reaction is neglected and, in particular, in which the horizon is unchanged by the passage of our object.

Following Ref. [3], let us consider then the thermal density matrix ρ_{R0} which results from tracing the Minkowski vacuum, $|0_M\rangle\langle 0_M|$, over the invisible[e] Rindler wedge:

$$\rho_{R0} = \mathrm{Tr}_{invisible}|0_M\rangle\langle 0_M|. \tag{10}$$

This describes all information that the Rindler observer can access in the Minkowski vacuum state. We wish to compare ρ_{R0} with another density matrix ρ_{R1} which provides the Rindler description of the state ρ_M above:

$$\rho_{R1} = \mathrm{Tr}_{invisible}\rho_M = \mathrm{Tr}_{invisible}\frac{1}{n}\sum_{i=1}^{n}|1_M;i\rangle\langle 1_M;i|. \tag{11}$$

We would like to compute the difference in energy

$$\delta E = \mathrm{Tr}[H(\rho_{R1} - \rho_{R0})], \tag{12}$$

and in entropy

$$\delta S = -\mathrm{Tr}[\rho_{R1}\ln\rho_{R1} - \rho_{R0}\ln\rho_{R0}], \tag{13}$$

where H is the Hamiltonian of the system and, in both cases, the sign has been chosen so that the change is positive when ρ_{R1} has the greater value of energy or entropy.

Now, the entropy of each state separately is well known to be divergent [33]. However, Ref. [3] points out that the change δS will still be well-defined. In particular, suppose the object has some moderately well-defined energy $E_{inertial}$ as described by the inertial observer and where the object is well localized. Then the energy measured by the Rindler observer will also be reasonably well-defined and the difference $\rho_{R1}\ln\rho_{R1} - \rho_{R0}\ln\rho_{R0}$ will have negligibly small diagonal entries at high energy. Thus, the above trace will exist. In other words, we may compute δS by first imposing a cutoff Λ, computing the entropy (S_0, S_1) and energy (E_0, E_1) of the two states (ρ_{R0}, ρ_{R1}) separately, subtracting the results, and removing the cutoff.

Now, the case which is most interesting for the second law is where $\ln n \gg E_{acc}/T$ so that the presumed violation of the second law is very

[e]The terms "visible" and "invisible" Rindler wedge will always be used in the context of what is in causal contact with our chosen Rindler observer. Of course, the entire spacetime is visible to the inertial observer.

large. Note that in this regime the expected number of similar objects in the thermal ensemble ρ_{R0} is already large; in particular, for $E_{acc} \gg T$ it is of order $ne^{-E_{acc}/T}$. The physical insight of Ref. [3] is that, in such a case, the difference between ρ_{R1} and ρ_{R0} is roughly that of adding *one*[f] object to a collection of $ne^{-E_{acc}/T}$ similar objects. Thus, one should be able to describe $\delta\rho = \rho_{R1} - \rho_{R0}$ as a small perturbation. One may therefore approximate δS through a first-order Taylor expansion around ρ_{R0}:

$$\delta S \approx -\mathrm{Tr}\left[\delta\rho \frac{\delta(\rho \ln \rho)}{\delta\rho}\bigg|_{\rho=\rho_{R0}}\right] = -\mathrm{Tr}[\delta\rho(1 + \ln \rho_{R0})] = -\mathrm{Tr}[\delta\rho \ln \rho_{R0}], \tag{14}$$

where in the last step we use the fact that $\mathrm{Tr}\rho_{R1} = \mathrm{Tr}\rho_{R0}$. Now recall that the initial density matrix is thermal, $\rho_{R0} = e^{-H/T}/(\mathrm{Tr}e^{-H/T})$. Thus, we have

$$\delta S \approx -\mathrm{Tr}[\delta\rho(-H/T)] = \frac{\mathrm{Tr}[H(\rho_{R1} - \rho_{R0})]}{T} = \frac{\delta E}{T}, \tag{15}$$

where we have again used $\mathrm{Tr}(\delta\rho) = 0$. As noted in Ref. [3], this key result is independent of any cut-off.

The above result can also be understood on the basis of classical thermodynamic reasoning. The initial configuration ρ_{R0} represents a thermal equilibrium. We wish to calculate the change in entropy during a process which increases the energy by an amount δE. Whatever the nature of the object that we add, this cannot increase the entropy by more than it would have increased had we added this energy as heat. Since we consider a small change in the configuration, the first law yields

$$\delta S \leq \delta S_{max} = \frac{\delta E}{T}. \tag{16}$$

We see that the process of adding a small object saturates this bound, at least to first order in small quantities. If on the other hand we considered finite changes, we would still expect the resulting δS to satisfy the bound (16), though it would no longer be saturated.

At this point the reader may wish to see more carefully how the above effect arises in an explicit example. In particular, the reader may question the step in which $\delta\rho$ is treated as a small perturbation, which has so far been argued only on physical grounds and without mathematical rigor[g]. As

[f] Actually, a number δ which depends on $E_{inertial}$ and T but is, at least in simple cases, independent of n; see, e.g., eq. (31) below and the surrounding discussion.

[g] One can do a bit better: An argument (due to Mark Srednicki) for this result was given in the appendix of Ref. [3] using a standard trick of statistical mechanics. However, this trick has also not been rigorously justified.

a result, it was recently claimed [13] that careful consideration of this issue would find large corrections over the analysis of Ref. [3] and invalidate the result. To study possible corrections in detail, we now turn in section 3.1 to a concrete example in the context of free bosons where all quantities of interest may be computed explicitly. We will find certain errors in Ref. [13], and show instead that the linear approximation is indeed highly accurate.

3.1. *An explicit example: Free Bosons*

Consider then a system of n free boson fields. Our "objects" will be merely the particles of such fields, with the understanding that the flavor of the particle is treated as unobservable. That is, each 1-particle state will correspond to a microstate of our object, but any two such states which differ only by change in the flavor of particles will correspond to the same macrostate of our object. Thus, we wish to calculate the entropy of "objects" described, from the inertial point of view, by a single-particle density matrix uniformly distributed over the different fields. The inertial observer therefore assigns the object an entropy $S_{inertial} = \ln n$. This example was briefly considered in Ref. [3], though only under the assumption that $\delta\rho$ could be treated perturbatively. An attempt was then made in Ref. [13] to analyze this key assumption in more detail by computing the result exactly. The claim of that work was that discrepancies were found that were sufficient to invalidate the conclusions of Ref. [3]. However, we will repeat this analysis below and, after correcting some elementary mistakes in the equations of Ref. [13], we will in fact find support for the above observer-dependence of entropy and in particular for the accuracy of the linearized treatment used above.

Our notation and set up essentially follow that of Ref. [3]. To begin, we write the Minkowski vacuum for a system of n free bosonic fields in terms of the Rindler Fock space. The result is [34]:

$$|0_M\rangle = \prod_{i=1}^{n}\prod_{j}(1 - e^{-\frac{\omega_j}{T}})^{\frac{1}{2}} e^{-\frac{\omega_j}{2T} a^\dagger_{ijL} a^\dagger_{ijR}} |0_{Rindler}\rangle. \tag{17}$$

Here i labels the different fields, j labels a complete set of modes u_{jL}, u_{jR} of positive Rindler frequency ω_j for each field, and the labels R and L refer respectively to the right and left Rindler wedges. Each mode u_{iL}, u_{jR} of the ith field has an associated annihilation operator a_{iL}, a_{iR} which annihilates the Rindler vacuum $|0_{Rindler}\rangle$ and satisfies a standard commutation relation of the form $aa^\dagger - a^\dagger a = 1$. The temperature T associated with the uniformly accelerating observer of interest is given by $T = \frac{a}{2\pi} = \frac{\kappa}{2\pi}$ where a is the observer's proper acceleration and κ is the surface gravity of the boost Killing field ξ normalized on our observer's worldline.

Similarly, the annihilation operator a_{iM} for a Minkowski mode u_M of the ith field can on general principles [34] be expressed in the form:

$$a_{iM} = \sum_j \Big[(u_M, u_{jR})(a_{ijR} - e^{-\frac{\omega_j}{2T}} a^\dagger_{ijL}) + (u_M, u_{jL})(a_{ijL} - e^{-\frac{\omega_j}{2T}} a^\dagger_{ijR}) \Big], \tag{18}$$

where (u, v) is the Klein-Gordon inner product. For simplicity we might suppose that we choose a mode u_M with no support on the invisible Rindler wedge (say, the left one) and for which (u_M, u_{jR}) is well modelled by a delta-function; we will return to the general case later in the subsection. In particular, this simplification means that the Rindler frequency ω of u_M is reasonably well defined[h]. For such a case the above Bogoliubov transformation becomes

$$a_{iM} = \frac{1}{\sqrt{1 - e^{-\frac{\omega}{T}}}} (a_{iR} - e^{-\frac{\omega}{2T}} a^\dagger_{iL}), \tag{19}$$

where the normalization of a_{iM} is fixed by the commutation relation. Here the Rindler operators refer to the one relevant pair of Rindler modes (and i continues to label flavors of fields). Note that since the we consider a free field, all other modes are decoupled and can be dropped from the problem. Thus, we simplify the discussion below by considering only the part of the Rindler Fock space associated with this pair of modes. Tracing over the invisible wedge yields

$$\rho_{R0} = (1 - e^{-\omega/T})^n e^{-N\omega/T}, \tag{20}$$

where $N = \sum_i a^\dagger_{iR} a_{iR}$ is the operator associate with the *total* number of particles without regard to flavor.

Since a_{iM} acts in the Hilbert space describing only the ith field, it is convenient to describe the action of a_{iM} on the vacuum $|0_{iM}\rangle$ for this particular field alone. One sees that the properly normalized Minkowski one-particle state is given in terms of an infinite number of Rindler excitations

$$a^\dagger_{iM} |0_{iM}\rangle = (1 - e^{-\frac{\omega}{T}}) \sum_k e^{-\frac{k\omega}{2T}} \sqrt{k+1} |k, k+1\rangle_i, \tag{21}$$

---

[h]If the object is well localized and located near our Rindler observer at some time, then ω is also the frequency measured by the inertial observer.

where $|k, k+1\rangle_i$ denotes the state of field i having k Rindler excitations in the right wedge and $k+1$ Rindler excitations in the left wedge. Here we consider only the factor in the Hilbert space that describes the modes appearing in (19).

Thus, if we return to our Minkowski density matrix

$$\begin{aligned}\rho_M &= \frac{1}{n}\sum_{i=1}^{n} |1_M; i\rangle\langle 1_M; i| \\ &= \frac{1}{n}\sum_{i=1}^{n} (a^\dagger_{iM}|0_M\rangle\langle 0_M|a_{iM}),\end{aligned} \tag{22}$$

tracing over the invisible Rindler wedge will give the desired result. As in Ref. [3], a short calculation gives

$$\begin{aligned}\rho_{R1} = \frac{1}{n}\sum_i \, (1-e^{-\omega/T})^n \times \\ \sum_{k_1,\dots,k_n} e^{-(\omega/T)\sum_j k_j}\, (e^{\omega/T}-1)k_i|k_1\dots k_n\rangle\langle k_1\dots k_n|,\end{aligned} \tag{23}$$

where again only the part of the state associated with the Rindler mode of interest has been displayed. Here the notation $|k_1\dots k_n\rangle$ denotes the state with k_j particles of type j in this mode of Rindler energy ω in the visible Rindler wedge. One can write this result in terms of the matrix elements between a complete set of modes as

$$\rho_{R1} = \sum_{\vec{k}} (\rho_{R1})_{\vec{k},\vec{k}}|k_1\dots k_n\rangle\langle k_1\dots k_n|, \tag{24}$$

where

$$(\rho_{R1})_{\vec{k},\vec{k}} = \left[(e^{\omega/T}-1)\frac{1}{n}\sum_{i=1}^{n} k_i\right](\rho_{R0})_{\vec{k},\vec{k}}, . \tag{25}$$

and $\vec{k} = \{k_1, \dots, k_n\}$. As a result, one sees that

$$\rho_{R1} = \frac{e^{\omega/T}-1}{n} N\rho_{R0}. \tag{26}$$

This expression is the one given in Ref. [3]. However, in Ref. [13] $\sum_i k_i$ has been replaced by k_i in what appears to be a typographic mistake (as the left hand side is clearly independent of the choice of any particular field i). Note that ρ_{R1} is diagonal in the standard basis. One can check explicitly that ρ_{R1} is properly normalized so that its trace is 1.

The change in the density matrix is thus $\delta\rho = \rho_{R1} - \rho_{R0}$. For non-interacting particles, the Hamiltonian is $H = \omega N$, so the average change in energy is

$$E_{acc} := \delta E = \omega \mathrm{Tr}(N\delta\rho) = \frac{\omega}{1 - e^{-\frac{\omega}{T}}}. \tag{27}$$

Note that (27) requires no approximation that $\delta\rho$ is in any sense "small."

As remarked in Ref. [3], this result is already of interest. Note that for $\omega \gg T$ and under the conditions of footnote (h), one finds $\delta E = E_{inertial} \approx \omega$. On the other hand, for $\omega \ll T$ the background of objects in the thermal bath leads to an amplification reminiscent of the effect of stimulated emission. In particular, for $\omega \ll T$ we have added (on average) far more than one object (in fact, $\frac{1}{1-e^{-\omega/T}}$ objects) to the thermal bath in constructing ρ_{R1} from ρ_{R0}.

However, the quantity of real interest is $\delta S = Tr(\delta\rho_{R0} \ln \rho_{R0}) - Tr(\delta\rho_{R1} \ln \rho_{R1})$. Let us therefore calculate the entropies S_{R0} and S_{R1} of ρ_{R0} and ρ_{R1}. Here we follow the calculation of Ref. [13], but correcting the mistake mentioned above. The entropy of ρ_{R0} is

$$S_{R0} = -Tr\rho_{R0} \ln \rho_{R0} = n\left(\frac{\omega/T}{e^{\omega/T} - 1} - \ln(1 - e^{-\omega/T})\right), \tag{28}$$

since ρ_{R0} is just a thermal ensemble of n harmonic oscillators. On the other hand, we have

$$\begin{aligned} S_{R1} &= -Tr\ \rho_{R1} \ln \rho_{R1} \\ &= Tr\left[\rho_{R1}\left(\ln \frac{n}{(e^{\omega/T} - 1)(1 - e^{-\omega/T})^n} + \frac{\omega}{T}N - \ln N\right)\right] \\ &= S_{R0} + \ln \frac{n}{e^{\omega/T} - 1} + \frac{\omega/T}{1 - e^{-\omega/T}} \\ &\quad -(1 - e^{-\omega/T})^n \frac{e^{\omega/T} - 1}{n} \times \\ &\quad \sum_{k_1,\ldots,k_n=0}^{\infty} (\sum_{i=1}^{n} k_i) e^{-(\sum_{j=1}^{n} k_j)\omega/T} \ln(\sum_{m=1}^{n} k_m), \end{aligned} \tag{29}$$

where the final expression has been computed using (26). The last term in the final expression is merely the last term in the second line written out in complete detail. This allows the reader to compare with and correct Ref. [13].

This last term is difficult to evaluate explicitly, but one may estimate the result based on the fact that it is the expectation value of $-\ln N$ in the state ρ_{R1}. Now, the expectation value of N in the thermal state ρ_{R0}

under these conditions is $\langle N\rangle_{R0} = \frac{n}{e^{\omega/T}-1} \gg 1$. But ρ_{R1} differs from ρ_{R0} only by adding one object (using a Minkowski creation operator). Since a Minkowski creation operator is a combination of a Rindler creation operator and a Rindler annihilation operator, the expectation value of N in ρ_{R1} will then lie in the range

$$\frac{n}{e^{\omega/T}-1} - \delta < \langle N\rangle_{R1} < \frac{n}{e^{\omega/T}-1} + \delta, \tag{30}$$

where δ is an unknown function of ω/T which measures the number of Rindler quanta created by the action of a Minkowski creation operator. Based on eq. (27), it is natural to expect that δ contains a similar "stimulated emission" effect, so that

$$\delta = \frac{\alpha}{1-e^{-\omega/T}}, \tag{31}$$

with α independent of ω/T. We will find further evidence for this form from numerical calculations below.

Furthermore, at large n the distribution of N will be sharply peaked about the mean $\langle N\rangle$, while we may approximate

$$\ln(\langle N\rangle_{R0} + \delta) \approx \ln\langle N\rangle_{R0} + \frac{\delta}{\langle N\rangle_{R0}} = \ln\left(\frac{n}{e^{\omega/T}-1}\right) + \alpha\frac{e^{\omega/T}}{n}. \tag{32}$$

Thus, one should be able to replace $-Tr[\rho_{R1}\ln N]$ with $\ln\left(\frac{n}{e^{\omega/T}-1}\right)$ up to a term of order $\frac{1}{ne^{-\omega/T}}$. Performing this substitution yields

$$S_{R1} - S_{R0} = \frac{\omega/T}{e^{\omega/T}-1} + O(\frac{1}{ne^{-\omega/T}}). \tag{33}$$

Note that (33) is indeed $E_{acc}/T = \frac{\langle H\rangle_{R0}-\langle H\rangle_{R1}}{T}$ to leading order, as predicted by the linearized calculation. In particular, for large n the Rindler observer assigns a much smaller entropy than does the inertial observer.

Lest the reader have some remaining doubt as to the accuracy of this approximation, some simple numerical calculations (performed using Mathematica) are included below. Figs. (1,2) each include two plots showing numerical computations of the error term $\Delta = S_{R1} - S_{R0} - E_{acc}/T$, which gives the discrepancy between the exact result and the linearized approximation. These computations are done by truncating the infinite sum in (29) at a fixed total number (N) of objects. In Figs. (1,2), this "truncation level" was taken to be $N = 250$. Fig. (3) shows the result of varying this truncation level and indicates that, for the parameters relevant to Figs. (1,2), the value of Δ has indeed stabilized by the time one reaches $N = 250$.

On the left side of Fig. (1), Δ has been plotted against the number of objects n for the case $\omega/T = 2$. One can see that it rapidly approaches zero *from below*, indicating not only that $\delta S = E_{acc}/T$ in the limit $n \to \infty$, but also that E_{acc}/T provides an upper bound on δS as is required for more general preservation of the second law. On the right in Fig. (1), an attempt has been made to gain better control over this residual error Δ. Since from (32) one expects the error to scale with $\frac{1}{ne^{-\omega/T}}$, we have plotted the ratio of Δ to this factor; i.e., the product $\Delta n e^{-\omega/T}$. This product does indeed appear to approach a constant in Fig. (1). In particular, it appears to approach the value $-1/2$, corresponding to $\alpha = +1/2$ in (31). For comparison, Fig. (2) shows the corresponding plot for $\omega/T = 4$ which shows similar results (and again suggests $\alpha = 1/2$).

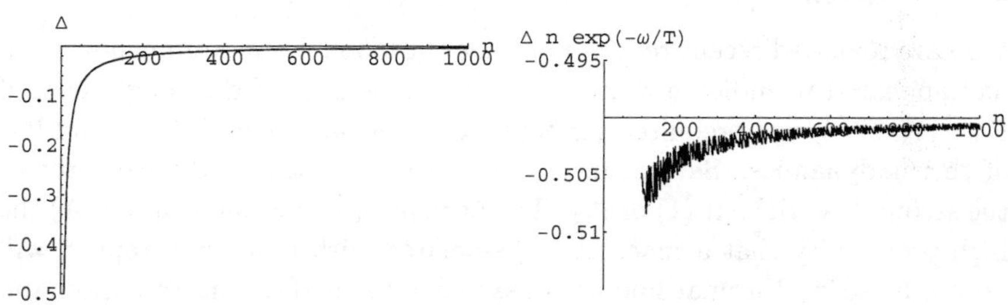

Figure 1. $\omega/T = 2$. On the left we plot Δ vs. the number of fields n up to $n = 1000$, so that $ne^{-\omega/T}$ ranges up to roughly 135. The result appears to converge to zero from below. On the right, Δ has been scaled by $ne^{-\omega/T}$. The product appears to converge to $-1/2$, further supporting the claim that Δ is $O([ne^{-\omega/T}]^{-1})$.

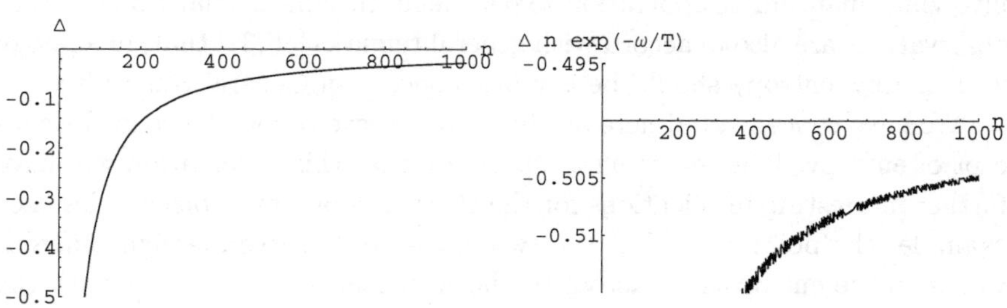

Figure 2. $\omega/T = 4$. On the left we plot Δ vs. the number of fields n up to $n = 1000$, so that $ne^{-\omega/T}$ ranges up to roughly 18. The result appears to converge to zero from below. On the right, Δ has been scaled by $ne^{-\omega/T}$. The product appears to converge to $-1/2$, further supporting the claim that Δ is $O([ne^{-\omega/T}]^{-1})$.

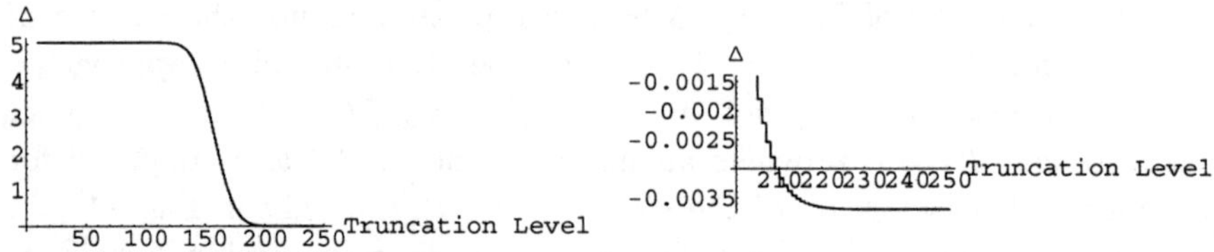

Figure 3. The accuracy of truncating the sum in (29) at a finite level is illustrated by plotting Δ against a range of truncation levels. The figure on the right shows an expanded view of the region between truncation levels 200 and 250. Note that a truncation level of 250 was used to generate Figs. (1) and (2).

4. Discussion

We have reviewed recent results pointing out loopholes in arguments which had appeared to indicate that novel entropy bounds of the rough form of Eqs. (1) or (2) were required for consistency of the generalized second law of thermodynamics. Several mechanisms were identified that can protect the second law without (1) or (2). The primary such mechanisms are A) the high probability that a macrostate associated with a large entropy S will be produced by thermal fluctuations, and thus in Hawking radiation and B) the realization that observers remaining outside a black hole associate a different (and, at least in interesting cases, smaller) flux of entropy across the horizon with a given physical process than do observers who themselves cross the horizon during the process. In particular, this second mechanism was explored using both analytic and numerical techniques in a simple toy model. We note that similar effects have been reported [35] for calculations involving quantum teleportation experiments in non-inertial frames. Our observations are also in accord with general remarks [36,37] that, in analogy with energy, entropy should be a subtle concept in General Relativity.

We have concentrated here on this new observer-dependence in the concept of entropy. It is tempting to speculate that this observation will have further interesting implications for the thermodynamics of black holes. For example, the point here that the two classes of observers assign different values to the entropy flux across the horizon seems to be in tune with the point of view (see, e.g., Refs. [38–42]) that the Bekenstein-Hawking entropy of a black hole does not count the number of black hole microstates, but rather refers to some property of these states relative to observers who remain outside the black hole. Perhaps the observations above will play some role in fleshing out this point of view?

In contrast, suppose that one remains a believer in, say, the Covariant Entropy Bound [8], which purports to bound the flux of entropy across a null surface. Since we have seen that this flux can depend on the choice of an observer, one may ask if this bound should apply to the entropy flux described by all observers, or only to one particular class of observers. We look forward to discussions of these, and other such points, in the future.

Acknowledgments

I am grateful to Djordje Minic, Simon Ross, and Rafael Sorkin for delightful collaborations leading to the results presented above and for other wonderful discussions. I would also like to thank Bob Wald and Ted Jacobson for many fascinating lessons on black hole thermodynamics and black hole entropy. Thanks also to Stefan Hollands for conversations relating to Refs. [16, 17]. Much of the research on the observer dependence of entropy was conducted at the Aspen Center for Physics. This work was supported in part by NSF grant PHY-0354978 and by funds from the University of California.

References

1. D. Marolf and R. Sorkin, *Phys. Rev. D* **66**, 104004 (2002) [arXiv:hep-th/0201255].
2. D. Marolf and R. Sorkin, *Phys. Rev. D* **69**, 024014 (2004), [hep-th/0309218].
3. D. Marolf, D. Minic and S. F. Ross, *Phys. Rev. D* **69**, 064006 (2004), [arXiv:hep-th/0310022].
4. J. D. Bekenstein, *Phys. Rev. D* **7**, 2333 (1973); **23**, 287 (1981); **49**, 1912 (1994).
5. G. 't Hooft, arXiv:gr-qc/9310026.
6. L. Susskind, *J. Math. Phys.* **36**, 6377 (1995) [arXiv:hep-th/9409089].
7. W. Fischler and L. Susskind, arXiv:hep-th/9806039.
8. R. Bousso, *JHEP* **9907**, 004 (1999) [arXiv:hep-th/9905177].
9. R. Brustein and G. Veneziano, 'A Causal Entropy Bound,' Phys. Rev. Lett. **84**, 5695 (2000) [arXiv:hep-th/9912055].
10. J. D. Bekenstein, arXiv:gr-qc/0107049.
11. W. G. Unruh and R. M. Wald, *Phys. Rev. D* **25**, 942 (1982).
12. W. G. Unruh and N. Weiss, *Phys. Rev. D* **29**, 1656 (1984).
13. I. Rumanov, arXiv:hep-th/0408153.
14. J. D. Bekenstein, *Phys. Rev. D* **70**, 121502 (2004), [arXiv:hep-th/0410106].
15. S. W. Hawking, *Commun. Math. Phys.* **43**, (1975) 199.
16. B. S. Kay and R. M. Wald, Phys. Rept. **207**, (1991) 49.

17. K. Fredenhagen and R. Haag, Commun. Math. Phys. **127**, (1990) 273.
18. J. M. Maldacena, *Adv. Theor. Math. Phys.* **2**, (1998) 231 [arXiv:hep-th/9711200].
19. L. Susskind and E. Witten, arXiv:hep-th/9805114.
20. E. E. Flanagan, D. Marolf and R. M. Wald, *Phys. Rev. D* **62**, (2000) 084035 [arXiv:hep-th/9908070].
21. R. Bousso, *Rev. Mod. Phys.* **74**, (2002) 825 [arXiv:hep-th/0203101].
22. R. Bousso, E. E. Flanagan and D. Marolf, *Phys. Rev. D* **68**, (2003) 064001 [arXiv:hep-th/0305149].
23. J. D. Bekenstein, *Ninth Marcel Grossmann Meeting on Recent Developments in Theoretical and Experimental General Relativity, Gravitation and Relativistic Field Theories*, eds. V. G. Gurzadyan, R. Jantzen and R. Ruffini, (World Scientific, Singapore 2002), pp. 553-559, [arXiv:gr-qc/0007062]; J. D. Bekenstein, *Phys. Lett. B* **481**, 339 (2000) [arXiv:hep-th/0003058].
24. R. Laflamme, *Nucl. Phys. B* **324**, 233 (1989).
25. A. O. Barvinsky, V. P. Frolov and A. I. Zelnikov, *Phys. Rev. D* **51**, 1741 (1995) [arXiv:gr-qc/9404036].
26. T. Jacobson, *Phys. Rev. D* **50**, 6031 (1994) [arXiv:gr-qc/9407022].
27. T. A. Jacobson, arXiv:hep-th/9510026.
28. T. Jacobson, *Class. Quant. Grav.* **15**, 251 (1998) [arXiv:gr-qc/9709048].
29. P. Kraus and F. Wilczek, *Nucl. Phys. B* **433**, 403 (1995) [arXiv:gr-qc/9408003]; P. Kraus and F. Wilczek, *Nucl. Phys. B* **437**, 231 (1995) [arXiv:hep-th/9411219]; E. Keski-Vakkuri and P. Kraus, *Nucl. Phys. B* **491**, 249 (1997) [arXiv:hep-th/9610045]; M. K. Parikh and F. Wilczek, *Phys. Rev. Lett.* **85**, 5042 (2000) [arXiv:hep-th/9907001].
30. S. Massar and R. Parentani, *Nucl. Phys. B* **575**, 333 (2000) [arXiv:gr-qc/9903027]; [arXiv:gr-qc/9801043].
31. R. D. Sorkin, in *Black Holes and Relativistic Stars*, ed. R.M. Wald (University of Chicago Press, Chicago, 1998) pp. 177-194, [arXiv:gr-qc/9705006].
32. T. Jacobson and R. Parentani, *Found. Phys.* **33**, (2003) 323 [arXiv:gr-qc/0302099].
33. R. D. Sorkin, "On The Entropy Of The Vacuum Outside A Horizon," Gen. Rel. Grav., proceedings of the GR10 Conference, Padova 1983, ed. B. Bertotti, F. de Felice, A. Pascolini (Consiglio Nazionale della Ricerche, Roma, 1983) Vol. 2; L. Bombelli, R. K. Koul, J. H. Lee and R. D. Sorkin, "A Quantum Source Of Entropy For Black Holes," Phys. Rev. D **34**, 373 (1986); M. Srednicki, "Entropy and area," Phys. Rev. Lett. **71**, 666 (1993) [arXiv:hep-th/9303048].
34. W. G. Unruh, *Phys. Rev. D* **14** (1976) 870.
35. P. M. Alsing, D. McMahon and G. J. Milburn, arXiv:quant-ph/0311096.
36. R. M. Wald, "Gravitation, thermodynamics, and quantum theory," Class. Quant. Grav. **16** (1999) A177 [arXiv:gr-qc/9901033].
37. R. M. Wald, "The thermodynamics of black holes," Living Rev. Rel. **4**, 6 (2001) [arXiv:gr-qc/9912119].
38. R. Parentani and T. Piran, Phys. Rev. Lett. 73 (1994) 2805 [hep-th/9405007].
39. T. Banks, Nucl. Phys. Proc. Suppl. **41** (1995) 21 [hep-th/9412131].

40. T. Jacobson, Phys. Rev. Lett. **75** (1995) 1260 [arXiv:gr-qc/9504004].
41. R. D. Sorkin, in Black Holes and Relativistic Stars, Chicago: (The University of Chicago Press, Chicago, 1998) ch. 9, pp. 177-194; [gr-qc/9705006].
42. T. Jacobson, in "General Relativity and Relativistic Astrophysics, Eighth Canadian Conference, Montréal, Québec, 1999," (AIP, Melville, NY, 1999, eds. C. P. Burgess and R. Myers) [arXiv:gr-qc/9908031].

ADVANCED GRAVITATIONAL-WAVE DETECTOR TECHNOLOGIES: FUTURE GENERATIONS OF INTERFEROMETERS

NERGIS MAVALVALA

LIGO Laboratory,
Massachusetts Institute of Technology,
NW17-161, 175 Albany Street,
Cambridge, MA 02139

This talk deals with plans for and progress on the design and construction of the Advanced LIGO facility. The strain sensitivity of the initial LIGO detector is recalled, and the astrophysical motivation for improvements is outlined. The dramatic increase in the number of potentially detectable sources motivate the sensitivity target for Advanced LIGO – a factor of 15 improvement over initial LIGO across the entire frequency band from about 40 Hz to 10 kHz. The goal of advanced detector technology is to reach sensitivities that are limited by fundamental noise sources (gravity gradients at very low frequencies, and quantum noise at higher frequencies, respectively). The talk outlines strategies for meeting these targets and the challenges that they present, and concludes with a discussion of the prospects for moving beyond the quantum limit.

The Advanced LIGO facility aims to improve the strain sensitivity of LIGO I by a factor of 10 - 15 and to improve on the lower frequency bound by a factor of 4. This will mean a greatly increased number of astrophysical and cosmological sources of gravitational waves will come into the detection range of the observatory: for example, sensitivity to neutron star binary mergers out to c. 350 Mpc, compared to 20 Mpc for initial LIGO. The enhanced performance is to be achieved through a reduction of noise across the three principal categories in the facility: seismic noise (both natural and man-made), thermal noise (arising from vibrations due to finite temperature) and optical/shot noise.

Seismic noise is reduced by the use of a multi-stage active isolation system, with arrays of sensors and actuators at each stage to measure and suppress vibrations. The mirrors of the interferometer are further suspended as the fourth stage of a quadruple pendulum. The reduction of thermal noise is principally effected by the use of more efficient suspension mechanisms for the test masses, and by the use of heavier test masses made from materials with higher mechanical quality. The main strategies for the reduction of optical noise are the factor of 20 increase of the power of the interferometer's laser source, and the use of signal recycling.

The final part of the talk deals with quantum noise in optical experiments and its removal, and the application of this to gravitational wave interferometry. Quantum noise manifests itself as radiation pressure noise (a photon-induced force on the test masses) and shot noise (statistical error in the photon counting by the photodetection process). Both sources of noise can be reduced by the injection of squeezing states into the antisymmetric port of the beamsplitter. These two quantum noise sources are associated with different quadratures of the laser field, and dominate at different frequencies, which requires squeezing with frequency-dependent quadrature (squeeze) angles for optimal sensitivity at all frequencies.

Other emerging detector technologies are briefly mentioned, including the use of cryogenic suspensions, broadband interferometers, all-reflective interferometers, reshaped laser beam profiles and quantum non-demolition techniques.

PROBING GENERAL RELATIVITY ON THE SCALES OF COSMOLOGY

P. J. E. PEEBLES

Joseph Henry Laboratories, Princeton University,
Princeton NJ 08544, USA
e-mail: pjep@princeton.edu

The cosmological tests are tight enough now to show that the Friedmann-Lemaître $\Lambda-$CDM cosmological model almost certainly is a useful approximation. This means general relativity theory passes significant tests of the extrapolation of some fifteen orders of magnitude from the length scales of the precision tests of gravity physics.

1. Introduction

Cosmology is tested by checking its consistency with observational constraints that are substantially more numerous than the freedom to adjust the theory and the interpretations of the observations. We are reaching this goal for some aspects of cosmology: particularly impressive is the abundance of evidence that the matter density parameter is in the range

$$0.15 \lesssim \Omega_m \lesssim 0.3. \tag{1}$$

This certainly is not precision cosmology, but the consistency of estimates that depend on many different aspects of the theory and a broad variety of astronomical methods show that it almost certainly is accurate cosmology, that is, that we have not been misled by systematic errors. That means general relativity theory passes significant empirical tests on the scale of the observable universe. The measurements supporting this remarkable advance are reviewed in Sec. 3. After brief comments on the more modest network of evidence for the detection of Λ, in Sec. 4, I conclude with a qualitative assessment of our understanding of the physics of spacetime and gravity on the scales of cosmology, and the case for developing a more systematic framework for judging the empirical evidence.

2. The Standard Cosmology

It will be useful for what follows to consider the elements of the present standard Friedmann-Lemaître $\Lambda-$CDM cosmology. In this model,

(1) gravity and spacetime are described by general relativity theory;
(2) the stress-energy tensor represents
   (a) textbook physics in the visible sector, with baryon density $\Omega_b = 0.04$ and thermal radiation at $T_o = 2.725$ K;
   (b) cold dark matter, with $\Omega_{CDM} + \Omega_b$ in the range of Eq. (1), and dark energy sufficient for flat space sections;
(3) initial conditions that describe
   (a) expansion from very high redshift;
   (b) growth of structure out of adiabatic, Gaussian, and scale-invariant departures from homogeneity and isotropy.

It is a measure of the durability of these concepts — and the slow progress of research in cosmology — that much of this picture was under discussion in the 1930s. Eddington disliked the idea that our universe expanded from a very different state, but Lemaître [1] celebrated it as a chance to learn new physics. Tolman showed that homogeneous and isotropic expansion cools a uniform sea of blackbody radiation while preserving the thermal spectrum. I know no evidence that he thought there might be an observable consequence; Gamow introduced that in the 1940s, with his proposal for the formation of deuterium and heavier elements. In the 1930s Zwicky and Hubble knew that the masses of nearby clusters of galaxies appear to be larger than the mass in the luminous parts of the galaxies, but I suspect it would have taken considerable explanation to convince them that Gamow's theory of the origin of the light elements requires that most of the missing mass — or dark matter — has to be nonbaryonic. The other component of the dark sector would have been easy to explain in the 1930s, and perhaps also the reason for its new name, because Lemaître, Eddington, and others were aware of the interpretation of Λ as an energy density with a curious equation of state (as reviewed in Sec. III.B.3 in Ref. 2). Pauli knew the problem with the quantum vacuum energy density, though perhaps not the possible relation to the issue of Λ. In his 1933 article [3] on quantum mechanics Pauli advised ignoring the zero-point energy of the electromagnetic field, because its energy density is absurd, but I expect he would have agreed that his prescription makes no sense. This is a good description of the present understanding of the vacuum energy.

In the 1930s Lemaître proposed that gravity drove structure formation. Four decades ago this was still debated. Two decades ago the gravitational instability picture was generally accepted but initial conditions were debated, whether set by explosions or cosmic strings or global monopoles, or to be stipulated as primeval isocurvature or adiabatic departures from homogeneity. The community has fixed on a variant of the last, in 3b, because it fits the observations. That is good science, but to be borne in mind when we consider the freedom to adjust the theory to fit the observations.

Einstein and de Sitter model proposed their model in 1932. The evidence that the mass density is lower than the Einstein-de Sitter value would not have surprised de Sitter. Would the evidence that the low mass density is accompanied by Λ rather than space curvature have disturbed or pleased Einstein? It was his invention, after all.

Table 1. Constraints on the matter density parameter.

	Measurement	Scale	Ω_m
1	peculiar velocities: relative rms	20 kpc $\lesssim r \lesssim$ 1 Mpc	$0.20e^{\pm 0.4}$
2	redshift space anisotropy	10 Mpc $\lesssim r \lesssim$ 30 Mpc	0.30 ± 0.08
3	mean relative velocities	10 Mpc $\lesssim r \lesssim$ 30 Mpc	$0.30^{+0.17}_{-0.07}$
4	numerical action solutions	$r \sim$ 1 Mpc	0.15 ± 0.08
5	virgocentric flow	$r \sim$ 20 Mpc	$0.20^{+0.22}_{-0.15}$
6	weak lensing: galaxy-mass	100 kpc $\lesssim r \lesssim$ 1 Mpc	$0.20^{+0.06}_{-0.05}$
7	mass-mass	300 kpc $\lesssim r \lesssim$ 3 Mpc	0.31 ± 0.08
8	angular size distance $r(z)$: SNe	$r \sim$ 3000 Mpc	$0.29^{+0.05}_{-0.03}$
9	cluster baryon fraction	$r \sim$ 3000 Mpc	$\lesssim 0.3$
10	local cluster baryon mass fraction	$r \sim$ 10 Mpc	0.27 ± 0.02
11	cluster mass function	$r \sim$ 10 Mpc	0.17 ± 0.05
12	mass fluctuation power spectrum	$r \sim$ 100 Mpc	0.23 ± 0.02
13	integrated Sachs-Wolfe effect	$r \sim$ 300 Mpc	~ 0.3

3. The Mean Mass Density

Table 1 lists measurements of the large-scale mean mass density. Many analyses nowadays employ joint fits to several of the observations listed in the table. That is efficient if we are working with the correct theory, but for the purpose of exploring what aspects of the theory are tested by the measurements it is best to separate the results from each physical effect or each significantly different means of observing an effect.

My comments, with sample references, may be too brief for those who do not follow this subject and redundant for those who do, but I hope the point is clear: quite different aspects of the theory and observations enter the thirteen measurements of nominally the same quantity, and the degree of consistency of the results checks the considerable variety of potential sources of systematic errors in the theory and observations.

3.1. *Mass Estimates*

The third column in the table indicates a characteristic range of length scales to which each measurement is sensitive. Here and throughout I adopt the distance scale represented by Hubble's constant

$$H_o = 70 \text{ km s}^{-1} \text{ Mpc}^{-1}. \tag{2}$$

The estimates of the mass density parameter Ω_m — the ratio of the mean mass density to the critical Einstein-de Sitter value — in the last column scale with the value of H_o in a variety of ways, but the effect of the uncertainty in H_o, perhaps 20 percent, on the scatter among results is smaller than the measurement errors.

The first seven estimates in Table 1 assume optically selected galaxies are useful mass tracers (where I mean by "useful" that the error introduced by this assumption is not much larger than the measurement error). I review the case for this assumption after commenting on these seven results.

The first two entries are based on the galaxy two-point correlation function in redshift space, where the distance of a galaxy with redshift z is cz/H_o. At separations $r \lesssim 1$ Mpc the random motions of galaxies elongate the clustering in redshift space along the line of sight. I have used an old measurement [4] of this small-scale effect in entry 1 because more recent surveys tend not to probe small scales. On larger scales the redshift space correlation function is flattened along the line of sight by the streaming motion accompanying the growth of clustering. Modern surveys beautifully demonstrate this effect [5]. The third entry probes the streaming by the measurement of the mean relative line-of-sight velocity as a function of separation [6]. The next entry grew out of the remark by Kahn and Woltjer [7] that the two large members of the Local Group of galaxies appear to have mass considerably larger than the observed star mass, in the approximation of a two-body system. Entry 4 comes from a more elaborate solution [8] that takes account of the mass distribution out to 40 Mpc distance, but I have entered a length scale characteristic of the Local Group because the motions of the galaxies in and near this system are the important aspect of the solution. The next entry is derived from the density contrast [9] and streaming flow [10] of galaxies in the Local Supercluster, which is centered on the Virgo Cluster, at about 20 Mpc distance, with us near the edge.

Entries 6 and 7 are based on weak gravitational lensing. The first uses the mean shear in the images of distant galaxies caused by the gravitational deflection by the masses in nearer galaxies, in the shear-galaxy cross correlation function measured by McKay *et al.* [11] and analyzed in Ref. 12. In entry 7 I quote the measurement and analysis of the shear autocorrelation function of galaxy images at redshifts $z \sim 1$ by Rhodes *et al.* [13]

The assumption that galaxies are useful mass tracers is tested by the consistency of the estimates of Ω_m from a considerable range of lengths, 100 kpc $\lesssim r \lesssim$ 30 Mpc, because if galaxies were biased mass tracers one might have thought the bias would vary with the length scale. An elegant and more direct test compares the galaxy three-point correlation function, scaled by the square of the two-point function, to the same expression for the mass correlation functions computed in lowest nontrivial order in perturbation theory [14]. The latter assumes Gaussian initial conditions and models the relation between the galaxy and mass density contrasts as

$$\delta_{\rm g} = b_1\delta_\rho + b_2\delta_\rho^2/2. \tag{3}$$

Feldman *et al.* [15] give a striking illustration of the consistency of functional forms of the galaxy and mass functions in this model. Their fit to the clustering of galaxies selected by the IRAS satellite detection at $\lambda \sim 100\mu$ requires constants $b_1 \sim 0.8$ and $b_2 \sim -0.4b_1^2$, consistent with the picture that IRAS galaxies are more smoothly distributed than the mass and particularly avoid the densest regions. This agrees with the observation that gas-rich galaxies that can be luminous in the far infrared avoid dense regions where interactions more readily deplete interstellar gas. For optically selected galaxies b_1 is quite close to unity and b_2 is close to zero [16], consistent with the picture that these galaxies trace mass on scales $\sim$ 5 to 30 Mpc. This test assumes Gaussian initial conditions, but if the result were an artifact of this assumption the agreement of functional forms would be a curious coincidence. The lessons are that galaxies need not trace mass, as illustrated by the IRAS sample, but that Nature seems to have been kind enough to have made optically selected galaxies useful mass tracers.

Entry 8 refers to measures of the angular size distance $H_oa_or(z)$ as a function of redshift from the redshift-magnitude relation for Type Ia supernovae. I quote the fit [17] for zero space curvature. A similar result is obtained when the supernovae data are supplemented by $r(z)$ estimates for radio galaxies [18]. Since this describes the general expansion of the observable universe I have entered the present Hubble length as a measure of the characteristic length scale.

The next two entries are based on the idea that rich clusters of galaxies likely are large enough to have gravitationally collected a close to fair sample of baryons and dark matter. The constraint on $r(z)$ from the condition that the cluster baryon mass fraction is independent of redshift indicates that Ω_m is close to zero if $\Lambda = 0$ and is about at the upper bound in Eq. (1) if space sections are flat [19]. Entry 10 is from the comparison of the mass fraction in clusters [20] to the mean baryon density [12] from the standard model for the light elements; it is not sensitive to Λ.

Entry 11 refers to the evolution of the number density of massive clusters. There are in effect two constraints: the number densities of clusters at low redshift and high. A fit [21] to the two free parameters offered by the standard cosmology with negligible space curvature yields the value of Ω_m in the table and an amplitude of the primeval scale-invariant mass density fluctuations that agrees with the fluctuations in galaxy counts, yet another indication that galaxies are useful mass tracers. The length scale here and in entry 10 is the radius from which the cluster mass is gathered.

Part 3b in the standard cosmology fixes in linear perturbation theory the functional form of the mass autocorrelation function $\xi(r)$ — for example, $\xi(r)$ passes through zero at $r = 30\Omega_m^{-1}$ Mpc (for H_o in Eq. (2)) — so if galaxies trace mass the shape of the large-scale galaxy two-point correlation function offers a measure of Ω_m. Entry 12 is from the 2dF galaxy redshift survey [22].

If space curvature or Λ make an appreciable contribution to the present expansion rate then in the standard model the 3 K CBR radiation is perturbed at low redshift by the evolving gravitational potential of the mass distribution. Boughn and Crittenden [23] show that the effect is detected with reasonable confidence in the cross correlation of the WMAP CBR anisotropy measurement [24] with the angular distributions of radio and X-ray sources at redshifts close to unity. The constraint is not tight — the effect is about what is expected if $\Omega_m \sim 0.3$ and space sections are flat — but it is an elegant convergence of theory and observation.

I have not entered the comparison of the expansion time from high redshift with stellar evolution and radioactive decay ages because the constraint on Ω_m is somewhat looser (as shown in Table 15 in Ref. 25), but it certainly is an important test of the cosmology.

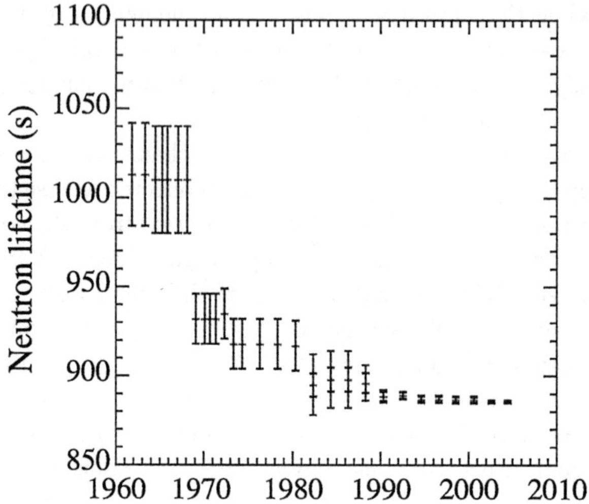

Figure 1. Evolution of the handbook value of the neutron mean life.

3.2. *Systematic Errors*

The considerable number of entries in Table 1 allows us the luxury of considering the considerable variety of potential sources of systematic error, under the headers of astronomy, physics and sociology.

To be confident about our deductions of aspects of fundamental physical reality from limited observations of processes operating on the far side of the visible universe we need to demonstrate consistency of evidence from a broad variety of astronomical methods that are subject to different uncertainties. I conclude that for the purpose of constraining the value of Ω_m at the modest precision of the entries in Table 1 this condition is satisfied: we can trust the astronomy.

We must consider that we have adjusted the theory to fit the observations: that is how we arrived at part 3b of the standard model. The case that this is more than curve fitting is supported by the magnificently successful fit of the standard model to the WMAP measurements of the CBR anisotropy spectrum, with parameters that agree with independent estimates of the baryon mass density and the slope and normalization of the mass fluctuation power spectrum [24]. We cannot know for sure that there is no viable modified — or alternative — model for structure formation, however, so it is important that most of the entries in Table 1 are not sensitive to the details in part 3b, and that the consistency with entries 11 and 12, which do depend on this part of the standard model, adds to the evidence that the ΛCDM cosmology is a useful approximation.

We also have to bear in mind that scrupulously careful measurements can be influenced by our respect for social norms. The reluctance to stray more than one or two standard deviations from what is generally accepted is illustrated by variations with time in the handbook values of physical constants, as in the example [26] in figure 1. One sees a similar effect in the history of estimates of Ω_m. In 1980, using the method in entry 1 in Table 1, I got [27] $\Omega_m = 0.65 \pm 0.25$. My prejudice then was no secret: I argued that the only reasonable case is the Einstein-de Sitter model, with $\Lambda = 0$ and $\Omega_m = 1$, and I was glad to see that my estimate was not significantly off the right answer. A few years later the inflation concept led a good part of the community to accept that the Einstein-de Sitter model very likely is the right answer. You can find in the literature of the following decade careful observational analyses that indicate consistency with Einstein-de Sitter. Different observational advances drove different groups away from this consensus. It happened to me in 1982, with the demonstration [4] of small relative velocity dispersion in the field (from the Center for Astrophysics redshift catalog), together with the absence of the segregation of normal and dwarf galaxies that I would have expected to see as a consequence of the biasing picture that was introduced to save this phenomenon. By the early 1990s the large scales of clustering of galaxies and clusters of galaxies [28], [29], [30] and the cluster mass function [30] and baryon mass fraction [31] had added to the evidence for low Ω_m. But the general — and unnervingly abrupt — swing of opinion away from the Einstein-de Sitter model was triggered in 1998 by the supernovae redshift-magnitude measurements [32, 33].

It takes nothing away from the enormous amount of careful work by many people to remark that that social pressure likely has narrowed the scatter in Table 1. Could the apparent consistency of evidence summarized in this table be a social construction, like the case for the Einstein-de Sitter model ten years ago? Each of us may decide whether there are enough checks to rule out significant systematic errors from this source as well as the astronomy and physics. The much broader variety of evidence developed since the 1990s, and the larger number of people who are probing for something new and interesting to do with the evidence, lead me to conclude that we have reached a compelling case for real consistency at the accuracy indicated by the general run of error flags in Table 1.

4. The Cosmological Constant

Is the Λ term in the standard cosmology a social construction? The accompanying evidence for flat space sections was welcomed as a prediction of inflation, but you can count that welcome as an effect of social pressure because inflation is a social construction, which is to say that it is a promising working hypothesis that awaits searching scientific tests. It is important that there are independent lines of evidence for the detection of Λ, mainly from measures of the angular size distance as a function of redshift [17], and from the WMAP measurement of the CBR anisotropy [24]. The latter is somewhat beclouded by its dependence on a structure formation model with anomalies that, if real, will drive adjustments of the model and maybe of the constraint on Λ, and it is beclouded also by the puzzle of the quantum vacuum energy density, which might drive adjustments of the world picture [34] or of the gravity theory [35, 36] and the interpretation of the cosmological tests. One can make a similar list of hazards for each estimate of Ω_m, of course; the big difference is in the lengths of the lists of independent evidence. The issue of Einstein's cosmological constant has been under discussion since 1917. I suggest we wait a few more years to see how the evidence develops — as from the cluster baryon mass fraction [20], the cluster mass function [21], and the expansion time scale [25] — before making a definite decision about the reality of this curious term.

5. Testing Gravity Physics on the Scales of Cosmology

In the 1960s, during the development of many of the precision tests of gravity physics on the scale of the Solar System and smaller, the PPN formalism offered a helpful way to judge how well the tests are doing: find the ranges of parameters allowed by the measurements, and see whether the allowed ranges include general relativity. We do not have an analogous formalism suitable for cosmology. The purpose of these qualitative remarks is to show why it is time to develop one.

The first seven entries in Table 1 make direct use of the inverse square law for gravitational acceleration and the assumption that galaxies are useful mass tracers. It would be absurd to imagine that a failure of both assumptions conspired to produce the apparent consistency of measurements over the range of scales from 100 kpc to 30 Mpc: we may conclude that the inverse square law passes this test, within the still substantial error budget.

The inverse square law has the special property that it preserves the functional form of the mass distribution in the growing mode of the departure from homogeneity in linear perturbation theory (because in this approximation the time derivative of the mass density contrast $\delta\rho/\rho$ is proportional to the divergence of the peculiar velocity field, which is proportional to the peculiar gravitational acceleration $\vec{g}$, and the inverse square law says the divergence of $\vec{g}$ is proportional to $\delta\rho/\rho$). A failure of the inverse square law would cause the rate of growth of the Fourier amplitudes of the mass distribution to depend on the wavelength [37, 38]. This means we have another check of the inverse square law from the apparent consistency of stories for the growth of clusters of galaxies, the shape of the large-scale mass fluctuation spectrum, the CBR anisotropy spectrum, and the detection of the contribution to the anisotropy by the ISW effect.

Entries 6 and 7 take account of the factor of two difference between the relativistic gravitational deflection of light and the naive Newtonian model. We know the factor of two is present on the scale of the Solar System. We have good reason to expect the factor of two is present on the scales of extragalactic astronomy, but I hope it is agreed that an actual observational check would be deeply satisfying.

Under the assumptions of standard local physics and a metric description of spacetime the angular size distance $r(z)$ fixes the redshift-magnitude relation, the expansion parameter $a(t)$ fixes the expansion time as a function of redshift, and both are used in the predictions of counts of conserved objects as a function of redshift and of the anisotropy of the CBR produced by the physics of decoupling. In the standard cosmology these two functions are determined by the Friedmann equations,

$$\left(\frac{\dot{a}}{a}\right)^2 = \frac{8}{3}\pi G\rho + \frac{1}{a^2R^2} + \frac{1}{3}\Lambda, \qquad \frac{\ddot{a}}{a} = -\frac{4}{3}\pi G(\rho + 3p) + \frac{1}{3}\Lambda, \tag{4}$$

where aR is the radius of curvature of space sections in the Robertson-Walker line element. The second expression — from the time derivative of the first — illustrates the contribution of the pressure to the active gravitational mass density. The angular size distance may be represented by the Sachs equation for the proper area $A = \mathrm{d}^2$ of a beam of light running from a distant object to an observer at $\mathrm{d} = 0$. In an exactly homogeneous and isotropic Friedmann-Lemaître model the Sachs equation is

$$\frac{d^2\mathrm{d}}{dt^2} - \frac{\dot{a}}{a}\frac{d\mathrm{d}}{dt} = -4\pi G(\rho + p)\mathrm{d}. \tag{5}$$

The cosmological constant Λ cancels here, but in models with evolving dark energy density, where $p_\Lambda \neq -\rho_\Lambda$, there is a contribution from the dark energy. The role of pressure in these examples, and in the dynamics of relativistic fluids, is elegant, and it is checked by the standard models for the light elements and the CBR anisotropy. A formal analysis of this check would be elegant too.

A theme of this essay is that we gain insight on what we are learning from the study of cosmology by considering how well we can constrain the fundamental elements of the subject. The present state of the tests draws special attention to the empirical evidence for the gravitational inverse square law and its relativistic generalizations. The evidence may be organized in three categories by the ranges of characteristic length scales,

$$\begin{aligned} 10^{-1}\ \mathrm{cm} &\lesssim r_{\mathrm{a}} \lesssim 10^{13}\ \mathrm{cm}, \\ 10^{13}\ \mathrm{cm} &\lesssim r_{\mathrm{b}} \lesssim 10^{21.5}\ \mathrm{cm}, \\ 10^{23.5}\ \mathrm{cm} &\lesssim r_{\mathrm{c}} \lesssim 10^{27}\ \mathrm{cm}. \end{aligned} \tag{6}$$

Class (a) represents the precision tests, from the laboratory and from measurements in the Solar System and binary pulsars [35]. The class (b) evidence is that the inverse square law continues to predict the gravitational acceleration, to a factor of two or so, to about 1 kpc, on the basis of the consistency of dynamical mass estimates in the inner parts of large galaxies with estimates of the seen masses in stars. In the factor ~ 100 gap to the class (c) results from cosmological tests, something happens that drives us to the postulate of dark matter with about five times the mass of the baryons. Since the standard theories of light elements and the CBR anisotropy require this dark matter we may expect to have a convincing closure of this gap. The modest accuracy in class (c) extends to ~ 30 Mpc, and to reasonable evidence of a detection, in the ISW effect, at ~ 300 Mpc. There are even better examples of the ability of standard physics to organize phenomena over enormous ranges of scales, but Eq. (6) is impressive enough, and cosmology is enjoying a growth spurt that will continue to improve the tests of gravity physics on really large scales.

Acknowledgments

I am grateful for advice from Steve Allen, Neta Bahcall, Masataka Fukugita, Will Percival, and Bharat Ratra.

References

1. G. Lemaître, *Nature* **127**, 706 (1931).
2. P. J. E. Peebles & B. Ratra, *Rev. Mod. Phys.* **75**, 559 (2003).
3. W. Pauli, *Handbuch der Physik* **24**, part 1, 83 (1933).

4. M. Davis, & P. J. E. Peebles, *ApJ* 267, 465 (1983).
5. E. Hawkins, *et al.*, *MNRAS* **346**, 78 (2003).
6. H. Feldman *et al.*, *ApJ* **596**, L131 (2003).
7. F. D. Kahn & L. Woltjer, *ApJ* **130**, 705 (1959).
8. E. J. Shaya, P. J. E. Peebles, & R. B. Tully, *ApJ* **454**, 15 (1995).
9. M. Davis & P. J. E. Peebles, *Ann. Rev. Astron. Ap.* **21**, 109 (1983).
10. M. Bureau, J. R. Mould, & L. Staveley-Smith, *ApJ* **463**, 60 (1996).
11. T. A. McKay, *et al.* astro-ph/0108013 (2001).
12. M. Fukugita & P. J. E. Peebles, *ApJ* **616**, 643 (2004).
13. J. Rhodes *et al.*, *ApJ* **605**, 29 (2004).
14. J. N. Fry, *Phys. Rev. Lett.* **73**, 215 (1994).
15. H. A. Feldman *et al.*, *Phys. Rev. Lett.* **86**, 1434 (2001).
16. L. Verde *et al.* *MNRAS* **335**, 432 (2002).
17. A. Riess *et al.*, *ApJ* **607**, 665 (2004).
18. R. A. Daly & S. G. Djorgovski, *ApJ* **597**, 9 (2003).
19. S. W. Allen, private communication (2004).
20. S. W. Allen *et al.*, *MNRAS* **353** 457, 2004
21. N. A. Bahcall & P. Bode, *ApJ* **588**, L1 (2003).
22. W. J. Percival and the 2dF GRS team, private communication (2004).
23. S. P. Boughn & R. G. Crittenden, *New Astron. Rev.* **49**, 75 (2005).
24. C. L. Bennett *et al.*, *ApJS* **148**, 1 (2003)
25. W. L. Freedman *et al.*, *ApJ* **553**, 47 (2001).
26. S. Eidelman *et al.*, *Phys. Lett.* **B 592**, 1 (2004).
27. P. J. E. Peebles, *New York Academy Sciences Annals* **375**, 157 (1981).
28. G. R. Blumenthal, A. Dekel, & J. R. Primack, *ApJ* **326**, 539 (1988).
29. G. Efstathiou, W. J. Sutherland, & S. J. Maddox, *Nature* **348**, 705 (1990).
30. N. A. Bahcall, & R. Cen, *ApJ* **398**, L81 (1992).
31. S. D. M. White, J. F. Navarro, A. E. Evrard, & C. S. Frenk, *Nature* **366**, 429 (1993).
32. A. G. Riess *et al.* *AJ* **116**, 1009 (1998).
33. S. Perlmutter *et al.* *ApJ* **517**, 565 (1999).
34. A. Vilenkin, astro-ph/0407586
35. E. G. Adelberger, B. R. Heckel, & A. E. Nelson, *Ann. Rev. Nuc. Particle Sci.* **53**, 77 (2003).
36. G. Dvali, G. Gabadadze, & M. Shifman, *Phys. Rev. D* **67** 044020 (2003).
37. P. J. E. Peebles, astro-ph/0208037.
38. C. Sealfon, L. Verde, & R. Jimenez, *Phys. Rev. D* **71** 083004 (2005).

RELATIVITY AND ASTROPHYSICS WITH LISA

STERL PHINNEY

California Institute of Technology , USA

We describe how LISA can be used for testing and discovery in astrophysics and relativity. Particular emphasis will be placed on how it can be used to develop an understanding of the evolution, properties, and static and dynamic space-time structure of black holes. We will also briefly describe what LISA, and a follow-on mission currently under study, can teach us about cosmology and the early universe.

THE GRAVITATIONAL SELF-FORCE

ERIC POISSON
Department of Physics, University of Guelph
Guelph, Ontario, Canada N1G 2W1

The self-force describes the effect of a particle's own gravitational field on its motion. While the motion is geodesic in the test-mass limit, it is accelerated to first order in the particle's mass. In this contribution I review the foundations of the self-force, and show how the motion of a small black hole can be determined by matched asymptotic expansions of a perturbed metric. I next consider the case of a point mass, and show that while the retarded field is singular on the world line, it can be unambiguously decomposed into a singular piece that exerts no force, and a smooth remainder that is responsible for the acceleration. I also describe the recent efforts, by a number of workers, to compute the self-force in the case of a small body moving in the field of a much more massive black hole. The motivation for this work is provided in part by the Laser Interferometer Space Antenna, which will be sensitive to low-frequency gravitational waves. Among the sources for this detector is the motion of small compact objects around massive (galactic) black holes. To calculate the waves emitted by such systems requires a detailed understanding of the motion, beyond the test-mass approximation.

1. Introduction: The Capra scientific mandate

This contribution describes how a body of mass m, supposed to be "small", moves in spacetime. In the test-mass approximation it is known that the body moves on a geodesic of a background spacetime whose metric $g_{\alpha\beta}$ does not depend on m. As m increases (but kept small) the test-mass description is no longer adequate. One might then say that the motion is still geodesic, but in a spacetime whose metric is perturbed with respect to the background metric. Alternatively, one might say that the motion is accelerated in the background spacetime. This is the point of view I shall take in this contribution: The body's motion will be described in the original (unperturbed) spacetime, and the gravitational influence of the body will be incorporated in its acceleration. The body will be said to move under the influence of its gravitational self-force.

The work described here was previously reviewed in a very long article published in *Living Reviews in Relativity* [1]. An abridged version of this article appeared in *Classical and Quantum Gravity* [2]. This contribution focuses on the highlights and presents the "big picture". The reader is warned that my presentation will be sketchy and incomplete, and is referred to the review articles for additional details. The scientific objectives behind the work described here have been pursued by a number of people; I like to refer to these objectives as the "Capra scientific mandate" and to these people as the "Capra posse".

The Capra scientific mandate includes three main objectives. The first is *to formulate the equations of motion of a small body in a specified background spacetime, beyond the test-mass approximation.* This first step was solved in 1997, first by Mino, Sasaki, and Tanaka [3], and then by Quinn and Wald [4]. The equations of motion are now known as the MiSaTaQuWa equations; I will sketch their derivation in this contribution. The second objective is *to concretely describe the motion of the small body in situations of astrophysical interest, including generic orbits of a Kerr black hole.* There has been much recent progress on this front, and I will describe some of the issues involved in this contribution. The third objective is *to properly incorporate the equations of motion into a wave-generation formalism.* This final objective is the most important, as the ultimate goal of this enterprise is to make detailed predictions toward eventual gravitational-wave measurements. This is the holy grail of the Capra program, and it has so far proved elusive. I will present a tentative outline of future work in the last section of this contribution.

The work reviewed in this contribution was shaped by a series of annual meetings named after the late movie director Frank Capra. The first of these meetings took place in 1998 and was held at Capra's ranch in Southern California; the ranch now belongs to Caltech, Capra's alma mater. Subsequent meetings were held in Dublin, Pasadena, Potsdam, State College PA, Kyoto, and Brownsville.

Members of the Capra posse include Paul Anderson, Warren Anderson, Leor Barack, Patrick Brady, Lior Burko, Manuella Campanelli, Steve Detweiler, Eanna Flanagan, Costas Glempedakis, Abraham Harte, Wataru Hikida, Bei Lok Hu, Scott Hughes, Sanjay Jhingan, Dong-Hoon Kim, Carlos Lousto, Eirini Messaritaki, Yasushi Mino, Hiroyuki Nakano, Amos Ori, Ted Quinn, Eran Rosenthal, Norichika Sago, Misao Sasaki, Takahiro Tanaka, Bob Wald, Bernard Whiting, and Alan Wiseman.

2. Astrophysical context

The motivation for the Capra program comes largely from the fact that solar-mass compact bodies moving around massive black holes have been identified as one of the promising sources of gravitational waves for the space-based interferometric detector LISA (Laser Interferometer Space Antenna). (The case for this identification is made in the contributions by Sterl Phinney and Sir Martin Rees.) These systems involve highly eccentric, nonequatorial, and relativistic orbits around rapidly rotating black holes. The waves produced by these orbits will be rich in information concerning the strongest gravitational fields in the Universe, and this information will be extractable from the LISA data stream. The extraction, however, will depend on sophisticated data-analysis strategies that will rely on a detailed and accurate modeling of the source. This modeling involves formulating the equations of motion for the small body in the field of the rotating black hole, *in a small-mass-ratio approximation that goes beyond the test-mass description.* And it involves a consistent incorporation of these equations of motion into a wave-generation formalism. In short, the extraction of this wealth of information relies on the successful completion of the Capra program.

The finite-mass corrections to the orbiting body's motion are important. Let m be the mass of the orbiting body, M the mass of the central black hole, and suppose that $m/M \ll 1$. For concreteness, assume that the orbiting body is a 10 $M_\odot$ black hole and that the central black hole has a mass of 10^6 $M_\odot$. Then $m/M = 10^{-5}$ is the order of magnitude of the correction to the equations of motion relative to the test-mass description. Simultaneously, $M/m = 10^5$ is the order of magnitude of the total number of wave cycles that will be received during a year's worth of LISA observation. This simplistic estimate illustrates that while the corrections to the equations of motion are small, in the course of a year they can accumulate and contribute a significant number of wave cycles.

Corrections to the equations of motion must incorporate both conservative and dissipative effects. Finite-mass corrections that are conservative in nature are familiar from Newtonian and post-Newtonian theory, and they occur also in a strong-gravity situation. These can accumulate over time. Imagine, for example, an eccentric orbit that undergoes periastron advance. A finite-mass correction to this effect will cause a steady drift in the phasing of the orbit, and this will directly be reflected in the phasing of the gravitational waves; after 10^5 orbital cycles the 10^{-5} correction will have grown into a sizable effect.

Dissipative effects, on the other hand, do not occur in Newtonian gravity, but they are familiar in post-Newtonian theory; they are also present in a strong-gravity situation. Dissipation is associated with the radiative loss of energy and angular momentum by the orbital system, and the resulting corrections to the motion also accumulate over time; this translates into a steady drift of the gravitational-wave signal in the LISA frequency band. These are finite-mass corrections, because the test-mass description makes no room for gravitational radiation and radiation reaction.

From a more theoretical point of view, the appeal of this work comes largely from the fact that while the motion of self-gravitating bodies has been studied extensively in the context of post-Newtonian theory (see, for example, Refs. [5,6]), very little is known in the case of strong fields and fast motions. To the relativist working in this area, this problem is irresistible: We have strong gravity, fast motion, and a smallness parameter in the form of m/M. We have a cool problem that can be solved by standard perturbative techniques.

3. Motion of a black hole

Let me restate the problem in its most general form: A body of mass m moves in an arbitrary (but empty of matter) spacetime whose radius of curvature (in the body's neighbourhood) is $\mathscr{R}$; what is the description of its motion when $m \ll \mathscr{R}$? This formulation of the problem is more general than the two-body version stated previously. When the small body is a member of a binary system, and is at a distance b from another body of mass M, then $\mathscr{R} \sim \sqrt{b^3/(M+m)}$ and

$$\frac{m}{\mathscr{R}} \sim \frac{m}{M+m} v^3,$$

where $v = \sqrt{(M+m)/b}$. For relativistic motion ($v \sim 1$) this is small whenever $m/M \ll 1$.

The clean separation of scales allows us to idealize the motion as following a world line in a spacetime whose metric $g_{\alpha\beta}$ will be specified below. While the region of spacetime occupied by the body is truly a world tube of finite extension, on a scale $\mathscr{R}$ — the only scale of relevance in the background spacetime — this extension is so small that little is lost by making this idealization. The body shall then follow a world line γ that will be described by parametric relations $z^\mu(\tau)$, in which τ is proper time measured in the metric $g_{\alpha\beta}$. We wish to determine this world line. It is understood that the world line loses all significance when the neighbourhood of the body is examined on the fine scale m; on this scale the finite extension of the world tube is fully revealed.

It is desirable to choose the body to have an internal structure that is as simple as possible. One thus eliminates an important source of technical complications, but one still feels confident that the resulting equations of motion will apply, to a good degree of accuracy, to a body of arbitrary structure. (This "effacement property" of the internal structure is well established in post-Newtonian theory [5].) This attitude often leads the researcher to assume that the body is a point particle. We shall refrain from doing so at this stage, but we will come back to this description in Sec. 4. We shall instead choose the body to have the simplest structure compatible with the laws of general relativity: it shall be a nonrotating black hole. This was one of the starting assumptions of Mino, Sasaki, and Tanaka [3] (see also Ref. [7]).

The motion of the black hole is determined by *matched asymptotic expansions*, a powerful technique that has known many useful applications in general relativity (see, for example, Refs. [8, 9]). In this approach the metric of the black hole perturbed by the tidal gravitational field supplied by the external universe is matched to the metric of the external universe perturbed by the moving black hole. Demanding that this metric be a valid solution to the Einstein field equations determines the motion without additional input.

The method of matched asymptotic expansions relies on the existence of (i) an *internal zone* in which gravity is dominated by the black hole, (ii) an *external zone* in which gravity is dominated by the conditions in the external universe, and (iii) an overlapping region known as the *buffer zone*, in which the black hole and the external universe have comparable influence. Let r be a meaningful measure of spatial distance from the black hole. Then the internal zone is the region of spacetime in which $r < r_i =$ constant, where $r_i \ll \mathscr{R}$; thus $r/\mathscr{R}$ is small throughout the internal zone and the tidal influence of the external universe is small. The external zone, on the other hand, is the region of spacetime in which $r > r_e =$ constant, where $r_e \gg m$; thus m/r is small throughout the external zone and the gravitational effects of the black hole are small. Finally, the buffer zone is the region of spacetime in which $r_e < r < r_i$. This region exists because $m \ll \mathscr{R}$, and $r/\mathscr{R}$ and m/r are simultaneously small in the buffer zone.

An expansion of the metric in the internal zone has the schematic form

$$\mathbf{g} = g(\text{Schwarzschild}) + O(r^2/\mathscr{R}^2) + \cdots, \tag{1}$$

where g(Schwarzschild) is the metric of a nonrotating black hole in isolation, and $O(r^2/\mathscr{R}^2)$ is the tidal field supplied by the external universe. (Terms that scale as $r/\mathscr{R}$ have been removed by a coordinate transformation.) On the other hand, an expansion of the metric in the external zone has the schematic form

$$\mathbf{g} = g(\text{background}) + O(m/r) + \cdots, \tag{2}$$

where g(background) is the metric of the external universe in the absence of the black hole, and $O(m/r)$ is the perturbation produced by the moving black hole. The key idea of matched asymptotic expansions is that the expansions of Eqs. (1) and (2) are both valid in the buffer zone, and that both forms of $\mathbf{g}$ represent the same metric (up to a coordinate transformation). Performing the matching returns the black hole's equations of motion.

External zone

To flesh out these ideas let us first examine the situation in the external zone. Let $g_{\alpha\beta}$ stand for g(background), the metric of the external universe in the absence of the small black hole. (In the astrophysical context of Sec. 2 this would be the metric of the massive black hole, in isolation.) Let γ be a fiducial world line in this background spacetime (later to be identified with the small hole's world line), and express the metric in normal coordinates centered on γ. It will have the form of $g = \eta + O(r/\mathscr{R}) + O(r^2/\mathscr{R}^2) + \cdots$, where η is the metric of flat spacetime. The normal coordinates do not extend beyond $r = \mathscr{R}$ and are therefore restricted to the buffer zone.

More specifically, let the normal coordinates be the retarded coordinates $(u, x^a = r\Omega^a)$, such that u is constant on each future light cone centered on γ (and u is equal to proper time on the world line), r is an affine parameter on the cone's null generators, and $\Omega^a \equiv x^a/r$ is constant on each generator. Then the time-time component of the background metric takes the form

$$g_{uu} = -\big(1 + 2ra_a\Omega^a + r^2\mathscr{E}_{ab}\Omega^a\Omega^b\big) + O(r^3/\mathscr{R}^3), \tag{3}$$

where $a_a(u)$ is the acceleration vector of the world line, and $\mathscr{E}_{ab}(u) = C_{uaub} = O(1/\mathscr{R}^2)$ are the electric components of the Weyl tensor evaluated on γ. (This three-tensor is symmetric and tracefree.) Our main goal is to eventually determine a_a; we can anticipate that $a_a = O(m/\mathscr{R}^2)$ since the motion must be geodesic in the test-mass limit.

Let $h_{\alpha\beta}$ be the perturbation produced by the moving black hole, so that the full metric is $\mathsf{g}_{\alpha\beta} = g_{\alpha\beta} + h_{\alpha\beta}$. A standard technique in perturbation theory is to introduce the trace-reversed potentials

$$\psi_{\alpha\beta} = h_{\alpha\beta} - \frac{1}{2} g_{\alpha\beta} (g^{\gamma\delta} h_{\gamma\delta}) \tag{4}$$

and to impose the Lorenz gauge condition

$$\psi^{\alpha\beta}{}_{;\beta} = 0. \tag{5}$$

Here and below, all indices are manipulated with the background metric, and covariant differentiation is taken to be compatible with this metric.

In the external zone the perturbation produced by the black hole cannot be distinguished from one produced by a point particle of mass m moving on γ, and $\psi_{\alpha\beta}$ must be a solution to the wave equation

$$\Box \psi^{\alpha\beta} + 2R^{\alpha}{}_{\gamma}{}^{\beta}{}_{\delta} \psi^{\gamma\delta} = -16\pi T^{\alpha\beta}, \tag{6}$$

where $T^{\alpha\beta}$ is the particle's stress-energy tensor. This equation can be solved by means of a retarded Green's function,

$$G^{\alpha\beta}{}_{\gamma'\delta'}(x,x') = U^{\alpha\beta}{}_{\gamma'\delta'}(x,x')\delta(\sigma) + V^{\alpha\beta}{}_{\gamma'\delta'}(x,x')\theta(-\sigma), \tag{7}$$

where $\sigma(x,x')$ is half the squared geodesic distance between the points x and x', and $U^{\alpha\beta}{}_{\gamma'\delta'}(x,x')$, $V^{\alpha\beta}{}_{\gamma'\delta'}(x,x')$ are smooth bitensors. The Green's function is decomposed into a singular "light-cone part" that has support on $\sigma = 0$ only, and a smooth "tail part" that has support on $\sigma < 0$ (so that x is in the chronological future of x'). (It should be noted that this representation of the Green's function is valid only when x is in the normal convex neighbourhood of x'. In the sequel I will simplify expressions by pretending that the decomposition holds globally. The reader is referred to LRR [1] for all the fine print.)

The solution to Eq. (6) is

$$\psi^{\alpha\beta}(x) = \frac{4m}{r} U^{\alpha\beta}{}_{\gamma'\delta'}(x,x') u^{\gamma'} u^{\delta'} + \psi^{\alpha\beta}_{\text{tail}}(x) \tag{8}$$

where $x' \equiv z^{\mu}(u)$ is the point on the world line that is linked to x by a null geodesic, u the value of the proper-time parameter at this retarded point, $u^{\gamma'}$ the four-velocity at the retarded point, and

$$\psi^{\alpha\beta}_{\text{tail}}(x) = 4m \int_{-\infty}^{u} V^{\alpha\beta}{}_{\mu'\nu'}(x,z') u^{\mu'} u^{\nu'} \, d\tau' \tag{9}$$

is the "tail" term. Here $z' \equiv z(\tau')$ stands for an arbitrary position on the world line, and the potentials of Eq. (8) depend on the particle's entire history prior to the retarded point $\tau = u(x)$.

Inverting Eq. (4) and expressing the results in the retarded coordinates $(u, r\Omega^a)$ returns

$$\begin{aligned} \mathsf{g}_{uu} = &-1 - r^2 \mathscr{E}_{ab}\Omega^a\Omega^b + O(r^3/\mathscr{R}^3) \\ &+ \frac{2m}{r} + h^{\text{tail}}_{uu} + r\big(2m\mathscr{E}_{ab}\Omega^a\Omega^b - 2a_a\Omega^a + h^{\text{tail}}_{uuu} + h^{\text{tail}}_{uua}\Omega^a\big) \\ &+ O(mr^2/\mathscr{R}^3) \end{aligned} \tag{10}$$

for the time-time component of the perturbed metric $\mathsf{g}_{\alpha\beta} = g_{\alpha\beta} + h_{\alpha\beta}$. The fields $h^{\text{tail}}_{\alpha\beta}$ are obtained from $\psi^{\text{tail}}_{\alpha\beta}$ by trace reversal, and $h^{\text{tail}}_{\alpha\beta\gamma} = \nabla_\gamma h^{\text{tail}}_{\alpha\beta}$.

Equation (10) gives the metric of the external universe perturbed by a black hole moving on a world line γ with an (as yet undetermined) acceleration $a_a(u)$. It is noteworthy that this metric appears to be singular when $r \to 0$. The metric, however, is valid only in the external zone $r > r_e \gg m$, and the limit $r \to 0$ is unattainable. The world line γ is outside the metric's domain of validity.

Internal zone

To obtain the metric of a nonrotating black hole that is slightly distorted by a tidal gravitational field is a standard application of black-hole perturbation theory. To leading order in the perturbation, which scales as $1/\mathscr{R}^2$, it is sufficient to integrate a set of time-independent perturbation equations, because each time derivative comes with an extra factor of $1/\mathscr{R}$, the inverse time scale associated with changes in the external universe. The perturbation must be well behaved on the hole's event horizon, and the asymptotic conditions for $r \gg 2m$ are such that the perturbation must behave as $-r^2\mathscr{E}_{ab}\Omega^a\Omega^b$, which has a quadrupolar form. These observations make solving the perturbation equations a very straightforward task.

In a set of internal light-cone coordinates $(\bar{u}, \bar{x}^a = \bar{r}\bar{\Omega}^a)$ the time-time component of the perturbed metric reads

$$\mathsf{g}_{\bar{u}\bar{u}} = -f - \bar{r}^2 f^2 \mathscr{E}_{ab}\bar{\Omega}^a\bar{\Omega}^b + O(\bar{r}^3/\mathscr{R}^3), \tag{11}$$

where $f = 1 - 2m/\bar{r}$ and $\mathscr{E}_{ab}(\bar{u})$ is the tidal gravitational field supplied by the external universe. In the limit $m/\bar{r} \to 0$ (keeping $\bar{r}/\mathscr{R}$ fixed) Eq. (11) becomes the metric of the external universe, expressed in coordinates for which $a_a = 0$. In the limit $\bar{r}/\mathscr{R} \to 0$ (keeping $m/\bar{r}$ fixed) Eq. (11) becomes the metric of a Schwarzschild black hole expressed in Eddington-Finkelstein coordinates. For small values of $\bar{r}/\mathscr{R}$ and arbitrary values of $m/\bar{r}$ Eq. (11) describes a tidally distorted black hole. The metric of Eq. (11) is valid everywhere in the internal zone, where $r < r_i \ll \mathscr{R}$.

Matching

Equation (10) gives the spacetime metric in the external zone, and the requirement that $r/\mathscr{R} \ll 1$ implies that the metric is in fact restricted to the buffer zone. Equation (11), on the other hand, gives the spacetime metric in the internal zone, and its specialization to $m/\bar{r} \ll 1$ implies also a restriction to the buffer zone. Since both metrics describe the geometry of the same region of the same spacetime, they must be related by a coordinate transformation.

The transformation from the external coordinates (u, x^a) to the internal coordinates $(\bar{u}, \bar{x}^a)$ is given in part by

$$\begin{aligned}\bar{u} = u - 2m\ln r - \frac{1}{2}\int^u h^{\rm tail}_{uu}\,du - \frac{1}{2}r\Big[h^{\rm tail}_{uu} + 2h^{\rm tail}_{ua}\Omega^a + h^{\rm tail}_{ab}\Omega^a\Omega^b\Big] \\ - \frac{1}{4}r^2\Big[h^{\rm tail}_{uuu} + (h^{\rm tail}_{uua} + 2h^{\rm tail}_{uau})\Omega^a + (h^{\rm tail}_{abu} + 2h^{\rm tail}_{uab})\Omega^a\Omega^b \\ + h^{\rm tail}_{abc}\Omega^a\Omega^b\Omega^c\Big] + O(mr^3/\mathscr{R}^3). \qquad (12)\end{aligned}$$

Applying this transformation to Eq. (10) gives

$$\begin{aligned}g_{\bar{u}\bar{u}} &= -1 - \bar{r}^2\mathscr{E}_{ab}\bar{\Omega}^a\bar{\Omega}^b + O(\bar{r}^3/\mathscr{R}^3) \\ &\quad + \frac{2m}{\bar{r}} + 4m\bar{r}\mathscr{E}_{ab}\bar{\Omega}^a\bar{\Omega}^b - 2\bar{r}\left(a_a - \frac{1}{2}h^{\rm tail}_{uua} + h^{\rm tail}_{uau}\right)\bar{\Omega}^a \\ &\quad + O(m\bar{r}^2/\mathscr{R}^3). \qquad (13)\end{aligned}$$

Comparison with Eq. (11) — linearized with respect to $m/\bar{r}$ — reveals that the acceleration vector must be given by

$$a_a = \frac{1}{2}h^{\rm tail}_{uua} - h^{\rm tail}_{uau}. \qquad (14)$$

As expected, the matching of the perturbed metrics in the buffer zone has returned the black hole's equations of motion.

MiSaTaQuWa equations

The tensorial form of Eq. (14) is

$$\frac{Du^\mu}{d\tau} = -\frac{1}{2}\big(g^{\mu\nu} + u^\mu u^\nu\big)\big(2h^{\rm tail}_{\nu\lambda\rho} - h^{\rm tail}_{\lambda\rho\nu}\big)u^\lambda u^\rho, \qquad (15)$$

and these are the MiSaTaQuWa equations of motion. Here, $z^\mu(\tau)$ gives the description of the black hole's motion in the background spacetime, $u^\mu = dz^\mu/d\tau$ is the normalized velocity vector, and $Du^\mu/d\tau$ is the covariant acceleration. All tensors, and all tensorial operations, refer to the background spacetime of the external universe; its metric is $g_{\alpha\beta}$, as it was introduced in Eq. (2).

The tail field on the right-hand side of Eq. (15) is obtained by differentiating Eq. (9); when expressed in terms of the original retarded Green's function of Eq. (7) it is given by

$$h^{\rm tail}_{\mu\nu\lambda} = 4m \int_{-\infty}^{\tau-\epsilon} \nabla_\lambda \Big(G_{\mu\nu\mu'\nu'} - \frac{1}{2} g_{\mu\nu} G^{\rho}{}_{\rho\mu'\nu'} \Big)(z,z') u^{\mu'} u^{\nu'}\, d\tau'. \tag{16}$$

It is understood that one takes the limit $\epsilon \to 0^+$ of this expression. Cutting the integration short excludes the (singular) light-cone part of the Green's function and isolates its tail part; the result is a smooth field on the world line. In Eq. (16), $z \equiv z(\tau)$ stands for the current position on the world line (at which the self-force is being evaluated) and $z' \equiv z(\tau')$ stands for a prior position.

4. Motion of a point mass

The derivation of the MiSaTaQuWa equations sketched in the preceding section is conceptually sound and devoid of any questionable assumptions. It is, however, technically involved and one wonders whether there may not be a shortcut to the final answer. Since the equations of motion refer to a body without internal structure, could they be derived on the assumption that the object is a point particle? The answer is in the affirmative, provided that one is willing to introduce additional assumptions and tolerate fields that are singular on the world line. The derivation sketched below was first presented by Mino, Tanaka, and Sasaki [3] and then by Quinn and Wald [4]; the approach described here relies heavily on concepts and techniques introduced by Detweiler and Whiting [10].

As we have seen in Sec. 3, the gravitational perturbation produced by a point particle of mass m is obtained from the potentials of Eq. (8) by trace reversal: $h_{\alpha\beta} = \psi_{\alpha\beta} - \frac{1}{2} g_{\alpha\beta}(g^{\gamma\delta}\psi_{\gamma\delta})$. In Sec. 3, Eq. (8) was assumed to hold in the external zone only ($r > r_e \gg m$), but we now accept its global validity. We also accept the fact that the perturbation is singular on the world line ($r = 0$). And we choose not to be bothered with the fact that while Eq. (8) was obtained by linearizing the Einstein field equations about the background metric $g_{\alpha\beta}$, the perturbation is obviously not small when r is smaller than m.

Our first additional assumption is that the particle will move on a geodesic of the metric $\mathsf{g}_{\alpha\beta} = g_{\alpha\beta} + h_{\alpha\beta}$. Writing down the geodesic equation in terms of tensorial quantities that refer to the background spacetime (such as the velocity vector $u^\mu = dz^\mu/d\tau$, which is normalized in the metric $g_{\alpha\beta}$) returns the equations of motion

$$\frac{Du^\mu}{d\tau} = -\frac{1}{2}\left(g^{\mu\nu} + u^\mu u^\nu\right)\left(2h_{\nu\lambda;\rho} - h_{\lambda\rho;\nu}\right)u^\lambda u^\rho. \tag{17}$$

These are superficially similar to Eq. (15), but the difference is important. While Eq. (15) involves tensor fields that are smooth on the world line, Eq. (17) involves highly singular quantities. These equations of motion are therefore meaningless as they stand, and $h_{\alpha\beta}$ must be regularized before Eq. (17) can be evaluated.

The regularization of a retarded field near its pointlike source is a problem that has a long history. This issue was encountered by Dirac [11] in the context of the electrodynamics of a point electric charge in flat spacetime. By appealing to energy-momentum conservation across a world tube surrounding the charge's world line, Dirac discovered that regularization could be achieved by decomposing the electromagnetic field into singular-symmetric "S" and regular-radiative "R" pieces; the "S" field would simply be removed from the retarded field and only the remainder — the "R" field — would be allowed to act on the particle. Dirac further discovered that in flat spacetime, the "S" field is given by the following combination of retarded and advanced solutions to Maxwell's equations: $F^{\rm S}_{\alpha\beta} = \frac{1}{2}(F^{\rm ret}_{\alpha\beta} + F^{\rm adv}_{\alpha\beta})$. The "R" field, on the other hand, is given by $F^{\rm R}_{\alpha\beta} = \frac{1}{2}(F^{\rm ret}_{\alpha\beta} - F^{\rm adv}_{\alpha\beta})$. Because the "S" field satisfies the same field equations as the retarded field (with a singular source term on the right-hand side), it is just as singular as the retarded field on the world line; removing it from the retarded field produces a smooth field. This is confirmed by the fact that the "R" field satisfies the sourcefree Maxwell equations.

Dirac's analysis was generalized to curved spacetime by DeWitt and Brehme [12], but the proper decomposition of the retarded electromagnetic field into "S" and "R" parts was identified only recently by Detweiler and Whiting [10]. For reasons of causality, the flat-spacetime prescription (a superposition of half retarded and half advanced fields) does not work in curved spacetime: The retarded field depends on the particle's entire past history, the advanced field depends on its future history, and a linear superposition would depend on the full history, both past and future (see Fig. 1). This would give rise to equations of motion with unacceptable causal properties.

The correct curved-spacetime prescription (identified by Detweiler and Whiting, and applied to the gravitational case) is to remove the "S" poten-

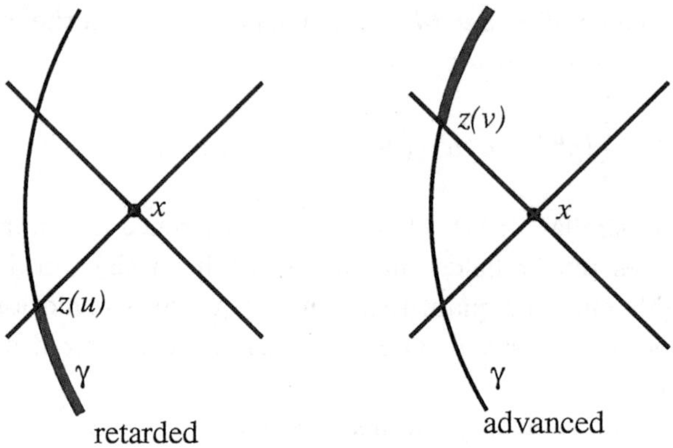

Figure 1. Causal properties of the retarded and advanced solutions. The retarded field at a point x depends on the particle's history prior to (and including) the retarded point $z(u)$. The advanced solution depends on the particle's history after (and including) the advanced point $z(v)$.

tial

$$\begin{aligned}\psi_{\rm S}^{\alpha\beta}(x) &= \frac{2m}{r} U^{\alpha\beta}{}_{\gamma'\delta'}(x,x') u^{\gamma'} u^{\delta'} + \frac{2m}{r} U^{\alpha\beta}{}_{\gamma''\delta''}(x,x'') u^{\gamma''} u^{\delta''} \\ &\quad - 2m \int_u^v V^{\alpha\beta}{}_{\mu\nu}(x,z) u^\mu u^\nu \, d\tau \end{aligned} \tag{18}$$

from the retarded potential. Here $x' \equiv z(u)$ stands for the retarded point on the world line associated with the field point x (and tensors with primed indices are evaluated at that point), and $x'' \equiv z(v)$ stands for the advanced point (with a similar meaning for doubly-primed indices); within the integral $z \equiv z(\tau)$ stands for an arbitrary point on the world line, and the integral extends from the retarded point to the advanced point. Detweiler and Whiting have shown that $\psi_{\rm S}^{\alpha\beta}(x)$ satisfies Eq. (6) — the same wave equation as the retarded potential — and comparison between Eqs. (8) and (18) reveals that the "S" potential is just as singular as the retarded potential on the world line. The "R" potential is then defined by

$$\psi_{\rm R}^{\alpha\beta}(x) = \psi^{\alpha\beta}(x) - \psi_{\rm S}^{\alpha\beta}(x), \tag{19}$$

and it satisfies the homogeneous form of Eq. (6); this is a smooth tensor field on the world line. The causal properties of the "S" and "R" fields are illustrated in Fig. 2.

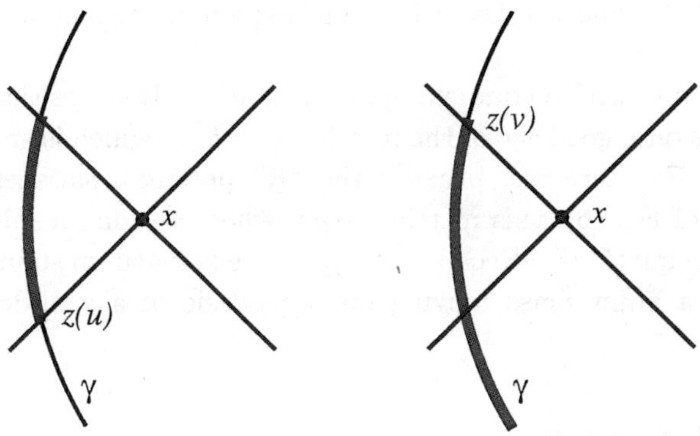

Figure 2. Causal properties of the "S" and "R" fields. The "S" field at a point x depends on the particle's history between (and including) the retarded point $z(u)$ and the advanced point $z(v)$. The "R" field depends on the particle's history up to the advanced point $z(v)$. When x is brought to the world line the "R" field stays smooth and $z(v)$ coincides with x; the field then depends on the particle's past history only.

Having introduced this unique decomposition of the retarded field into "S" and "R" pieces, Detweiler and Whiting postulate that *the "S" field exerts no force on the particle; the entire self-force arises from the action of the "R" field.* This axiom can be seen as another unavoidable assumption that must be introduced in order to make sense of the motion of point particles. Alternatively, it can be motivated by showing that the average of the "S" field on a spherical surface (as seen in the particle's instantaneous rest frame) surrounding the particle is zero. (This calculation is performed in LRR [1].) In any event, the postulated equations of motion are

$$\frac{Du^\mu}{d\tau} = -\frac{1}{2}\left(g^{\mu\nu} + u^\mu u^\nu\right)\left(2h^{\mathrm{R}}_{\nu\lambda;\rho} - h^{\mathrm{R}}_{\lambda\rho;\nu}\right)u^\lambda u^\rho, \tag{20}$$

and one finds that the "R" field reduces to

$$h^{\mathrm{R}}_{\mu\nu;\lambda} = -4m\Big(u_{(\mu}R_{\nu)\rho\lambda\xi} + R_{\mu\rho\nu\xi}u_\lambda\Big)u^\rho u^\xi + h^{\mathrm{tail}}_{\mu\nu\lambda} \tag{21}$$

on the world line, where $h^{\text{tail}}_{\mu\nu\lambda}$ is the tail field of Eq. (16). Substituting Eq. (21) into Eq. (20) eliminates the terms involving the Riemann tensor, and we end up with the MiSaTaQuWa equations of Eq. (15). We conclude that the equations of motion can indeed be derived on the basis of a point mass, at the price of introducing the geodesic postulate (which was not needed for a black hole) and the Detweiler-Whiting axiom (which also was not needed).

Equation (20) comes with a compelling interpretation. It states that the point particle moves on a geodesic of the metric $g_{\alpha\beta} + h^{\text{R}}_{\alpha\beta}$, which is smooth on the world line. Furthermore, because the "R" potential satisfies the homogeneous form of Eq. (6), this metric is everywhere a vacuum solution to the Einstein field equations. (Recall that $g_{\alpha\beta}$ is itself a vacuum solution.) We therefore have a point mass moving on a geodesic of a well-defined vacuum spacetime.

5. Newtonian self-force

The notion of a self-force exists also in the simple setting of Newtonian theory. And the Newtonian potential also can be decomposed into "S" and "R" pieces. This pedagogical illustration was first presented to me by Steve Detweiler. The following presentation draws heavily from Ref. [13].

Consider, in Newtonian theory, a large mass M at position $\boldsymbol{\rho}(t)$ relative to the centre of mass, and a small mass m at position $\boldsymbol{R}(t)$. We assume that $m \ll M$ and the centre of mass condition reads $m\boldsymbol{R} + M\boldsymbol{\rho} = \boldsymbol{0}$. We denote the position of an arbitrary field point by $\boldsymbol{x}$, and $r \equiv |\boldsymbol{x}|$ is its distance from the centre of mass. We shall also let $R \equiv |\boldsymbol{R}|$ and $\rho \equiv |\boldsymbol{\rho}|$.

We begin with a test-mass description of the situation, according to which the smaller mass moves in the gravitational field of the larger mass, which is placed at the origin of the coordinate system. The background Newtonian potential is

$$\Phi_0(\boldsymbol{x}) = -\frac{M}{r} \tag{22}$$

and the background gravitational field is $\boldsymbol{g}_0 = -\boldsymbol{\nabla}\Phi_0 = -M\boldsymbol{x}/r^3$. In this description, the smaller mass m moves according to $d^2\boldsymbol{R}/dt^2 = \boldsymbol{g}_0(\boldsymbol{x} = \boldsymbol{R})$. If the motion is circular, then m possesses a uniform angular velocity given by ${\Omega_0}^2 = M/R^3$, where R is the orbital radius. These results are in close analogy with a relativistic description in which the smaller mass is taken to move on a geodesic of the background spacetime, in a test-mass approximation.

We next improve our description by incorporating the gravitational effects produced by the smaller mass. The exact Newtonian potential is

$$\Phi(\boldsymbol{x}) = -\frac{M}{|\boldsymbol{x}-\boldsymbol{\rho}|} - \frac{m}{|\boldsymbol{x}-\boldsymbol{R}|}, \tag{23}$$

and for $m \ll M$ this can be expressed as $\Phi(\boldsymbol{x}) = \Phi_0(\boldsymbol{x}) + \delta\Phi(\boldsymbol{x})$, with a perturbation given by

$$\delta\Phi(\boldsymbol{x}) = -\frac{M}{|\boldsymbol{x}-\boldsymbol{\rho}|} + \frac{M}{r} - \frac{m}{|\boldsymbol{x}-\boldsymbol{R}|}. \tag{24}$$

This gives rise to a field perturbation $\delta\boldsymbol{g}$ that exerts a force on the smaller mass. This is the particle's "bare" self-acceleration, and the correspondence with the relativistic problem is clear.

An examination of Eq. (24) reveals that the last term on the right-hand side diverges at the position of the smaller mass. But since the gravitational field produced by this term is isotropic around $\boldsymbol{R}(t)$, we know that this field will exert no force on the particle. We conclude that the last term can be identified with the singular "S" part of the perturbation,

$$\Phi_{\rm S}(\boldsymbol{x}) = -\frac{m}{|\boldsymbol{x}-\boldsymbol{R}|}, \tag{25}$$

and that the remainder makes up the regular "R" potential,

$$\Phi_{\rm R}(\boldsymbol{x}) = -\frac{M}{|\boldsymbol{x}-\boldsymbol{\rho}|} + \frac{M}{r}. \tag{26}$$

The full perturbation is then given by $\delta\Phi(\boldsymbol{x}) = \Phi_{\rm S}(\boldsymbol{x}) + \Phi_{\rm R}(\boldsymbol{x})$, and only the "R" potential affects the motion of the smaller mass. Once more the correspondence with the relativistic problem is clear.

It is easy to check that to first order in m/M, Eq. (26) simplifies to

$$\Phi_{\rm R}(\boldsymbol{x}) = m\frac{\boldsymbol{R}\cdot\boldsymbol{x}}{r^3}; \tag{27}$$

this simplification occurs because of the centre-of-mass condition, which implies that $\boldsymbol{\rho}$ is formally of order $m/M \ll 1$. The "R" part of the field perturbation is then

$$\boldsymbol{g}_{\rm R}(\boldsymbol{x}) = m\frac{3(\boldsymbol{R}\cdot\boldsymbol{x})\boldsymbol{x} - r^2\boldsymbol{R}}{r^5}, \tag{28}$$

and evaluating this at the particle's position yields a correction to the background field $\boldsymbol{g}_0(\boldsymbol{x}=\boldsymbol{R}) = -M\boldsymbol{R}/R^3$ given by $\boldsymbol{g}_{\rm R}(\boldsymbol{x}=\boldsymbol{R}) = 2m\boldsymbol{R}/R^3$; the force still points in the radial direction but the active mass has been shifted from M to $M-2m$. For circular motion the angular velocity becomes $\Omega^2 = (M - 2m)/R^3$. This can be cast in a more recognizable form if we express the angular velocity in terms of the total separation $s \equiv R + \rho = (1 + m/M)R$ between the two masses. To first order in m/M we obtain $\Omega^2 = (M + m)/s^3$, which is just the usual form of Kepler's third law. The "R" part of the field perturbation is therefore responsible for the finite-mass correction to the angular velocity.

Notice that the "R" potential of Eq. (27) has a pure dipolar form. By contrast, the potentials $\delta\Phi$ and $\Phi_{\rm S}$ contain an infinite number of multipole moments. This observation should be kept in mind as the reader proceeds through the remaining sections of this contribution.

6. Concrete evaluation of the self-force

I turn next to the practical issues involved in a concrete evaluation of the gravitational self-force — the right-hand side of Eq. (20), which shall be denoted $a^\mu[h^{\rm R}]$ — for a particle moving in the field of a Schwarzschild or Kerr black hole.

The first sequence of steps are concerned with the computation of the (retarded) metric perturbation $h_{\alpha\beta}$ produced by a point particle moving on a specified geodesic of the Kerr spacetime. A method for doing this was elaborated by Lousto and Whiting [14] and Ori [15], building on the pioneering work of Teukolsky [16], Chrzanowski [17], and Wald [18]. The procedure consists of (i) solving the Teukolsky equation for one of the Newman-Penrose quantities ψ_0 and ψ_4 (which are complex components of the Weyl tensor) produced by the point particle; (ii) obtaining from ψ_0 or ψ_4 a related (Hertz) potential Ψ by integrating an ordinary differential equation; (iii) applying to Ψ a number of differential operators to obtain the metric perturbation in a radiation gauge that differs from the Lorenz gauge of Eq. (5); and (iv) performing a gauge transformation from the radiation gauge to the Lorenz gauge. For a Schwarzschild black hole one can rely instead on the formalism of metric perturbations [19, 20], and the procedure simplifies.

It is well known that the Teukolsky equation separates when ψ_0 or ψ_4 is expressed as a multipole expansion, summing over modes with (spheroidal-harmonic) indices l and m. In fact, the procedure outlined above relies heavily on this mode decomposition, and the metric perturbation returned at the end of the procedure is also expressed as a sum over modes:

$$h_{\alpha\beta} = \sum_\ell h^\ell_{\alpha\beta}. \tag{29}$$

(For each l, m ranges from $-l$ to l, and summation of m over this range is henceforth understood. Indices on the metric perturbation and other tensors will now be omitted for ease of notation.) From the modes h_ℓ, mode contributions to the self-acceleration can be computed: $a[h_\ell]$ is obtained from Eq. (20) by substituting h_ℓ in place of $h^{\rm R}$. These mode contributions are finite on the world line, but $a[h_\ell]$ is discontinuous at the radial position of the particle. The sum over modes does not converge, because the "bare" acceleration (constructed from the retarded field h) is formally infinite.

The next sequence of steps is concerned with the regularization of each $a[h_l]$ by removing the contribution from $h^{\rm S}$. Starting with Eq. (18), the singular field can be constructed locally in a neighbourhood of the particle, and then decomposed into modes of multipole order ℓ. This gives rise to modes $a[h^{\rm S}_\ell]$ for the singular part of the self-acceleration; these are also finite and discontinuous, and their sum over ℓ also diverges. But the true modes $a[h^{\rm R}_\ell] = a[h_\ell] - a[h^{\rm S}_\ell]$ of the self-acceleration are continuous at the radial position of the particle, and their sum does converge to the particle's acceleration.

The general structure of $a[h^{\rm S}_\ell]$, as worked out in Refs. [21–27], is given by

$$a^\mu[h^{\rm S}_\ell] = (\ell + \tfrac{1}{2})A^\mu + B^\mu + \frac{C^\mu}{\ell + \frac{1}{2}} + \text{convergent terms}. \tag{30}$$

The "regularization parameters" A^μ, B^μ, and C^μ are independent of ℓ but they depend on the details of the spacetime and on the particle's state of motion. It is evident that the term involving A^μ in Eq. (30) will diverge quadratically when summed over ℓ, that the term involving B^μ will diverge linearly, and that the term involving C^μ will diverge logarithmically; all other terms converge, and keeping additional terms produces faster convergence of the sum $\sum_\ell a[h^{\rm R}_\ell]$.

The structure of Eq. (30) can easily be recovered in the Newtonian setting of the preceding section. The multipolar decomposition of the singular potential of Eq. (25) is given by

$$\Phi_{\rm S}(\boldsymbol{x}) = -m \sum_\ell \frac{(r_<)^\ell}{(r_>)^{\ell+1}} P_\ell(\hat{\boldsymbol{n}} \cdot \hat{\boldsymbol{R}}), \tag{31}$$

where $r_< = \min(r, R)$, $r_> = \max(r, R)$, $\hat{\boldsymbol{n}} = \boldsymbol{x}/r$, $\hat{\boldsymbol{R}} = \boldsymbol{R}/R$, and $P_\ell(\mu)$ are Legendre polynomials. Taking the gradient of Eq. (31) and then the limit

$\boldsymbol{x} \to \boldsymbol{R}$ returns

$$g_{\rm S}^r = \frac{m}{R^2} \sum_\ell \left[\mp(\ell + \tfrac{1}{2}) - \frac{1}{2} \right] \tag{32}$$

for the radial component of the singular field (the angular components all vanish). This has the same form as Eq. (30), with $A^r = \mp m/R^2$, $B^r = -m/(2R^2)$, and $C^r = 0$. The choice of sign in front of the $(\ell + \frac{1}{2})$ term in Eq. (32) comes from the two ways in which the limit $r \to R$ can be taken: The upper (minus) sign corresponds to $r \to R^+$ while the lower (plus) sign corresponds to $r \to R^-$. This Newtonian calculation reproduces the leading-order term in a post-Newtonian expansion of the relativistic regularization parameters [21–27].

The self-acceleration is obtained by first computing $a[h_\ell]$ from the retarded metric perturbation, then computing the counterterms $a[h_\ell^{\rm S}]$ by mode-decomposing the singular field, and finally summing over all $a[h_\ell^{\rm R}] = a[h_\ell] - a[h_\ell^{\rm S}]$. This procedure is lengthy and involved, and thus far it has not been brought to completion, except for the special case of a particle falling radially toward a nonrotating black hole [28]. Another evaluation of the self-force, which did not rely on a mode decomposition, was carried out in the context of weak fields and slow motions [29]; this reproduced the standard post-Newtonian result.

The procedure described above is lengthy and involved, but it is also incomplete when the background spacetime is that of a Kerr black hole. The reason is that the metric perturbations $h^\ell_{\alpha\beta}$ that can be recovered from ψ_0 or ψ_4 do not by themselves sum up to the complete gravitational perturbation produced by the moving particle. Missing are the perturbations derived from the other Newman-Penrose quantities: ψ_1, ψ_2, and ψ_3. While ψ_1 and ψ_3 can always be set to zero by an appropriate choice of null tetrad, ψ_2 contains such important physical information as the shifts in mass and angular-momentum parameters produced by the particle [30]. Because the mode decompositions of ψ_0 and ψ_4 start at $l = 2$, one might say that the "$\ell = 0$ and $\ell = 1$" modes of the metric perturbations are missing. It is not currently known how the procedure can be completed so as to incorporate *all modes* of the metric perturbations. Specializing to a Schwarzschild spacetime eliminates this difficulty, and in this context the low multipole modes have been studied for the special case of circular orbits [13, 31]. In view of the fact that the Newtonian self-force is purely $\ell = 1$ (refer back to the last paragraph of Sec. 5), it is clear that these low multipoles cannot be ignored.

Taking into account these many difficulties (and I choose to stay silent on others, for example, the issue of relating metric perturbations in different gauges when the gauge transformation is singular on the world line), it is perhaps not too surprising that such a small number of concrete calculations have been presented to date. But progress in dealing with these difficulties has been steady, and the situation should change dramatically in the next few years.

7. Beyond the self-force

The successful computation of the gravitational self-force is not the end of the story. After the difficulties reviewed in the preceding section have all been dealt with and the motion of the small body is finally calculated to order m, it will still be necessary to obtain gauge-invariant information associated with the body's corrected motion. Because the MiSaTaQuWa equations are not by themselves gauge-invariant [they are formulated in the Lorenz gauge of Eq. (5)], this step will necessitate going beyond the self-force.

To see how this might be done, imagine that the small body is a pulsar, and that it emits light pulses at regular proper-time intervals. The motion of the pulsar around the central black hole modulates the pulse frequencies as measured at infinity, and information about the body's corrected motion is encoded in the times-of-arrival of the pulses. Because these can be measured directly by a distant observer, they clearly constitute gauge-invariant information. But the times-of-arrival are determined not only by the pulsar's motion, but also by the propagation of radiation in the perturbed spacetime. This example shows that to obtain gauge-invariant information, one must properly combine the MiSaTaQuWa equations of motion with the metric perturbation.

In the astrophysical context of the Laser Interferometer Space Antenna, reviewed in Sec. 2, the relevant observable is the instrument's response to a gravitational wave, which is determined by gauge-invariant waveforms, h_+ and $h_\times$. To calculate these is the ultimate goal of the Capra program, and the challenges that lie ahead go well beyond what I have described thus far. To obtain the waveforms it will be necessary to solve the Einstein field equations to *second order* in perturbation theory.

To understand this, consider first the formulation of the first-order problem. Schematically, one introduces a perturbation h that satisfies a wave equation $\Box h = T[z]$ in the background spacetime. Here $T[z]$ is the stress-energy tensor of the moving body, which is a functional of the world line $z(\tau)$. In first-order perturbation theory, the stress-energy tensor must be conserved in the background spacetime, and $z(\tau)$ must describe a geodesic. It follows that in first-order perturbation theory, the waveforms constructed from the perturbation h contain no information about the body's corrected motion.

The first-order perturbation, however, can be used to correct the motion, which is now described by the world line $z(\tau) + \delta z(\tau)$. In a naive implementation of the self-force, one would now re-solve the wave equation with a corrected stress-energy tensor, $\Box h = T[z + \delta z]$, and the new waveforms constructed from h would then incorporate information about the corrected motion. This implementation is naive because this information would not be gauge-invariant. In fact, to be consistent one would have to include *all* second-order terms in the wave equation, not just the ones that come from the corrected motion. Schematically, the new wave equation would have the form of $\Box h = (1+h)T[z+\delta z]+(\nabla h)^2$, and this is much more difficult to solve than the naive problem (if only because the source term is now much more singular than the distributional singularity contained in the stress-energy tensor). But provided one can find a way to make this second-order problem well posed, and provided one can solve it (or at least the relevant part of it), the waveforms constructed from the second-order perturbation h will be gauge invariant. In this way, information about the body's corrected motion will have properly been incorporated into the gravitational waveforms.

8. Conclusion

There has been significant progress toward solving the self-force problem over the last several years. The foundations (reviewed in Secs. 3 and 4) are now solid, thanks to the work of Mino, Sasaki, and Tanaka [3], Quinn and Wald [4], and Detweiler and Whiting [10]. The regularization parameters A^μ, B^μ, and C^μ (introduced in Sec. 6) have been calculated for arbitrary motion in the Schwarzschild and Kerr spacetimes by Barack and Ori [21–24] and Mino, Nakano, and Sasaki [21, 25], and their expressions were independently verified [26, 27]. And finally, the reconstruction of the metric perturbation from the Teukolsky function (described in Sec. 6) is now better understood, thanks to the work of Lousto and Whiting [14] and Ori [15].

While this progress is encouraging, the challenges that lie ahead are still numerous. Among the outstanding issues are (i) the question of defining and calculating the "$\ell = 0$ and $\ell = 1$" contributions to the self-force in the Kerr spacetime (as was discussed in Sec. 6), (ii) the design of useful gauge-invariant quantities that could be computed from the self-force and the metric perturbation (as was discussed in Sec. 7), and (iii) the incorporation of the self-force into a consistent wave-generation formalism (as was also discussed in Sec. 7). These issues are fascinating, and in the next few years they will be vigourously pursued by the Capra posse.

Acknowledgments

I wish to thank the organizers of GR17 for inviting me to talk about the gravitational self-force at this prestigious international conference. And I wish to thank the Capra posse (listed by name at the end of Sec. 1) for numerous discussions on this topic. The work presented in this contribution was supported by the Natural Sciences and Engineering Research Council of Canada.

References

1. E. Poisson, *The motion of point particles in curved spacetime*, Living Rev. Relativity **7** (2004), 6. [Online article]: cited on July 29, 2005, http://www.livingreviews.org/lrr-2004-6.
2. E. Poisson, *Radiation reaction of point particles in curved spacetime*, Class. Quantum Grav. **21**, R153 (2004).
3. Y. Mino, M. Sasaki, and T. Tanaka, *Gravitational radiation reaction to a particle motion*, Phys. Rev. D **55**, 3457 (1997).
4. T. C. Quinn and R. M. Wald, *Axiomatic approach to electromagnetic and gravitational radiation reaction of particles in curved spacetime*, Phys. Rev. D **56**, 3381 (1997).
5. T. Damour, *The problem of motion in Newtonian and Einsteinian gravity*, in *300 Years of Gravitation*, edited by S. W. Hawking and W. Israel (Cambridge University Press, Cambridge, England, 1987).
6. L. Blanchet, *Gravitational radiation from post-Newtonian sources and inspiralling compact binaries*, Living Rev. Relativity **5** (2002), 3. [Online article]: cited on July 29, 2005, http://www.livingreviews.org/lrr-2002-3.
7. E. Poisson, *Retarded coordinates based at a world line and the motion of a small black hole in an external universe*, Phys. Rev. D **69**, 084007 (2004).
8. K. S. Thorne and J. B. Hartle, *Laws of motion and precession for black holes and other bodies*, Phys. Rev. D **31**, 1815 (1985).
9. P. D. D'Eath, *Black holes: Gravitational interactions* (Clarendon, Oxford, 1996).
10. S. Detweiler and B. F. Whiting, *Self-force via a Green's function decomposition*, Phys. Rev. D **67**, 024025 (2003).

11. P. A. M. Dirac, *Classical theory of radiating electrons*, Proc. Roy. Soc. London **A167**, 148 (1938).
12. B. S. DeWitt and R. W. Brehme, *Radiation damping in a gravitational field*, Ann. Phys. (N.Y.) **9**, 220 (1960).
13. S. Detweiler and E. Poisson, *Low multipole contributions to the gravitational self-force*, Phys. Rev. D **69**, 084019 (2004).
14. C. O. Lousto and B. F. Whiting, *Reconstruction of black hole metric perturbations from the Weyl curvature*, Phys. Rev. D **66**, 024026 (2002).
15. A. Ori, *Reconstruction of inhomogeneous metric perturbations and electromagnetic four-potential in Kerr spacetime*, Phys. Rev. D **67**, 124010 (2003).
16. S. A. Teukolsky, *Perturbations of a rotating black hole. I. Fundamental equations for gravitational, electromagnetic, and neutrino-field perturbations*, Astrophys. J. **185**, 635 (1973).
17. P. L. Chrzanowski, *Vector potential and metric perturbations of a rotating black hole*, Phys. Rev. D **11**, 2042 (1975).
18. R. M. Wald, *Construction of solutions of gravitational, electromagnetic, or other perturbation equations from solutions of decoupled equations*, Phys. Rev. Lett. **41**, 203 (1978).
19. T. Regge and J. A. Wheeler, *Stability of a Schwarzschild singularity*, Phys. Rev. **108**, 1063 (1957).
20. F. J. Zerilli, *Gravitational field of a particle falling in a Schwarzschild geometry analyzed in tensor harmonics*, Phys. Rev. D **2**, 2141 (1970).
21. L. Barack, Y. Mino, H. Nakano, A. Ori, and M. Sasaki, *Calculating the gravitational self-force in Schwarzschild spacetime*, Phys. Rev. Lett. **88**, 091101 (2002).
22. L. Barack and A. Ori, *Regularization parameters for the self-force in Schwarzschild spacetime: Scalar case*, Phys. Rev. D **66**, 084022 (2002).
23. L. Barack and A. Ori, *Regularization parameters for the self-force in Schwarzschild spacetime. II. Gravitational and electromagnetic cases*, Phys. Rev. D **67**, 024029 (2003).
24. L. Barack and A. Ori, *Gravitational self-force on a particle orbiting a Kerr black hole*, Phys. Rev. Lett. **90**, 111101 (2003).
25. Y. Mino, H. Nakano, and M. Sasaki, *Covariant self-force regularization of a particle orbiting a Schwarzschild black hole*, Prog. Theor. Phys. **108**, 1039 (2003).
26. S. Detweiler, E. Messaritaki, and B. F. Whiting, *Self-force of a scalar field for circular orbits about a Schwarzschild black hole*, Phys. Rev. D **67**, 104016 (2003).
27. D. Kim, *Regularization parameters for the self-force of a scalar particle in a general orbit about a Schwarzschild black hole* (2004), http://xxx.lanl.gov/abs/gr-qc/0402014.
28. L. Barack and C. O. Lousto, *Computing the gravitational self-force on a compact object plunging into a Schwarzschild black hole*, Phys. Rev. D **66**, 061502 (2002).
29. M. J. Pfenning and E. Poisson, *Scalar, electromagnetic, and gravitational self-forces in weakly curved spacetimes*, Phys. Rev. D **65**, 084001 (2002).

30. R. M. Wald, *On perturbations of a Kerr black hole*, J. Math. Phys. **14**, 1453 (1973).
31. H. Nakano, N. Sago, and M. Sasaki, *Gauge problem in the gravitational self-force: First post Newtonian force under Regge-Wheeler gauge*, Phys. Rev. D **68**, 124003 (2003).

COSMIC SUPERSTRINGS REVISITED

JOSEPH POLCHINSKI
Kavli Institute for Theoretical Physics
University of California
Santa Barbara, CA 93106-4030

It is possible that superstrings, as well as other one-dimensional branes, could have been produced in the early universe and then expanded to cosmic size today. I discuss the conditions under which this will occur, and the signatures of these strings. Such cosmic superstrings could be the brightest objects visible in gravitational wave astronomy, and might be distinguishable from gauge theory cosmic strings by their network properties.

Seeing superstrings of cosmic size would be a spectacular way to verify string theory. Witten considered this possibility in the context of perturbative string theory, and found that it was excluded for several reasons [1]. Perturbative strings have a tension close to the Planck scale, and so would produce inhomogeneities in the cosmic microwave background far larger than observed. The scale of this tension also exceeds the upper bound on the energy scale of the inflationary vacuum, and so these strings could not have been produced after inflation, and any strings produced earlier would have been diluted beyond observation. Reference [1] also identified instabilities that would prevent long strings from surviving on cosmic time scales.

In recent years we have understood that there are much more general possibilities for the geometry of the compact dimensions of string theory, including localized branes, and this allows the string tension to be much lower, anything between the Planck scale and the weak scale. Also, we have found new kinds of extended object in string theory. Thus the question of cosmic superstrings (and branes) must be revisited, and this has been done beginning in Refs. [2,3]. A necessary set of conditions is:

(1) The strings must be *produced* after inflation.
(2) They must be *stable* on cosmological timescales.
(3) They must be *observable* in some way, but not already excluded.

Reference [1] thus showed that perturbative strings fail on all three counts. If we do find models that satisfy these three conditions then there is one more that would also be important:

(4) Cosmic superstrings should be *distinguishable* from other kinds of cosmic string, in particular gauge theory solitons.

We will see that each of these issues is separately model-dependent, but that there are simple models, including the most fully-developed models of inflation in string theory, in which all these conditions are met.

The list above provides an outline for this paper, after an introductory section which reviews the story of cosmic strings in grand unified field theories.

0. Cosmic String Review

Cosmic strings might also arise as gauge theory solitons, and for some time these were a candidate for the source of the gravitational perturbations that produced the galaxies. It is useful to give a brief summary here to set a context for the superstring discussion, but for a more complete discussion and references see the reviews [4,5]; for a recent overview see Ref. [6].

In any field theory with a broken $U(1)$ symmetry, there will be classical solutions that are extended in one dimension [7,8]. These are topological solitons: as one traverses a circle around the string core, the Higgs field winds around the manifold of vacua, which is also a circle. The broken $U(1)$ can be either a global or a gauge symmetry. In the global case the Goldstone boson field has a $1/r$ gradient at long distance, giving a logarithmic potential between strings. In the gauge case this gradient is pure gauge and the physical fields all fall off exponentially. Most of the discussion of cosmic strings has focused on gauge strings, though one should keep the global case in mind for later reference.

These solutions exist whenever there is a broken $U(1)$ symmetry, and whenever a $U(1)$ symmetry *becomes* broken during the evolution of the universe, a network of strings must actually form. This is the Kibble argument [9]: the Higgs field starts at zero, and then rolls down to one of the vacua. By causality, it cannot roll in the same direction everywhere in the universe, as it cannot be correlated on distances greater than the horizon scale (in practice, the correlation length is usually less than this, being the microphysical correlation length of the field theory). Thus it chooses random directions in different places, and inevitably there will be

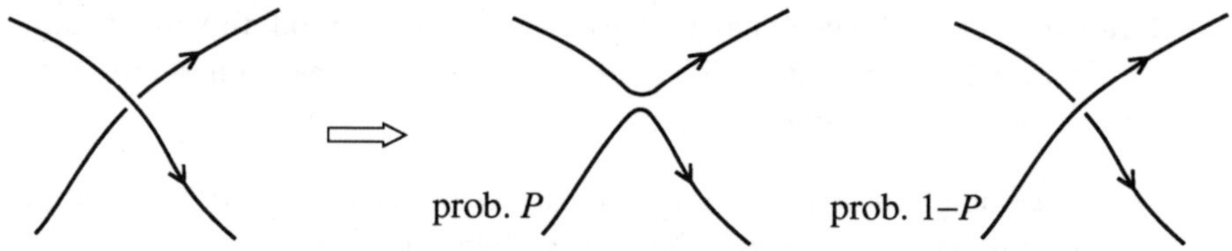

Figure 1. When two strings of the same type collide, they either reconnect, with probability P, or pass through each other, with probability $1 - P$. For classical solitons the process is deterministic, and $P = 1$ for the velocities relevant to the string network.

some trapped winding, so that strings are left over at the end. This is indeed what simulations show. A fraction of the string is in the form of finite sized loops, and a fraction is in the form of infinite strings; the latter enter and leave the boundaries of the simulation volume no matter how large this is taken to be. These populations are cleanly separated because the distribution of lengths of finite loops falls rapidly for long loops, giving a convergent integral for the total amount of string in loops. Presumably the existence of the infinite strings is implied by the causality argument. They are important because they begin to stretch with the expansion of the universe, while the small loops quickly decay away.

In the subsequent evolution of the network, the important processes in addition to the expansion of the universe are collisions of strings and the emission of gravitational radiation.[a] When two strings collide they can either pass through each other or they can reconnect (intercommute), as in Fig. 1. In an adiabatic collision, they will always reconnect, $P = 1$, because reconnection allows the flux in the string core to take an energetically favorable shortcut. Simulations show that this remains true at the moderately relativistic velocities that are present in string networks [11-13].

During the radiation and matter dominated eras, the horizon distance grows linearly in t while the comoving distance grows more slowly, as $t^{1/2}$ or $t^{2/3}$ respectively. Thus as time goes on we see more of the universe, and if no decay processes were operating we would see more strings. The relevant decay process is the combined effect of reconnection and gravitational radiation. Reconnection turns long straight strings into long kinked strings and then allows loops of string to break off, and these loops decay

[a]For reasons of space I am focusing on strings whose only long-distance interactions are gravitational. Global strings would also emit Goldstone bosons. There are also superconducting strings [10], which have strong couplings to gauge fields. These are perhaps less likely to arise in the superstring case, for reasons that I will explain.

by gravitational radiation. Simulations show that this occurs at the maximum rate allowed by causality, so that the string network always looks the same as viewed at the horizon scale, with a few dozen strings spanning the horizon volume and a gas of loops of various sizes. This is known as the *scaling solution.* The scaling solution is an attractor: if we start with too much string, the higher collision rate will reduce it, while if we start with too little then there will be few collisions until the amount of string per horizon volume approaches the scaling form. It is possible that the strings have a short-distance structure that does not scale with the horizon size.

Since we are considering strings whose only interactions are gravitational, all of their effects are controlled by the dimensionless product of Newton's constant and the string tension, $G\mu$. This is the string tension in Planck units, and it sets the size of the typical metric perturbation produced by a string. For example, the geometry around a long straight string is conic, with a deficit angle $8\pi G\mu$.

The string network produces inhomogeneities proportional to $G\mu$, and because of the scaling property of the string network these are scale invariant. A value $G\mu \sim 10^{-5.5}$ would give scale-invariant perturbations of the right magnitude to produce the galaxies and CMB fluctuations.[b] This was a viable theory for some time [14,15], but it is now excluded. For example, the CMB power spectrum is wrong: the actual spectrum shows a pattern of peaks and dips, whereas the spectrum from strings would be smooth. There is a simple reason for this. The fluctuations produced in inflation have a definite phase. This phase is maintained from the end of inflation until the perturbations go nonlinear, and is imprinted as oscillations of the power spectrum. Strings, on the other hand, each keep their own time, there is no common phase. Fitting cosmological constant plus cold dark matter *plus* strings to the CMB power spectrum gives an upper limit $G\mu < 10^{-6}$ [16,17]. Beyond the power spectrum, strings will produce nongaussianities in the CMB. Recent limits are somewhat stronger than those from the power spectrum, around $G\mu < 10^{-6.5}$ [18,19].

The strongest current limit on $G\mu$ comes from pulsar timing; this was on the verge of ruling out the cosmic string structure scenario even before

[b]I am going to quote values of $G\mu$ to the nearest half order of magnitude. This is less precise than most numbers given in the literature, but it is all I will need and it roughly reflects the uncertainties in the understanding of the string network. To give more precise numbers would require a more detailed discussion, and so the reader should consult the references. Incidentally, some of the bounds given here are stronger than those that I was aware of when this talk was given.

the CMB, and has recently gotten much stronger. Because the energy in the strings eventually goes into gravitational waves, strings produce a large stochastic gravitational wave background. The precision of pulsar timing puts a limit on such a background, $\Omega_{\rm GW} < 4\times10^{-9}$ in the relevant frequency range [20] (0.1 inverse year, set by the observation time). This translates into $G\mu < 10^{-7}$ [4].

Thus far we have quoted upper bounds, but there are possible detections of strings via gravitational lensing. A long string will produce a pair of images symmetric about an axis, very different from lensing by a point mass. Such an event has been reported recently [21,22]. The separation of around two arc-seconds corresponds to $G\mu$ equal to 4×10^{-7} times a geometric factor that is at least 1. This appears to exceed the upper limit from pulsars, unless there exist fewer strings than in the usual network simulations. I have heard varying opinions on how seriously to take such an observation, as similar pairs in the past have turned out to be coincidental. There is further discussion in the review [6], which also discusses a possible time-dependent lens, as from an oscillating loop, with $G\mu \sim 10^{-7.5}$ [23].

Network simulations show that the total energy density in the string network is of order $60G\mu$ times the matter density (during the matter-dominated era), so the upper bound on $G\mu$ implies that the strings contribute only a small fraction to the total energy density. In particular, the strings are not the dark matter, if they have the usual network properties.

1. Production of Cosmic F- and D-Strings

In order to discuss inflationary cosmology, one needs a fairly clear understanding of the scalar field dynamics. This has been an area of recent progress in string theory, and a nice geometric idea has emerged for obtaining a slow-roll potential [24-27]. That is, the early universe could have contained an extra brane-antibrane pair, separated in the compact directions. The potential energy of these branes would drive inflation. The inflaton is then the separation between the branes: this has a potential which is rather flat when the branes are separated and steepens as they approach, until at some point a field becomes tachyonic and the brane-antibrane annihilate rapidly.

If these branes are D-branes, then there is a $U(1)$ gauge symmetry on each of the brane and antibrane, and this $U(1) \times U(1)$ disappears when the branes annihilate. One linear combination of the $U(1)$'s is Higgsed. The Kibble argument then applies, so that a network of strings must be left over when the branes annilate [2,3]. These are D-strings, as one can see by studying the conserved charges [28,29]. More precisely, if the branes that annihilate are $D(3+k)$ branes, extended in the three large dimensions and wrapped on k small dimensions, then the result is $D(1+k)$ branes that extend in one large dimension and are wrapped on the same small dimensions. The simplest case is $k = 0$, where D3/$\overline{\text{D3}}$ annihilation produces D1-branes.

It is important that this process produces only strings, and not monopoles or domain walls [2,3]. We have seen that the existence of strings is consistent with observation, provided that the scale of their tension is a few orders of magnitude below the string scale. By contrast, zero- and two-dimensional defects are observationally ruled out because if they were produced then their energy density would quickly come to dominate the universe (unless they were removed by instabilities, inflation, or further symmetry breaking). The point here is that Π_1 topology from the breaking of a $U(1)$ produces defects of codimension two, and the Kibble argument requires that the codimension be in the large directions: the small directions are in causal contact.[c]

The second linear combination of $U(1)$'s is confined. We can think of confinement as dual Higgsing, by a magnetically charged field, and so we would expect that again the Kibble argument implies production of strings [30,31]. These are simply the F-strings, the 'fundamental' superstrings whose quantization defines the theory, at least perturbatively.

It is striking that what appears to be the most natural implementation of inflation in string theory produces strings and not dangerous defects, but we should now ask how generic this is. Even these models are not, in their current form, completely natural: like all models of inflation they require tuning at the per cent level to give a sufficiently flat potential [32]. There might well be other flat regions in the large potential energy landscape of string theory. An optimistic sign is that there are arguments entirely independent of string theory to indicate that inflation terminates with a symmetry-breaking transition: this is known as hybrid inflation [33], and

[c]Reference [30] gives arguments that the Kibble process cannot produce D1-branes; we disagree. The argument in Sec. IV of that paper is based on the energetics of the Kaluza-Klein RR fields. These are massive and should be integrated out, and so cannot affect the phase transition (the point is that one is not producing real KK particles, rather these fields are just adjusting in response to the light fields). Also, the model in Sec. V of that paper has its Π_1 disorder partly in the compact directions, which is not the case for the D-strings under consideration [2,3].

leads to efficient reheating as well as production of strings [34-36]. However, not every symmetry-breaking pattern produces strings. For example, in strongly coupled heterotic string theory [37], there are M5-branes and M2-branes. It would be natural to use the M5-branes, with two wrapped dimensions, to implement brane inflation. Since the M2-branes have codimension three relative to the M5-branes, it will not be so easy to produce strings; this is currently under investigation.[d] It may still be that our vacuum is well-described by weakly coupled heterotic string theory. In that case inflation, and cosmic strings, might simply arise from the low energy effective field theory [1].

2. Stability

Reference [1] identified two instabilities that would prevent superstrings from growing to cosmic size. Actually, these same two instabilities exist for field theory soliton strings [44] — one for global strings and the other for gauge strings — so let us first discuss them in this context.

In the case of global strings, we have noted that the long-ranged Goldstone boson has gradient energy. It does not have potential energy at long distance as long as the broken $U(1)$ symmetry is exact: the broken vacua are then exactly degenerate. However, there are general arguments in string theory that there are no exact global symmetries [45,46]. More generally, the no-hair theorems imply that black holes can destroy global charges, so in any theory of gravity these can not be exactly conserved. Thus the degeneracy of the vacua will not be exact, and there will be a potential energy cost at long distance from the global string. This takes the form of a domain wall, with energy proportional to its area, bounded by the string. The wall exerts a transverse force on the string and forces it to collapse, as in Fig. 2. This is clear for a loop of string bounding a domain wall, but is less intuitive for a network with infinite strings. One can picture the network, with the domain walls, as a complicated shape formed from strips whose width is the typical transverse separation between strings. The timescale for the collapse of the strips, and the disappearance of the strings, is set by the width of the strips, not their (potentially unbounded) length [47].

For gauge strings, the $U(1)$ symmetry is exact because it is a gauge symmetry, and all energies fall exponentially with distance from the string.

[d] There has also been recent discussion of a more exotic symmetry-breaking pattern in D-brane inflation [38-40]. For other discussions of string production in brane inflation see Refs. [41-43].

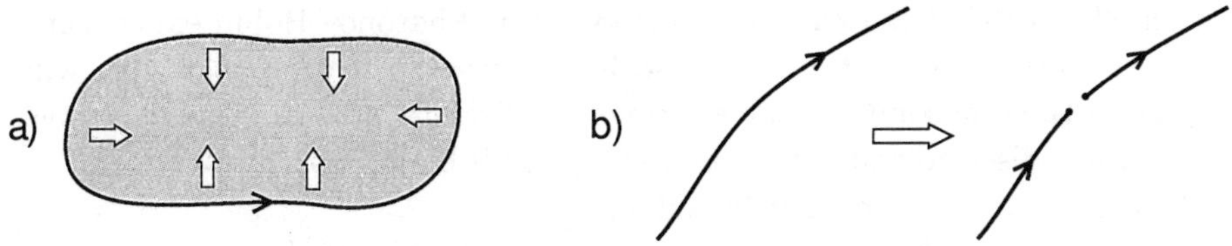

Figure 2. Instabilities of macroscopic strings: a) Confinement by a domain wall. b) Breakage.

A magnetic flux runs along the core of the string, and it is the conservation of this flux that prevents the string from breaking. However, in any unified theory one expects that there will be electric and magnetic sources for every flux [48], so that the string can break by creation of a monopole/antimonopole pair. If this is possible, it will happen not only once but everywhere along the length of the string, and so the string breaks up into short segments rather than growing to cosmic length.

These two instabilities are rather generic, but there are two ways that strings might evade them. First, they might turn out to be slow on cosmic timescales because they are suppressed by large ratios of scales. For monopole pair creation, for example, the classic Schwinger calculation gives the pair production rate as $e^{-\pi M^2/\mu}$, where M is the monopole mass. If the monopole mass is an order of magnitude larger than the scale of the string tension, then the decay will be slow on cosmological time scales. One can think of a succession of symmetry breakings $SU(2) \to U(1) \to 1$. There is no $U(1)$ at the beginning so there are no stable strings, but if the scale of the first symmetry breaking is higher than that of the second then there will be string solutions in the effective field theory describing the second breaking, and they can be long-lived.

Second, there is way that strings can be exactly stable. Consider the Abrikosov flux tubes in an ordinary superconductor. The Higgs field there is an electron pair and has charge $2e$, so the tube has total flux $2\pi/2e$. However, because there do exist singly charged electrons, Dirac quantization gives the minimum monopole charge as twice this, $2\pi/e$. Thus a flux tube cannot end on a monopole, though two can; equivalently one can think of the monopole as a bead on a string, at which the flux reverses. So the Abrikosov flux tubes (in an infinite system) are absolutely stable. One can think of this in terms of an unbroken discrete gauge symmetry $(-1)^{Q/e}$, which acts as -1 on the electron and $+1$ on the BCS condensate. As one circles the string, fields come back to themselves only up to this transformation. Thus the string can be detected by an Aharonov-Bohm experiment at arbitrarily long distance, and so it can never just disappear. We will refer to these absolutely stable strings as 'discrete' strings. Gauge strings without a discrete gauge charge are truly invisible at long distance, and so there is no obstacle to their breaking.

Now let us turn to string theory. Consider first perturbative strings [1], compactified for example on a torus or a Calabi-Yau manifold. The heterotic and type II strings are effectively global strings, because they couple to long-distance form fields $B_{\mu\nu}$ and $C_{\mu\nu}$, which are dual to scalars. The confining force is then produced by instantons, which are magnetic sources for these fields (thus the completeness of the magnetic sources [48] again enters). It seems difficult to suppress the confining force enough for the strings to survive to today — in particular, QCD instantons contribute in the heterotic case and give a lower bound on the force. The type I string couples to no form field, and of course it can break. The time scale for the breaking is the string scale unless one tunes the string coupling essentially to zero.

The most fully developed model of inflation in string theory is the **K**LMT model [32], in which inflation is due to a D3/$\overline{\text{D3}}$ pair. This model has F- and D1-strings, and these are effectively gauge strings. In ten dimensions these strings couple to form fields, but there are no massless form fields in four dimensions; they are removed by the orientifold or F theory monodromy that the model requires. Thus the strings are unstable to breakage, and this can occur in several ways.

First, the projection that removes the form field produces an oppositely oriented image string on the covering space of the compactification, and breakage occurs through a segment of the string annihilating with its image. If the image is coincident with the string, breakage will be rapid. If the image is not coincident, then the string must fluctuate to find its image. At this point an important feature of the **K**LMT model enters, the warping of the compact dimensions. That is, the metric is

$$ds^2 = e^{2\Delta(y)}\eta_{\mu\nu}dx^\mu dx^\nu + \dots \; . \tag{1}$$

The factor $e^{\Delta(y)}$ is a gravitational redshift which varies strongly as a function of the compact coordinates y. Generally $e^{\Delta(y)}$ is around 1 over most of the compact space, but falls to much smaller values in a few small regions known as 'throats.' The effective tension of a string as seen by a four-dimensional physicist depends on its ten-dimensional tension μ_0 and the local gravitational redshift at its position:

$$\mu = e^{2\Delta(y)}\mu_0 \ . \tag{2}$$

This is the Randall-Sundrum (RS) idea [49], that different four-dimensional scales arise from a single underlying scale through gravitational redshifting. This means that the strings feel a strong potential with local minima in the throats. The same is true for the inflationary D3/$\overline{\text{D3}}$: these sit near the bottom of some throat, and the strings produced in their annihilation sit near the bottom of the same throat. As we will discuss further in the next section, the depth of this throat is at least of order $e^{\Delta(y)} \sim 10^{-3}$. Any process that requires the strings to tunnel out of the inflationary well thus involves a ratio of scales that is this large, and pays a penalty of at least e^{-10^6} in the Schwinger calculation. Thus the annihilation is completely suppressed if the image is not in the inflationary throat [31]. Equivalently, it is suppressed if there is no orientifold fixed point in the throat. There is no particular reason for the throat to be coincident with a fixed point; their relative positions are fixed by the complex structure moduli, which depend on flux integers, and these are expected to take rather generic values [50-52].

The strings can also break on a brane. The model must include branes on which the Standard Model (SM) fields live. If these are D3-branes in the inflationary throat the strings will break; if they are D7-branes that pass through the inflationary throat then all but the D1-string will break [31]. If they are outside the throat then the strings are stable for the same reason as above. In the simplest implementation of the RS idea the SM branes must be in a different throat: the depth of the inflationary throat is something of order the GUT scale, while the depth of the SM throat should be of order the weak scale. One must ask whether other branes might still be in the inflationary throat, but this is not possible: these would have low energy degrees of freedom which would receive most of the energy during reheating, rather than the SM fields. So this gives a simple scenario in which the strings are stable, but much more work is needed to see whether it is viable, and whether it is generic.

The stabilization of moduli has not been as fully developed in models based on large compact dimensions, but many of the same considerations apply. The warp factor will depend at least weakly on the compact dimensions, because of the symmetry breakings required in realistic models, and this will localize the various branes and strings. The annihilation of objects that are physically separated in the compact directions is then suppressed due to their separation.

In summary, it is encouraging to see that strings can be stabilized as a side effect of certain generic properties such as warping and/or large dimensions, which are needed to lower the inflationary scale below the Planck scale in these models. Incidentally, there are examples of exactly stable discrete strings, for example by wrapping higher-dimensional branes on torsion cycles, but it is not clear how a network of these strings would be produced. It is not sufficient to have one kind of string that is produced, and a different kind that is stable![e]

3. Seeing Cosmic Strings

We will focus on strings that have only gravitational interactions. Light matter fields live on branes; in most cases stability requires that the branes and strings be physically separated so the light fields will interact with the strings only through bulk interactions. In particular there will then be no charged zero modes, as there are in superconducting strings [10]. We have noted at least one way that stable strings can coincide with a brane, namely D1-branes with a D7-brane, but the zero modes will become massive due to symmetry breaking effects unless protected by a chiral symmetry. It seems difficult to obtain such a symmetry in the models that we are considering; it is an interesting question whether there are models in which stable superconducting strings are produced. In the absence of chiral symmetries (noting that supersymmetry must ultimately be broken), the *only* light or zero modes on cosmic superstrings will be their bosonic coordinates in the noncompact directions.

Thus we must ask what is the likely range for $G\mu$. In models with large compact dimensions [54,55] this is suppressed by some power of $R/L_{\rm P}$. In models with large warping [49] it is suppressed by $e^{2\Delta}$ as in Eq. (2). Thus the tension is essentially a free parameter; for example in the warped models it is the exponential of a ratio of flux quanta [56]. In models of brane inflation, the value of $G\mu$ can be deduced from the observed value of the CMB fluctuations $\delta T/T$. That is, one assumes that $\delta T/T$ arises from the quantum fluctuations of the inflaton; this is natural given the flat form of the inflaton potential. For any given brane geometry the inflaton potential has a definite functional form. For example, in the D3/$\overline{\rm D3}$ system it is $V \sim V_0 - O(\phi^{-4})$. Fitting the observed $\delta T/T$ then determines the normalization

[e]In this connection, the possible stable strings discussed in Ref. [53] involve additional degrees of freedom beyond those involved in brane annihilation, and so would not be produced in brane inflation.

of V_0, and this in turn determines the string tension. For example, in the **K**LMT model V_0 is the Planck scale times $e^{4\Delta}$ and the string tension is just the square root of this (times $g_s^{1/2}$ for F-strings and $g_s^{-1/2}$ for D-strings). This puts the tensions in the likely range $10^{-10} \lesssim G\mu \lesssim 10^{-9}$ [32]. For models based on large dimensions, for various geometries, Refs. [3,57] find values in the range $10^{-12} \lesssim G\mu \lesssim 10^{-6}$.

The CMB and pulsar bounds on $G\mu$ quoted in the review are at the upper end of the brane inflation range, ruling out the highest-tension models. Both bounds will improve in the coming decade, at least by one or two orders of magnitude, due to improved data. However, the more exciting prospect comes from LIGO [58,59]. Under most circumstances LIGO is at a disadvantage looking for cosmological backgrounds because these fall with increasing frequency: LIGO is looking at frequencies that are 10^{10} times those of the pulsar measurements (100 Hz versus years^{-1}). However, something unexpectedly nice happens. When a loop of string in three space dimensions oscillates, typically it forms a cusp several times per oscillation [60]. The instantaneous shape is $y = |x|^{2/3}$, with the tip moving at the speed of light in the x-direction. Like the crack of a whip, a great deal of energy is concentrated in the tip, but this whip is perhaps hundreds of light-years long, with tension not so far below the Planck scale, and so it emits an intense beam of gravitational waves in the direction of its motion [58]. The Fourier transform of such a singularity is much larger at high frequency than for a smooth function, large enough that it is within reach of LIGO.

This is shown in Fig. 3, reproduced from [59]. Under optimistic assumptions (but not, I think, too optimistic), even LIGO I is close to discovery sensitivity of one event per year over much of the range of interesting tensions, including the narrow range of the **K**LMT model. This is remarkable: cosmic superstrings might be the brightest objects in gravitational wave astronomy, and the first discovered! LIGO I to date has around 0.1 design-year of data, but it is supposed to begin a new science run in January 2005 at close to design sensitivity and with a good duty cycle. Advanced LIGO is sensitive over almost the whole range, and with a higher event rate. It has not yet been funded, but it is likely to be in operation around five years from now. LISA, which may follow a few years after that, is even more sensitive. In magnitude of h it is comparable to LIGO I, but it is looking at a frequency 10,000 times lower and so the typical strains are 1000 times greater [59]. The cusp events might be seen in a search for unmodeled bursts. The shape is not as complex as for stellar and black hole inspirals,

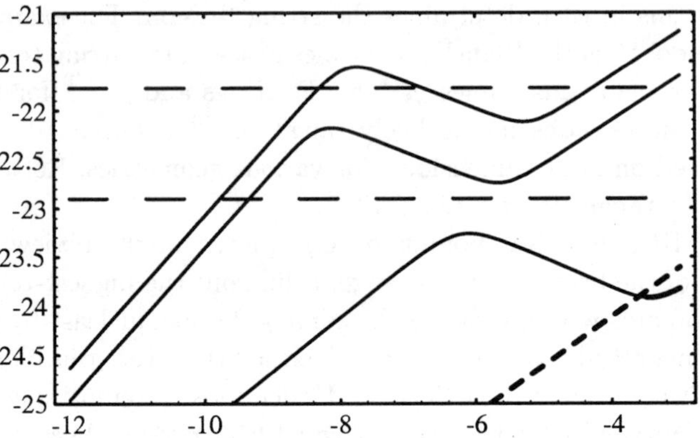

Figure 3. Gravitational wave cusp signals, taken from Damour and Vilenkin [59]. The horizontal axis is $\log_{10}\alpha$ where $\alpha = 50G\mu$. Thus the brane inflation range $10^{-12} \lesssim G\mu \lesssim 10^{-6}$ becomes $-10.3 < \log_{10}\alpha < -4.3$. The vertical axis is $\log_{10} h$ where h is the gravitational strain in the LIGO frequency band. The upper and lower dashed horizontals are the sensitivities of LIGO I and Advanced LIGO at one event per year. The upper two curves are the cusp signal under optimistic and pessimistic network assumptions. The lowest solid curve is the signal from kinks, which form whenever strings reconnect. The dashed curve is the stochastic signal.

but modeling the specific frequency dependence will increase the signal-to-noise ratio. The power-law frequency dependence of the cusp is distinctive. See also Ref. [61] for a discussion of the form of the cusp signal.

The dependence of the sensitivity on string tension in Fig. 3 is interesting because it is not monotonic. This comes about as follows [59]. As the string tension decreases, the coupling to gravity becomes weaker and so does the gravitational wave burst from a given cusp. However, since gravitational radiation is the only decay channel for string loops they will live longer.[f] Thus as we decrease $G\mu$ there are more but smaller cusps. Seeing smaller intrinsic events requires that the events be closer to us. Thus, the three regimes that are evident in Fig. 3 correspond at the smallest tensions to cusps that took place at redshifts less than one, at the intermediate tensions to cusps that are at redshifts greater than one but in the matter-dominated era, and at the largest tensions to cusps that occurred during the radiation-dominated era. The rise in the event rate with decreasing tension in the middle range comes about because the signals from smaller,

[f]There is another effect as well, the dependence of the short-distance network structure on the gravitational radiation, but we can overlook this for simplicity.

later, cusps suffer less from redshifting during the relatively rapid matter-dominated expansion. The pulsar bounds are also strengthed by taking into account the cusps. This requires careful treatment of statistics, but these might also reach most of the interesting range of tensions [59].

4. Distinguishing Superstrings

Let us now imagine the best case, that LIGO has observed some cusp events. Can we hope to distinguish a network of F- and/or D-strings from a network of gauge theory soliton strings?

The microscopic structure of the string core does not affect the evolution of strings that are light-years in length, except when two strings cross and their cores interact. We have noted that gauge theory solitons will always reconnect. For F-strings, reconnection is a quantum process, and takes place with a probability P of order g_s^2. The numerical factors are worked out in Ref. [62]. To be precise, P is a function of the relative angle and velocity in the collision, but it is simplified to the value averaged over collision parameters.

An important issue is the motion of the string in the compact dimensions. For many supersymmetric compactifications, strings can wander over the whole compact space. Thus they can miss each other, leading to a substantial suppression of P [57,30]. However, we have noted that in realistic compactifications strings will always be confined by a potential in the compact dimensions. Even if the scale of the potential is low, the fluctuations of the strings are only logarithmic in the ratio of scales (this is characteristic of one-dimensional objects) [31]. Thus there is no suppression by powers of the size of the compact dimensions, but the logarithm can be numerically important — it tends to offset powers of π that appear in the numerator. The value of g_s, and the scale of the confining potential, are not known, but in a variety of models Ref. [62] finds $10^{-3} \lesssim P \lesssim 1$. For D-D collisions the situation is more complicated, and in the same models one finds $10^{-1} \lesssim P \lesssim 1$. For F-D collisions, P can vary from 0 to 1.

Given the value of P, to determine the observational effect one must feed this into the network simulations. A simple argument suggests that the signatures scale as $1/P$: the amount of string in the network must be increased by this factor in order for an increased number of collisions (per unit length of string) to offset the reduced P in each collision [63]. This is a bit oversimplified, because there are issues connected with the sub-horizon scale structure in the string network [64,65] that can work in either direction.

If we take $1/P$ as a model, we see that for the smaller values of P discussed above there can be a substantial increase in the signal even above the encouraging values found in the last section, so that LIGO might soon begin to see *many* cusps. Of course, the existing bounds become stronger, e.g. $G\mu < 10^{-7}/P$ from pulsars. If P is only slightly less than one, say 0.5, then it will require precision simulation of the networks and good statistics on the signatures to distinguish this from 1.0. It should be noted that even with given values of μ and P there are still substantial uncertainties in the understanding of the behavior of string networks. This has recently been discussed in Ref. [63], which concludes that the sensitivities given in Fig. 3 are only weakly dependent on the unknowns.

To first approximation there are two relevant parameters, μ and P. Each individual cusp event has only a single parameter to measure, its overall strength h: because it is a power law there is no characteristic frequency scale. (There is a high frequency cutoff, determined by the alignment of the cusp with the detector [59], but this gives no information about the cusp itself.) After $O(10)$ cusps are seen one can begin to plot a spectrum, $dN \sim Ah^{-B}dh$, and from the two parameters A and B fix μ and P. There are degeneracies — B depends primarily on the epoch in which the cusp took place — but with a more detailed spectrum, and ultimately with data from kink events and pulsars, this degeneracy will be resolved. Thus μ and P will be overdetermined, and nonstandard network behavior (such as we are about to discuss) will be detectable.

The second potentially distinguishing feature of the superstring networks is the existence of both F- and D-strings [31,30], and moreover bound states of p F-strings and q D-strings with a distinctive tension formula

$$\mu = \mu_0 \sqrt{p^2 + q^2/g_s^2}\ . \tag{3}$$

In this case, when strings of different types collide, rather than reconnecting they form more complicated networks with trilinear vertices. It is then possible that the network does not scale, but gets into a frozen phase where it just stretches with the expansion of the universe [9,66]. If so, its density would come to dominate at the tensions that we are considering. The F-D networks have not yet been simulated, but simulations of comparable networks suggest that they scale, possibly with an enhanced density of strings [67-69]. From the discussion above, it follows that one will not directly read off the spectrum (3) from the observations, but there should eventually be enough information to distinguish F-D string networks from other types.

Networks with multiple types of string can also arise in field theories, though I do not know any classical field theory that gives the particular spectrum (3). However, because of duality there will be gauge theory strings that are very hard to differentiate from the F- and D-strings that we are discussing. In particular, in the **K**LMT model the strings exist in a Klebanov-Strassler [70] throat which has a dual description as a cascading gauge theory.[g] Thus the (p, q) spectrum, and well as the property $P \ll 1$, do arise in this theory. The point is that the dual description of the F-strings is as electric flux tubes, which are quantum mechanical objects rather than classical solitons, whereas almost all work on field theory cosmic strings has been in the context of perturbative unification. The possibility that electric flux tubes could be cosmic strings was discussed in [1].

Indeed, the existence of dualities between string theories and field theories raises the issue, what really is string theory? This is beyond our current scope, but I note that in the present case there is a quantitative question. The **K**LMT model has a parameter gM; when this is large the string description is the valid one, and when it is small the gauge description is the valid one. To fantasize about the maximum possible information that might be extracted from the string network, let us suppose that we could map it out in detail by lensing. In this case we could see the spectrum (3) in the deficit angles, and by measuring various correlations in the network we might deduce the actual function $P(v, \theta)$. This is obtained as a piece of the Virasoro-Shapiro amplitude [78,62], and so in this best case we might truly see string theory written in the sky.

5. Conclusions

As we have seen, each of the four conditions that we discussed at the beginning is independently model-dependent.[h] However, quite a number of things have worked out surprisingly well: the production of strings in brane inflation, the possible stabilization of the strings as a side effect of other

[g]For recent discussions of relations between field theory strings and F- and D-strings see Refs. [71-77].

[h]I think that I am presenting this in an overly pessimistic way. When I hear talks on new physics (especially in the context of cosmology!) there is much less of a sense of an a priori measure of what is likely. I am quite sure that many of the ideas that are prominently discussed have an a priori probability very much less than cosmic superstrings. (I had started to include a list, but decided that it would be distracting and inflammatory. It was rather long.)

properties of the models (in particular, of the stabilization of the vacuum itself), the possibility to see strings over many interesting orders of magnitude of tension, and the existence of properties that distinguish different kinds of string so that after the strings are discovered we can do a lot of science with them. In any case, searching for cosmic strings is a tiny marginal cost on top of experiments that will already be done, and it is great that string theorists will have a stake in these experiments over the coming decade or more.

Acknowledgments

I would like to thank E. Copeland, M. Jackson, N. Jones, and R. Myers for collaborations, and N. Arkani-Hamed, L. Bildsten, G. Dvali, A. Filippenko, T. Kibble, A. Lo, A. Lommen, J. Preskill, H. Tye, T. Vachaspati, and A. Vilenkin for discussions and communications. This work was supported by National Science Foundation grants PHY99-07949 and PHY00-98395.

References

1. E. Witten, *Phys. Lett.* **B153**, 243 (1985).
2. N. Jones, H. Stoica, and S. H. H. Tye, *JHEP* **07**, 051 (2002), hep-th/0203163.
3. S. Sarangi and S. H. H. Tye, *Phys. Lett.* **B536**, 185 (2002), hep-th/0204074.
4. A. Vilenkin and E. Shellard, *Cosmic strings and other topological defects* (Cambridge Univ. Press, 1994), p. 517.
5. M. B. Hindmarsh and T. W. B. Kibble, *Rept. Prog. Phys.* **58**, 477 (1995), hep-ph/9411342.
6. T. W. B. Kibble, talk at COSLAB 2004, Ambleside (unpublished).
7. A. A. Abrikosov, *Sov. Phys. JETP* **5**, 1174 (1957).
8. H. B. Nielsen and P. Olesen, *Nucl. Phys.* **B61**, 45 (1973).
9. T. W. B. Kibble, *J. Phys.* **A9**, 1387 (1976).
10. E. Witten, *Nucl. Phys.* **B249**, 557 (1985).
11. E. P. S. Shellard, *Nucl. Phys.* **B283**, 624 (1987).
12. R. A. Matzner, *Computers in Physics* **2**, 51 (1988).
13. K. J. M. Moriarty, *E. Myers, and C. Rebbi, Phys. Lett.* **B207**, 411 (1988).
14. Y. B. Zeldovich, *Mon. Not. Roy. Astron. Soc.* **192**, 663 (1980).
15. A. Vilenkin, *Phys. Rev. Lett.* **46**, 1169 (1981).
16. L. Pogosian, S. H. H. Tye, I. Wasserman, and M. Wyman,*Phys. Rev.* **D68**, 023506 (2003), hep-th/0304188.
17. L. Pogosian, M. C. Wyman, and I. Wasserman, *JCAP* **09** 008 (2004), astro-ph/0403268.
18. E. Jeong and G. F. Smoot, *Astrophys. J.* **624**, 21 (2005), astro-ph/0406432.
19. A. Lo and E. Wright, (2005), astro-ph/0503120.
20. A. N. Lommen, (2002), astro-ph/0208572.

21. M. Sazhin *et al.*, *Mon. Not. Roy. Astron. Soc.* **343**, 353 (2003), astro-ph/0302547.
22. M. V. Sazhin *et al.*, (2004), astro-ph/0406516.
23. R. E. Schild, I. S. Masnyak, B. I. Hnatyk, and V. I. Zhdanov, *Astron. and Astrophys.* **422**, 477 (2004), astro-ph/0406434.
24. G. R. Dvali and S. H. H. Tye, *Phys. Lett.* **B450**, 72 (1999), hep-ph/9812483.
25. S. H. S. Alexander, *Phys. Rev.* **D65**, 023507 (2002), hep-th/0105032.
26. C. P. Burgess *et al.*, *JHEP* **07**, 047 (2001), hep-th/0105204.
27. G. R. Dvali, Q. Shafi, and S. Solganik, (2001), hep-th/0105203.
28. A. Sen, *JHEP* **09**, 023 (1998), hep-th/9808141.
29. E. Witten, *JHEP* **12**, 019 (1998), hep-th/9810188.
30. G. Dvali and A. Vilenkin, *JCAP* **0403**, 010 (2004), hep-th/0312007.
31. E. J. Copeland, R. C. Myers, and J. Polchinski, *JHEP* **06**, 013 (2004), hep-th/0312067.
32. S. Kachru *et al.*, *JCAP* **0310**, 013 (2003), hep-th/0308055.
33. A. D. Linde, *Phys. Rev.* **D49**, 748 (1994), astro-ph/9307002.
34. J. Yokoyama, *Phys. Rev. Lett.* **63**, 712 (1989).
35. L. Kofman, A. D. Linde, and A. A. Starobinsky, *Phys. Rev. Lett.* **76**, 1011 (1996), hep-th/9510119.
36. I. Tkachev, S. Khlebnikov, L. Kofman, and A. D. Linde, *Phys. Lett.* **B440**, 262 (1998), hep-ph/9805209.
37. E. Witten, *Nucl. Phys.* **B471**, 135 (1996), hep-th/9602070.
38. J. Urrestilla, A. Achucarro, and A. C. Davis, *Phys. Rev. Lett.* **92**, 251302 (2004), hep-th/0402032.
39. T. Watari and T. Yanagida, *Phys. Lett.* **B589**, 71 (2004), hep-ph/0402125.
40. K. Dasgupta *et al.*, *JHEP* **08**, 030 (2004), hep-th/0405247.
41. E. Halyo, (2004), hep-th/0402155.
42. T. Matsuda, *Phys. Rev.* **D70**, 023502 (2004), hep-ph/0403092.
43. T. Matsuda, *Phys. Lett.* **B611**, 129 (2005), hep-ph/0406064.
44. J. Preskill and A. Vilenkin, *Phys. Rev.* **D47**, 2324 (1993), hep-ph/9209210.
45. T. Banks and L. J. Dixon, *Nucl. Phys.* **B307**, 93 (1988).
46. J. Polchinski, *String theory. Vol. 2: Superstring theory and beyond* (Cambridge, UK: Univ. Pr., 1998), p. 531.
47. A. Vilenkin and A. E. Everett, *Phys. Rev. Lett.* **48**, 1867 (1982).
48. J. Polchinski, *Int. J. Mod. Phys.* **A19S1**, 145 (2004), hep-th/0304042.
49. L. Randall and R. Sundrum, *Phys. Rev. Lett.* **83**, 3370 (1999), hep-ph/9905221.
50. R. Bousso and J. Polchinski, *JHEP* **06**, 006 (2000), hep-th/0004134.
51. M. R. Douglas, *JHEP* **05**, 046 (2003), hep-th/0303194.
52. S. Ashok and M. R. Douglas, *JHEP* **01**, 060 (2004), hep-th/0307049.
53. L. Leblond and S. H. H. Tye,*JHEP* **03**, 055 (2004), hep-th/0402072.
54. N. Arkani-Hamed, S. Dimopoulos, and G. R. Dvali, *Phys. Lett.* **B429**, 263 (1998), hep-ph/9803315.
55. I. Antoniadis, N. Arkani-Hamed, S. Dimopoulos, and G. R. Dvali, *Phys. Lett.* **B436**, 257 (1998), hep-ph/9804398.
56. S. B. Giddings, S. Kachru, and J. Polchinski, *Phys. Rev.* **D66**, 106006 (2002),

hep-th/0105097.
57. N. T. Jones, H. Stoica, and S. H. H. Tye, *Phys. Lett.* **B563**, 6 (2003), hep-th/0303269.
58. T. Damour and A. Vilenkin, *Phys. Rev. Lett.* **85**, 3761 (2000), gr-qc/0004075.
59. T. Damour and A. Vilenkin, *Phys. Rev.* **D64**, 064008 (2001), gr-qc/0104026.
60. N. Turok, *Nucl. Phys.* **B242**, 520 (1984).
61. X. Siemens and K. D. Olum, *Phys. Rev.* **D68**, 085017 (2003), gr-qc/0307113.
62. M. G. Jackson, N. T. Jones, and J. Polchinski, (2004), hep-th/0405229.
63. T. Damour and A. Vilenkin, Phys. Rev. **D71**, 063510 (2005), hep-th/0410222.
64. D. P. Bennett and F. R. Bouchet, (1990), invited talks given at Cosmic String Workshop, Cambridge, England, Jul 2-7, 1989.
65. D. Austin, E. J. Copeland, and T. W. B. Kibble, *Phys. Rev.* **D48**, 5594 (1993), hep-ph/9307325.
66. A. Vilenkin, *Phys. Rev. Lett.* **53**, 1016 (1984).
67. T. Vachaspati and A. Vilenkin, *Phys. Rev.* **D35**, 1131 (1987).
68. D. Spergel and U.-L. Pen, *Astrophys. J.* **491**, L67 (1997), astro-ph/9611198.
69. P. McGraw, *Phys. Rev.* **D57**, 3317 (1998), astro-ph/9706182.
70. I. R. Klebanov and M. J. Strassler, *JHEP* **08**, 052 (2000), hep-th/0007191.
71. G. Dvali, R. Kallosh, and A. Van Proeyen, *JHEP* **01**, 035 (2004), hep-th/0312005.
72. E. Halyo, *JHEP* **03**, 047 (2004), hep-th/0312268.
73. P. Binetruy, G. Dvali, R. Kallosh, and A. Van Proeyen, *Class. Quant. Grav.* **21**, 3137 (2004), hep-th/0402046.
74. S. S. Gubser, C. P. Herzog, and I. R. Klebanov, *JHEP* **0409**, 036 (2004), hep-th/0405282.
75. A. Achucarro and J. Urrestilla, *JHEP* **08**, 050 (2004), hep-th/0407193.
76. S. S. Gubser, C. P. Herzog, and I. R. Klebanov, *Comptes Rendus Physique* **5**, 1031 (2004), hep-th/0409186.
77. A. Lawrence and J. McGreevy, *JHEP* **0410** 056 (2004), hep-th/0409284.
78. J. Polchinski, *Phys. Lett.* **B209**, 252 (1988).

QUANTUM COMPUTATION

JOHN PRESKILL
California Institute of Technology, USA

A quantum computer, which processes a quantum state rather than classical bits, could easily perform calculations that would take far longer than the age of the universe on todays supercomputers. But constructing practical quantum computers will be tremendously challenging. A particularly daunting difficulty is that quantum computers are far more susceptible to making errors than conventional digital computers. The theory of fault-tolerant quantum computation prescribes how a robust subsystem of a larger quantum system can be constructed and reliably manipulated. A large-scale fault-tolerant quantum computer will be a highly unusual quantum many-body system, having much in common with topological quantum field theory and loop quantum gravity.

BLACK HOLES IN ACTIVE GALACTIC NUCLEI

MARTIN REES

Institute of Astronomy, Cambridge University, UK

By the early 1970s the astrophysically-relevant theoretical properties of black holes – the Kerr metric, the 'no hair' theorems, and so forth – were well established. But it took much longer for the observers to discover well-authenticated candidates. By now, the evidence points insistently towards the existence of dark objects, with deep potential wells and 'horizons' through which matter can pass into invisibility. However, observers cannot yet definitively confirm the form of the metric in the strong-gravity region, though there is exciting progress towards this goal.

1. Evidence for Supermassive Compact Objects

There are two compellingly convincing cases of massive collapsed objects in nearby galaxies. The first, in the peculiar spiral NGC 4258, has been revealed by amazingly precise mapping of gas motions via the 1·3 cm maser emission line of H_2O. This putative hole has an inferred mass of $3{\cdot}6 \times 10^7$ solar masses. The second utterly convincing candidate lies in our own Galactic Centre. Fast moving stars have been 'tracked' in orbits around a mass of around 3×10^6 solar masses, associated with a variable source of radio, infrared and x-ray emission. Compact dark objects, with deep gravitational potential wells, are implicated in the power output from active galactic nuclei (AGNs), and in the production of relativistic jets that energise strong radio sources. The demography of these massive holes has been clarified by studies of relatively nearby galaxies (mainly with quiescent nuclei): most seem to harbour dark central masses. There is a remarkably close proportionality between the hole's mass and the velocity dispersion of the central bulge or spheroid in the stellar distribution.

2. How did they form?

We still cannot decide which formation route for supermassive black holes is most likely. The main options are:

(i) Monolithic formation. Some of the gas settling towards the centre of a galaxy may undergo monolithic collapse into a supermassive star (which is then unstable for 'post-Newtonian' reasons) rather fragmenting into ordinary stars.

(ii) Coalescence in a dense stellar system. Coalescence leads to heavier stars; these settle by dynamical friction towards the centre; runaway coalescence then ensues.

(iii) Mergers of binary holes. Most present-day galaxies have been assembled via a series of mergers. After each merger event, the central holes already present in each galaxy would themselves coalesce, via dynamical friction, and then (when they get close) via gravitational radiation.

(iv) Growth via accretion. This is inevitable during the 'active' phase of a galactic nucleus. Moreover, comparison of the total mass of holes in the present universe with the integrated radiative output from all AGNs suggests that much of the mass must have been acquired via radiatively-efficient accretion.

The role of processes (i) and (ii) is unclear. However, (iii) and (iv) are almost certainly important.

3. Gravitational Wave Recoil

A single big galaxy can be traced back to the stage when it was in hundreds of smaller components with individual internal velocity dispersions as low as 20 km/sec. Did black holes form with the same efficiency in small galaxies (with shallow potential wells), or did their formation had to await the buildup of substantial galaxies with deeper potential wells? This issue is important because it determines the expected event rate detected by LISA, and whether there is a population of high-z miniquasars. Of particular interest is whether the merger history can be traced back to 'seed' holes that are the remnants of an early generation of 'Very Massive Objects (VMOs)' with helium cores above 130 solar masses

When the members of a black hole binary coalesce, there is a recoil due to the non-zero net linear momentum carried away by gravitational waves in the coalescence. If the holes have unequal masses, a preferred longitude in the orbital plane is determined by the orbital phase at which the final plunge occurs. For spinning holes there may be a rocket effect perpendicular to the orbital plane, since the spins break the mirror symmetry with respect to the orbital plane. This recoil could be so violent that the merged hole breaks loose from shallow potential wells.

4. Do the Holes obey the Kerr Metric?

So much for the astrophysics of black holes. What are the prospects of using them to test relativistic gravity? The flow patterns close to the hole offer, in principle, a probe of the metric. There is of course no hope (until X-ray interferometry is developed) of actually 'imaging' the inner discs. However, the large frequency-shifts reveal themselves spectroscopically — substantial gravitational redshifts would be expected, as well as large doppler shifts. There are already convincing cases of broad asymmetric emission lines indicative of a relativistic disc. The value of a/m can in principle be constrained too, because the emission is concentrated closer in, and so displays larger shifts, if the hole is rapidly rotating, and there is some evidence that this occurs. The spin of a hole affects the efficiency of 'classical' accretion processes; the value of a/m in a Kerr hole also determines how much energy is in principle extractable from the hole itself. (And the importance of such energy-extraction processes is still unclear). Moreover, the orientation of the spin axis may be important in relation to jet production, etc.

Spin-up is a natural consequence of prolonged disc-mode accretion: any hole that has (for instance) doubled its mass by capturing material that is all spinning the same way would end up with a/m being at least 0.5. A hole that is the outcome of a merger between two of comparable mass would also, generically, have a substantial spin. On the other hand, if it had gained its mass from capturing many low-mass objects (holes, or even stars) in randomly-oriented orbits, a/m would be small. The Chandra and XMM X-ray satellites are extending these studies; they also find rapid quasi-periodic variations which may be caused by 'blobs' on very close orbits. These instruments may offer enough sensitivity, in combination with time-resolution, to study flares, and even to follow a 'hot spot' on a plunging orbit.

The swing in the polarization vector of photon trajectories near a hole was long ago suggested as another diagnostic; but this is still not feasible because X-ray polarimeters are far from capable of detecting the few per-cent polarization expected. Interpretation of these studies is impeded by the inherent 'messiness' of gas dynamics. A far cleaner probe would be a compact mass in a precessing and decaying orbit around a massive hole. Detecting and monitoring such a system may have to await the launch of LISA.

THE IMPLICATIONS FOR COSMOLOGY OF WMAP DATA

LICIA VERDE
University of Pennsylvania, USA

The primary goal of the WMAP mission is to produce a high-fidelity all-sky polarization-sensitive map of the CMB radiation to determine the cosmology of our Universe. WMAP has produced a full sky map of the microwave sky in 5 frequencies, with a resolution a factor 30 higher than the previous full sky map as produced by the COBE satellite in 1992. This is the cleanest picture of the early Universe; the structures on the CMB – the pattern of hot and cold spots – carry information about the composition, geometry, age, etc. of our Universe. We will review the implications for cosmology of WMAP data. The highlight is that cosmology now has a standard cosmological model. With only 6 parameters the model fits not only WMAP data remarkably well, but also a host of other astronomical observations.

WORKSHOP REPORTS

A1: EXACT SOLUTIONS AND THEIR INTERPRETATION

MARC MARS
Facultad de Ciencias, Universidad de Salamanca,
Plaza de la Merced s/n, 37008 Salamanca, Spain.
e-mail: marc@usal.es

1. Introduction

Exact solutions have played a very important role in the development of general relativity and other gravitational theories. For many years, the lack of solutions and the great difficulty of solving the Einstein field equations meant that, generally, new solutions were published almost automatically. As the field matured, and the number of solutions increased substantially, it became gradually clear that finding new solutions for their own sake was not the way forward. Instead, the new solution should come with some added bonus, be it either an interesting physical interpretation, an innovative method of solving the equations, the classification of a whole set of metrics satisfying suitable conditions, etc. Moreover, the analysis and physical interpretation of existing solutions was perceived to be a fundamental issue in the field. Also studying geometrical properties of the solutions which could have interest in other fields became more and more important, for instance as test beds for numerical simulations, as counterexamples of general conjectures, or even as sources of new general conjectures via induction. This shift in the point of view, which started some ten years ago, has in my opinion definitely settled in the field. This means, in particular, that some works which years ago would not have belonged naturally to the field of exact solutions, now fit perfectly well there. This fact was clearly noticeable from the type of abstracts sent to the workshop and it is also reflected in the selection made for oral presentations. In addition to many contributions devoted to presenting new exact solutions, there were works dealing with geometrical and/or physical studies of classes of solutions, in particular of black holes, radiative spacetimes and cylindrical models. Other contributions emphasized geometrical properties of general spacetimes, others dealt with existence or non-existence of solutions within some class, etc. Thus, the range of questions treated is nowadays much broader than pure solution finding, something which I personally consider very positive.

As a summary, over seventy-five abstracts were received for the A1 workshop, which meant a time allocation of seven and a half hours. The talks were distributed uniformly in thirty presentations of fifteen minutes, including questions. Most of the abstracts not selected for oral presentation were given the opportunity of a poster presentation. For the sake of brevity, only the oral presentations will be reported here. For posters, the interested reader is addressed to the book of abstracts of the conference. Regarding the oral presentations, I tried to group them together into several topics and the presentations themselves were scheduled, as much as possible, to follow these groupings. In this report I will devote one section to each topic.

2. Black holes in four dimensions

The contribution of **E. Ruiz** (reporting on work done in collaboration with **V.S. Manko**) was concerned with the so-called Kramer-Neugebauer spacetime [1], which is a vacuum stationary and axially symmetric solution of the Einstein field equations. This spacetime has been considered since its publication as a non-linear superposition of two Kerr solutions. A more careful analysis shows that in general this spacetime also contains a strut between the two objects (see [2] for a review), which can be removed by requiring the spacetime to have a well-defined symmetry axis [3]. In this contribution it was shown that, even when this condition is satisfied, there is still enough freedom [4] in the parameters to allow for equilibrium configurations of (i) two super-extreme Kerr objects, (ii) a normal and a super-extreme Kerr object or, surprisingly, (iii) two normal Kerr objects (perhaps two black holes). Imposing further that each object has a positive Komar mass, case (iii) was shown to be impossible. This settles the so-called Hoenselaers conjecture [5]. Although valid only for the particular Kramer-Neugebauer solution, this result may have more profound implications when combined with the results of G. Weinstein [6] or G. Neugebauer [7] on general asymptotically flat stationary and axially symmetric vacuum solutions describing a black hole with two connected components. In particular, the results of this contribution suggest that the Kerr metric uniqueness theorem is still true even for non-connected horizons (at least for the two-component case).

The contribution of **W.S. Bonnor** (reporting on work [8] done in collaboration with **B.R. Steadman**) dealt also with the Kramer-Neugebauer solution. A class of metrics containing five parameters was analyzed from the physical point of view by using a perturbation scheme in the mass parameters m_1 and m_2 of the two spinning objects. It was found that, in general, the metric is not truly asymptotically flat but, instead, contains NUT sources which makes it unphysical. These NUT charges are absent in two different cases: (I) whenever $a_1/m_1 = a_2/m_2$ where a_i are the specific angular momenta of the objects or (II) whenever $a_1 + a_2 = 0$. In the latter case the NUT charge is removed by a suitable unitary transformation of the metric. In case (I) it was shown that there are closed timelike curves (CTC) nearby the axis of symmetry between the two spinning objects. Thus, the axis of symmetry is not regular between the objects. The source of this singularity is a strut between the two Kerr like objects plus a spinning rod of finite length. In case (II) there are no closed timelike curves near the axis of symmetry and the source between the two bodies is just a strut. The authors conclude that, in view of the existence of CTC, general relativity does not give a satisfactory account of this physical system.

C. Klein discussed explicit solutions describing disks of matter around black holes. Such configurations are of considerable astrophysical interest in the context of accretion disks around black holes or supermassive black holes in the nucleus of certain galaxies. A relativistic description of such systems is necessary since black holes are genuinely relativistic objects. If the disks are massive, an exact relativistic description is in order. An explicit treatment is in principle possible in the case of stationary axisymmetric settings with an infinitesimally thin disk. In this case the Einstein equations reduce to the completely integrable Ernst equation. A large class of solutions to this equation in terms of hyperelliptic θ functions due to Korotkin [9] can be used in this context. In this contribution it was shown [10] that a subclass of these solutions describing black holes with a surrounding disk of pressureless matter, so called dust, can be obtained in terms of θ functions on partially degenerated Riemann surfaces of genus g. In all cases the metric is explicit and, in the case of vanishing genus, the metric coefficients even become elementary functions.

J. Louko (reporting on work done in collaboration with **R.B. Mann**) discussed the possibility of constructing geons with angular momentum and gauge charges. An eternal black hole spacetime with a nondegenerate Killing horizon and suitable discrete isometries has a variant, constructed by suitable identifications, in which the spatial hypersurfaces have only one asymptotic region. These unconventional geon-type black holes provide an arena for probing our understanding of both classical (e.g. [11]) and quantum mechanical (e.g. [12]) properties of black holes. In this contribution, geons with angular momenta and gauge charges in four and more dimensions were constructed. It was shown, in particular, that four-dimensional spherically symmetric $SU(2)$ black holes have a geon variant.

3. Vacuum metrics with extra curvature restrictions

The contribution of **V. Pravda** discussed the general solution of the Einstein vacuum field equations with all curvature invariants equal to zero, i.e. the so-called VSI spacetimes. A curvature invariant of order n is a scalar constructed from the Riemann tensor and its covariant derivatives up to order n. In Riemannian geometry the vanishing of all zeroth order invariants implies flat space whereas in Lorentzian geometry there are non-trivial examples (e.g. pp-waves) of spacetimes with all invariants of all orders vanishing. In this contribution all such spacetimes were obtained. It turns out that they represent sixteen classes within Kundt's class and all metrics can be given explicitly [13]. Besides general relativity, VSI spacetimes have also applications in string theory [13, 14] because many of these solutions, like pp-waves, are exact vacuum solutions to string theory to all orders in the coupling constant. It is therefore of interest to generalise the VSI theorem to higher dimensions, which was done in [15]. This required generalising the Petrov classification for higher dimensions [16, 17].

The contribution of **A. Barnes** dealt with vacuum spacetimes having the electric and magnetic parts of the Weyl tensor (with respect to some timelike congruence $\vec{u}$) proportional to each other with a proportionality *constant*. Such spacetimes are referred to as *constant argument* spacetimes and obviously generalise the purely magnetic spacetimes. Since long ago, the latter are conjectured to be non-existent (except for trivial cases). Such a general result has proven elusive and the conjecture has been shown only in special cases (type D, shear-free congruence, hypersurface orthogonal $\vec{u}$ or geodesic congruence, see references in [18]). In this contribution, the conjecture was extended to constant argument spacetimes [18] and a number of special cases where the result can be proven were presented. The general case is under investigation by analyzing the integrability conditions of the Bianchi+vacuum field equations using the Newman-Penrose formalism.

L. Wylleman (reporting on work done in collaboration with **N. Van den Bergh**) discussed inhomogeneous silent universes of Petrov type I. Silent universes [19] are dust spacetimes such that the fluid velocity vector $\vec{u}$ is irrotational (and, of course, geodesic from the Euler equations) and the

Weyl tensor is purely electric with respect to $\vec{u}$. The Petrov type is then restricted to be either O, D or I [20]. The first two cases are well known, the type D solutions corresponding to the Szekeres family of spacetimes [21]. Thus, only type I solutions remain to be understood. After discussing some developments in classifying them [22, 23], the author discussed the natural but unfortunate omitting of the cosmological constant Λ in the consistency problem of Petrov type I silent universes [24]. In this setting, the surviving dynamical variables and differential equations within the Ellis-MacCallum formalism were introduced. The autonomous dynamical subsystem of evolution equations was clearly shown to be consistent under the identical vanishing of one of the eigenvalues of the expansion tensor, but this only for $\Lambda > 0$. It was pointed out that the remaining equations then give some additional constraints but can be closed, leading to two new families of inhomogeneous type I silent universe metrics. Finally, a new suggestion based on [23] was made to prove the conjecture that the only $\Lambda = 0$ Petrov type I silent universes are the orthogonally spatially homogeneous Bianchi type I metrics. Both after this and after A. Barnes' talk, an interesting discussion took place where J. Carminati announced new, still unpublished, results solving some of the problems mentioned in the talks.

4. Higher or lower dimensional solutions

H.S. Reall reviewed recent progress in the field of exact solutions of the Einstein equations in more than four dimensions. Three different topics were covered. 1. Black rings: these are stationary, asymptotically flat black hole solutions [25] of the five-dimensional vacuum Einstein equations with horizons of topology $S^1 \times S^2$. The fact that the topology of the horizon is not spherical implies in particular that the black hole uniqueness theorems valid in four dimensions are no longer true in five dimensions. 2. Generalised Weyl solutions: these are higher dimensional generalisations [26] of the Weyl metrics, which are static and axially symmetric solutions of the Einstein vacuum field equations. The generalisation consists in D-dimensional vacuum spacetimes admitting $D-2$ orthogonal, commuting and non-null Killing vector fields. 3. Supersymmetric solutions: solutions admitting a super-covariantly constant spinor. These solutions play an important role in string theory. Years ago K.P. Tod [27] showed how to obtain all such solutions in four dimensions. In this contribution it was explained how one can obtain a general classification of such solutions in five dimensions [28].

R. Mann (reporting on work done in collaboration with **C. Stelea**) presented exact solutions of the vacuum Einstein field equations with cosmological constant in n dimensions [29], for $n \geq 5$. The spacetimes are locally asymptotically flat, de Sitter or anti-de Sitter and have non-trivial topology. These solutions are NUT charged generalisations of topological black hole solutions in higher dimensions. Both the Lorentzian and the Euclidean cases may be considered, the relation between them being suitable analytic continuation of the coordinates and parameters. The existence of such NUT charged spacetimes in odd dimensions was explicitly demonstrated, and examples were given both in dimensions five and seven. The NUT charges (and therefore the non-trivial topology) is closely related to the cosmological constant. Making the latter zero makes, in general, the NUT charges also to vanish and therefore the topology to become trivial. The structure of the fixed points of the Killing vector along the S^1 fiber defining the NUT charge (or charges) can be either of nut type (i.e. of dimension less than $n-2$) or of bolt type (i.e. of dimension $n-2$). The global structure of these solutions was also analyzed and possible applications to string theory, in particular regarding the AdS/CFT correspondence was discussed.

The contribution by **A.N. Aliev** dealt with a charged rotating black hole in five dimensions obtained to first order in perturbation theory. The metric generalises the Schwarzchild-Tangherlini solution [30], which corresponds to a static spherically symmetric vacuum black hole, to a situation when there is charge present and two small independent rotation parameters. In this sense, the metric can be considered a charged generalisation of the Myers-Perry black hole [31] in the approximation of slow rotation. This solution should be considered as a first step towards obtaining the full charged generalisation of the Myers-Perry vacuum black hole. Another first step was obtained in [32] where the Maxwell equations in the Myers-Perry solution were solved in the case of a purely magnetic field aligned to the two angular momenta of the black hole.

G. Clément (reporting on work done in collaboration with **K. Ait Moussa** and **C. Leygnac**) discussed black holes in topologically massive gravity [33]. Topologically massive gravity is obtained [34] from Einstein gravity in 2+1 dimensions by the addition of a gravitational Chern-Simons term. It was shown that the Vuorio [35] solution of this theory may be continued to a two-parameter family of non-asymptotically flat black hole solutions. A particularity of these black holes is that their ergosphere extends to infinity, so that no observers can remain static. However locally stationary observers are allowed. These spacetimes are geodesically complete but admit, within a certain parameter range, closed timelike curves hidden behind the two horizons. The superangular momentum approach in (2+1)-dimensional gravity with two Killing vectors [36] was used to evaluate the mass and angular momentum of these black holes.

5. Gravitational waves

P. Krtouš discussed accelerated black holes in an anti-de Sitter background, based on the C-metric with cosmological constant, which is a classic solution of the Einstein-Maxwell equations with Λ. These spacetimes play an important role in studies of radiative properties of gravitational fields and cosmological pair creation of black holes, among others. However, a deeper understanding of their physical and global properties, including the character of radiation, has been missing. Recently, the C-metric with positive [37] and negative [38] Λ have been analyzed. For positive Λ conformal infinity has a spacelike character and the spacetime contains asymptotically de Sitter domains each of them with two accelerated black holes. Different de Sitter domains are then connected through the interior of black holes. For negative Λ, infinity has a timelike character and the global structure of the spacetime is more complex. For small accelerations, $a < \sqrt{-\Lambda/3}$, the metric describes a single uniformly accelerated black hole, whereas for $a > \sqrt{-\Lambda/3}$ it represents a sequence of pairs of accelerated black holes [39]. The global structure of these spacetimes was illustrated by animated 3D conformal-like diagrams (see [40]). The behavior of the radiation in these spacetimes was also investigated. It was found that the radiative field component depends on a direction along which conformal infinity is approached. Moreover, the asymptotic directional structure is given by the algebraic structure of the field, namely by the degeneracy of the principal null directions and their orientation with respect to infinity [41].

The contribution of **M. Ortaggio** dealt with exact solutions representing impulsive gravitational waves which propagate either in the Bertotti-Robinson family of electrovacuum direct product geometries [42] or in the static, cylindrically symmetric "fluxtube" of Bonnor-Melvin [43]. These spacetimes belong to Kundt's class and are, in general, of Petrov type II. They can be conveniently visualized as submanifolds of impulsive pp-waves in a higher dimensional flat space, and also admit a simple four-dimensional line-element. The metrics discussed in this contribution are generalisations of well known impulsive solutions in Minkowski space [44, 45] which were previously already extended [46] to anti-de Sitter (see [47] for reviews).

M.P. Machado Ramos (reporting on work done in collaboration with **S.B. Edgar**) described a method for obtaining solutions of Einstein's equations by using the generalised invariant formalism (GIF) and applied the method to a particular class of Petrov type N pure radiation spaces [48,49] . The basic ideas of this approach were developed by Edgar and Vickers [50] who found, as an application, the general conformally flat pure radiation metric. The procedure consists of applying the GIF operators to the non-zero quantities arising at each step of the Karlhede algorithm [51], and therefore constructing involutive tables of these derivatives. From these tables, invariant scalar fields with zero weight should be extracted. If four of them (functionally independent) can be determined, a local coordinate system is obtained. Finally, the tables for these scalars enables one to write the complex null tetrad, and hence the metric. A big advantage of the method is that there is no need of solving differential equations. The difficulty consists in obtaining the simplest possible tables so that finding the four functionally independent scalars with zero weight is viable. This is not always an easy task, since often it becomes hard in practice to obtain the coordinate candidates. In this contribution the procedure was applied to obtain the pure radiation metric, first found by Edgar and Ludwig [52]

J. Griffiths discussed the stability of the Killing–Cauchy horizon which occurs in some colliding plane wave spacetimes. The spacetimes describing the collision and subsequent interaction of two plane gravitational waves traveling in a Minkowski background always become incomplete within the interaction region after some finite time into the future. In most cases, this is because of the formation of a curvature singularity [53] but there also exist classes of solution depending on an infinite number of parameters where a Killing-Cauchy horizon appears. It is an interesting problem whether such a horizon is stable or not. Previous analysis on this problem showed instability. In this contribution, it was pointed out that the perturbations used to claim the instability result were such that the initial waves which interact are themselves singular as they also become unbounded prior to the collision. The difficulty arises because colliding plane wave solutions are not traditionally derived from initial data. By reformulating the situation as a characteristic initial value problem, the solutions in the case of initial waves having constant aligned polarizations can be obtained using Abel transforms [54]. Using this approach, a proof that horizons in these spacetimes are unstable relative to bounded perturbations of the initial waves was presented.

6. Cylindrically and spherically symmetric solutions

J. Bičák (reporting on work done in collaboration with **T. Ledvinka, M. Žofka** and **B.G. Schmidt**) discussed several aspects of spacetimes generated by static and infinite cylinders of perfect fluid. The emphasis was laid on their global aspects. It was shown that for any smooth monotonic equation of state, or for incompressible fluid, and for any value of the density at the axis there exists a unique global solution [55]. If the pressure vanishes at finite radius then a unique exterior Levi-Cività solution can be joined to the inner solution. Its mass and conicity parameters are determined uniquely by the density/pressure at the axis. In the case of a fluid admitting a non-vanishing density for zero pressure the radius of the cylinder has to be finite. For incompressible fluid, the field equations were solved analytically for nearly Newtonian cylinders and numerically in fully relativistic situations. In fully relativistic cases, various remarkable phenomena were shown as, e.g., that the proper radius of the cylinders increases with increasing central pressure but the circumferential radius starts to decrease from a certain value of the central pressure. Properties of cylindrical shells made of various types of matter were also investigated. Assuming Minkowski spacetime inside the shell, the Israel formalism was used to study the nature of the matter within the shell (including massive particles, perfect fluids and counter-spiraling streams of photons). Relations between mass per unit length of the shell, its radius and the parameters of the Levi-Cività exterior were found [56]. As a particular case, it was shown that cylinders of counter-rotating photons can have Minkowski space on both sides [57].

J. Geršl discussed a spacetime describing an infinitely thin cylindrical shell constituted by a charged perfect fluid as a source [58]. The interior of the shell is described by the Bonnor-Melvin solution [59], which describes the gravitational field of a homogeneous magnetic field. Thus, it corresponds to an interior of a rotating charged hollow cylinder in stationary rotation. As an exterior the Datta-Raychaudhuri solution [60] was chosen. This metric contains both a radial electric and an axial magnetic field. Some properties of the spacetime which arises from solving the Israel junction conditions with a perfect fluid energy-momentum on the shell were discussed. In particular, ranges of free parameters for which energy conditions in the shell are satisfied, trajectories of charged test particles inside the cylinder or the expression for their angular velocity were discussed. It was also shown that the spacetime must contain closed timelike curves outside the shell. Timelike curves must extend far enough to the shell in order to become closed. The dependence of this minimum radial distance in terms of the charge of the shell was analyzed.

The contribution of **D.A. Konkowski** dealt with singularities of Levi-Cività spacetimes. Both "classical" and "quantum" type singularities were addressed. Quantum singularities in general relativistic spacetimes are determined by the behavior of quantum test particles whereas classical singularities are determined by incomplete geodesics or incomplete paths of bounded acceleration in maximal spacetimes. A static spacetime with a timelike singularity is quantum mechanically singular if the spatial portion of the wave operator is not essentially self-adjoint [61]. Weyl's limit point-limit circle criterion is a convenient way to determine whether a wave operator is essentially self-adjoint [62]. This test was applied to scalar wave packets in Levi-Cività spacetimes with (LCC) and without (LC) cosmological constant to help elucidate their physical properties. These spacetimes can be generally interpreted as cylindrically symmetric spacetimes with an infinite line mass along the axis [63, 64]; LCC spacetimes may also have a singular hypersurface. For the LC and LCC spacetimes a scalar curvature singularity was found along the axis and/or hypersurface except for three special cases some of which contain quasiregular singularities (see e.g. [63] and references therein). Both classes of spacetimes have two independent parameters: σ which is a mass parameter and C which is a topological parameter measuring a conical deficit angle. All LC and LCC spacetimes that are classically singular (with quasiregular or scalar curvature singularities) are also quantum mechanically singular and all that are classically nonsingular are also quantum mechanically nonsingular [62, 65]. The following interesting result was also presented: both for LC and LCC spacetimes, the enigmatic $\sigma = 1/2$ flat spacetime present a quantum wave packet behavior that clearly illustrates their flat, planar symmetry structure and sets them apart from the other LC and LCC spacetimes [62, 65]. This result illustrates how studying quantum wave behavior on spacetimes can help us understand the spacetimes' metric parameters together with the classical singularity structure of the spacetime.

F.C. Mena (reporting on work done in collaboration with **K.P. Tod**) discussed spacetimes obtained by joining an infinite cylinder made of a collapsing spatially homogeneous dust with a non-stationary vacuum exterior preserving the cylindrical symmetry. The use of an an evolving metric outside the cylinder is motivated by the result [66] that the matching of a spatially homogeneous cylindrically symmetric interior to a static exterior, also with cylindrical symmetry, leads to very restrictive interior solutions with no evolution along one spacelike direction. Thus the Oppenheimer-Snyder model of collapse cannot be generalised (in a physically reasonable way) to cylindrical symmetry with spatially homogeneous interiors and static exteriors. The fact that static exteriors restricts strongly the possible interiors was also observed for homogeneous and isotropic perfect fluid cylinders [67] and for arbitrary homogeneous and isotropic perfect fluids [68]. In order to analyze a collapsing situation for homogeneous (not isotropic) interiors, the authors considered LRS spatially homogeneous collapsing cylinders. Given an interior with diagonal metric, existence and uniqueness results for the exterior was proven. The matched solutions cannot be asymptotically flat and contain both trapped surfaces and singularities [69].

G. Fodor's contribution dealt with static and spherically symmetric perfect fluid solutions of the Einstein field equations. A method was presented which can be used to generate all such solutions without the need to solve differential equations. The equation of state of the fluid is not imposed a priori but calculated from the generating function just by differentiation. The same holds for the radial pressure profile. The positivity and monotonicity of pressure and the fulfillment of the energy conditions can be ensured by imposing simple conditions on the generating function. The full form of the metric can be obtained by simple differentiation and just one quadrature, making the method useful for producing a large number of physically interesting exact solutions. A specific choice of generating function was presented in detail. In that case the metric could be given in terms of elementary functions. This configuration represents a two parameter family of finite bodies surrounded by vacuum.

7. Einstein-Maxwell solutions

A.A. García (reporting on work done in collaboration with **E. Ayón–Beato** and **M. Cataldo**) discussed spacetimes in $2+1$ dimensions describing stationary and cyclically symmetric electromagnetic fields. The general structure of the electromagnetic field in such a situation can be given, in terms of its dual $*\mathbf{F}$, by

$$*\mathbf{F} = a\mathbf{dt} + b\mathbf{d}\phi + c\frac{g_{rr}}{\sqrt{-g}}\mathbf{dr}, \tag{1}$$

where the constants a, b and c are subject, by virtue of the Ricci circularity conditions, to the equations $a\,c = 0 = b\,c$. Consequently, it follows that there exist three branches of solutions $c \neq 0$, and $c = 0$, $a \neq 0, b = 0$ or $c = 0$, $a = 0, b \neq 0$. The corresponding solutions were explicitly given and their physical properties were studied in detail.

I. Eriksson (reporting on work done in collaboration with **G. Bergqvist** and **J.M.M. Senovilla**) presented new conservation laws for Einstein-Maxwell spacetimes [70]. The derivation was based on the so-called Chevreton tensor [71] which is defined as a suitable symmetrization of the super-energy tensor [72] constructed from the first covariant derivative of the electromagnetic field. It is therefore quadratic in these derivatives and has strong similarities with the Bel-Robinson tensor (in fact, it was introduced as an electromagnetic counterpart to the Bel-Robinson tensor for the gravitational field). In this contribution it was shown that the trace of the Chevreton tensor is symmetric, traceless and *divergence-free* whenever the electromagnetic field satisfies the source-free Maxwell equations and the spacetime metric is either an Einstein space (so that electromagnetic field is a test field) or the electromagnetic field is self-gravitating and the Einstein-Maxwell equations are satisfied. This tensor is also related to the Bach tensor. The existence of a divergence-free tensor immediately introduces new conservations laws, which are, in general, different and independent from the usual conservation law corresponding to the Maxwell energy-momentum tensor [70,73]. There is, however, a differential relationship between both which is explored in [74] for Ricci-flat spacetimes and in [75] for general spacetimes.

8. Symmetries and exact solutions

J.D. Finley (reporting on work done in collaboration with **J.K. McIver**) described the complete (non-Abelian) algebra of generalised symmetries for the $SDiff(2)$ Toda equation, $\partial_x \partial_y \Omega + {\partial_s}^2 e^{\Omega} = 0$ [76]. This equation is interesting for several reasons, in particular because all solutions of the Einstein field equations with self-dual Weyl tensor and one Killing vector, with non-zero Killing bivector, are determined by this equation. The algebra of generalised symmetries is two copies of the algebra $SDiff(2)$, along with two more generators with derivation-like properties on both of the copies. It contains as a subalgebra a doubly-infinite, Abelian algebra, equivalent to the infinite hierarchy of higher flows first shown by Takasaki and Takebe [77]. The explicit presentation requires a prolongation of the jet bundle to include infinitely many new potentials, only the first two of which are linear. Being non-Abelian, a recursion relation is obtained, which generates all these symmetries. It seems likely that the knowledge of the full algebra of generalised symmetries of the $SDiff(2)$ Toda equation, will allow the (future) determination of large classes of new exact solutions of this equation and, in particular, new vacuum solutions of Einstein's equations.

E.G.L.R. Vaz (reporting on work done in collaboration with **J. Carot** and **M.P. Machado Ramos**) discussed double warped spacetimes. These are Lorentzian manifolds admitting a local decomposition as the product of two manifolds M_1 and M_2 such that the metric is, locally, the direct sum of a metric on M_1 conformally rescaled with a function on M_2 plus a metric on M_2 conformally rescaled with a function on M_1. The definition obviously generalises that of a warped spacetime in which one of the functions is constant (or, equivalently, one). Depending on the dimensionality of M_1 and M_2, two classes arise corresponding respectively to $1+3$ and to $2+2$ decompositions. In each case a characterization of double warped spacetimes in terms of the Newman-Penrose spin coefficients in suitable canonical null bases was described [78]. Finally, a practical procedure to determine whether a given metric belongs to the double warped class was discussed.

The contribution of **G.S. Hall** dealt with the structure of the orbits of the generators of local groups of symmetries in a spacetime. More specifically, the talk discussed the mathematics of the orbits arising from a finite-dimensional Lie algebra A of vector fields on a spacetime (M, g) where A could be composed of Killing, homothetic, conformal, affine or projective vector fields on M. The orbits were not assumed to be of the same dimension everywhere on M and so the resulting (generalised) distribution is not necessarily of the Frobenius type. Thus, the usual theorems about Frobenius distributions cannot be used. It was explained how the associated flow of diffeomorphisms arising from A define the orbits in a natural way [79] and how they turn out to be integral manifolds of the (generalised) distribution [80]. The submanifold structure also turns out to be well behaved [81]. The importance of this for studying spacetime symmetries was stressed. A brief explanation of the corresponding isotropies was given and a method of classifying them mentioned [82]. Finally it was pointed out that some problems still exist and to get round these, the concept of "orbit stability" was introduced [83]. This defines an orbit O to be dimensionally stable if for each $p \in O$ there is a neighbourhood U of p such that the orbits through each point of U have the same dimension as O. Such orbits turn out to be "well-behaved". The theory was briefly applied when A was of the possible symmetry types mentioned above.

J.A. Valiente Kroon showed in his contribution that the Kerr spacetime, and other stationary asymptotically flat spacetimes, do not admit any conformally flat slice. The interest in determining whether such slices exist or not comes from various reasons, ranging from numerical to mathematical relativity. Some partial negative answers were already given by Garat and Price [84], and the author himself [85]. However, these results assumed several a priori conditions on the slices, which made the results less conclusive than desired. In this talk a new argument was put forward [86] which allowed treating much more general slices. The main idea is to make use of the multipolar expansions of the asymptotically flat stationary solutions (which are defined in the quotient manifold) to calculate an asymptotic expansion for the Cotton-York tensor of the slices. It was shown that for the slices to be locally conformally flat (i.e. with vanishing Cotton-York tensor) a certain quantity of quadrupolar nature must vanish. This quantity is, for a given stationary spacetime, universal in the sense that it does not depend on the choice of slice. Thus, the non-vanishing of this object can be viewed as an obstruction of the existence of conformally flat slices. As an application, it was shown that this quantity is non-zero for the Kerr solution.

9. Cosmology

The contribution of **R. Vera** (reporting on work done in collaboration with **B. Nolan**) aimed at studying finite rotating objects in equilibrium in cosmological backgrounds. The general framework is the construction of global models via the matching of spacetimes. The studies of rotating objects in equilibrium (i.e. stars, planets, galaxies, clusters) in general relativity have usually focused on the case when they are completely isolated so that the spacetime can be assumed to be asymptotically flat. However, in a more realistic astrophysical situation, the spacetime far away from the object should approach not Minkowski spacetime but a FLRW cosmological model. Thus, the boundary conditions must be modified accordingly. To that aim, the general matching of vacuum stationary and axisymmetric spacetimes with (necessarily) dust FLRW cosmological models was studied. Two main results were presented. First of all, it was shown that the matching conditions impose that the slicing of the matching hypersurface with respect to the cosmological time at the FLRW side must coincide with the slicing with respect to a canonical time in the stationary region. The latter is any function defining spacelike hypersurfaces containing both the surfaces orthogonal to the orbits of the isometry group and the orbits of the axial symmetry. Secondly, and more importantly, it was shown that the surfaces corresponding to the intersection of the matching hypersurface and the time slicing above must be geometric 2-spheres. This means that the surface of the vacuum region inside FLRW as seen by the cosmological observer must be an (evolving) sphere. This result generalises to the stationary setting a previous result obtained in the static case [68].

V.M. Villalba (reporting on work done in collaboration with **A. di Teodoro**) defined asymptotic symmetries in Bianchi cosmological models. Both asymptotic isometries and asymptotic collineations were considered. The method can be used to analyze under which condition a Bianchi model approaches a homogeneous and isotropic FLRW cosmological model at infinity. In particular, it was shown that asymptotically FLRW cosmologies can be covariantly characterized in terms of scalars constructed from the Killing equations. The results were applied also to a perturbative setting in order to define asymptotic symmetries associated with scalar cosmological perturbations.

R. Lazkoz (reporting on work done in collaboration with **L.P. Chimento**) discussed form invariance transformations of the Friedman-Lemaître equations for homogeneous and isotropic cosmologies [87]. These transformations can be used as solution generating methods, and it was stressed that they are *not* Lie point symmetries of the equations, because the equation of state of the perfect fluid does not remain invariant. The main motivation for the construction lies in the generation of new so-called phantom cosmologies [88] which correspond to FLRW cosmologies with $\gamma \equiv 1 + p/\rho < 0$. These models have attracted considerable attention recently because according to several observations (including that of distant supernovae) the universe at present is not only expanding but also accelerating. A possible explanation to this current state is dark energy with a phantom equation of state. One of the active areas in the theoretical research of phantom models lies on exact solutions. The approach of this contribution provides a method to construct a phantom version of FLRW scalar field cosmologies by applying form-invariance transformations . Such transformation maps the Hubble "constant" $H \to -H$ and $\rho+p \to -(\rho+p)$, which in turn give $\gamma \to -\gamma$ and $a \to a^{-1}$. The discussion was presented in a general setup, and then illustrated using power-law solutions.

In **L. Fernández-Jambrina**'s contribution, the issue of existence of geodesically complete cosmological perfect fluid spacetimes was addressed. The first non-singular cosmological model was presented by Senovilla [89] in 1990. Since then some other non-singular cosmologies were found. However, except for the class found in [90] which generalised the metric in [91], the solutions always contained a finite number of parameters, so that they never represented an open set of solutions (even within their symmetry class). The Bičák-Griffiths solution above [90] did contain arbitrary functions and no singularities, although no explicit proof of geodesic completeness was given. Nevertheless, the general belief was that the class of singularity-free cosmologies is probably negligible within the whole class of solutions. In this contribution, it was shown that among inhomogeneous stiff fluid cosmologies with an orthogonally-transitive, mutually orthogonal, 2-dimensional group of isometries there is a family of regular (i.e. geodesically complete) solutions depending on two nearly arbitrary functions [92]. The conditions for regularity are very easy to check and to implement (see also [93]). It is the largest family of non-singular cosmological perfect fluid spacetimes presented so far and includes the Bičák-Griffiths family. This result poses the question of the amount of regular cosmologies in the set of solutions of Einstein equations, which may be much larger than previously thought.

References

1. D. Kramer, G. Neuegebauer, "Superposition of 2 Kerr Solutions" *Phys. Lett. A* **75**, 259-261 (1980).
2. W. Dietz, C. Hoenselaers, "2 Mass solutions of Einstein Vacuum equations - The double Kerr solutions" *Ann. Phys.* **165**, 319-383 (1985).
3. A. Tomimatsu, M. Kihara, "Conditions for regularity on the symmetry axis in a superposition of 2 Kerr-NUT solutions" *Prog. Theor. Phys.* **67**, 1406-1414 (1982).
4. V.S. Manko, E. Ruiz, J.D. Sanabria-Gómez, "Extended multi-soliton solutions of the Einstein field equations: II. Two comments on the existence of equilibrium states" *Class. Quantum Grav.* **17**, 3881-3898 (2000).
5. V.S. Manko, E. Ruiz, "Exact solution of the double-Kerr equilibrium problem" *Class. Quantum Grav.* **18**, L11-L15 (2001).
6. G. Weinstein, "The stationary axisymmetrical 2-body problem in general-relativity" *Comm. Pure Appl. Math.* **45**, 1183-1203 (1992).
7. G. Neugebauer, R. Meinel, "Progress in relativistic gravitational theory using the inverse scattering method" *J. Math. Phys.* **44**, 3407-3429 (2003).
8. W.B. Bonnor, B.R. Steadman, "The double-Kerr solution" *Class. Quantum Grav.* **21**, 2723-2732 (2004).
9. D.A. Korotkin, "Finite-gap solutions of the stationary axisymetric Einstein equations in vacuum" *Theor. Math. Phys.* **77**, 1018-1031 (1989).
10. C. Klein, "Exact relativistic treatment of stationary black-hole disk systems" *Phys. Rev. D* **68**, 27501 (2003).
11. J.L. Friedman, K. Schleich, D.M. Witt, "Topological Censorship" *Phys. Rev. Lett.* **71**, 1486-1489 (1993); Erratum-ibid. **75**, 1872 (1995).
12. J. Louko, D. Marolf "Inextendible Schwarzschild black hole with a single

exterior: How thermal is the Hawking radiation?" *Phys.Rev. D* **58**, 024007 (1998).
13. V. Pravda, A. Pravdová, A. Coley, R. Milson, "All spacetimes with vanishing curvature invariants" *Class. Quantum Grav.* **19**, 6213-6236 (2002).
14. A. Coley, "A Class of Exact Classical Solutions to String Theory" *Phys. Rev. Lett.* **89**, 281601 (2002).
15. A. Coley, R. Milson, V. Pravda, A. Pravdová, "Vanishing scalar invariant spacetimes in higher dimensions" submitted to *Class. Quantum Grav.* (2004).
16. R. Milson, A. Coley, V. Pravda, A. Pravdová, "Alignment and algebraically special tensors in Lorentzian geometry" *Int. J. Geom. Meth. Mod. Phys.* **2**, 41-61 (2005).
17. A. Coley, R. Milson, V. Pravda, A. Pravdová, "Classification of the Weyl tensor in Higher Dimensions" *Class. Quantum Grav.* **21**, L35-L41 (2004).
18. A. Barnes, "Purely magnetic spacetimes" in Proceedings of the 27th Spanish Relativity Meeting. (Eds. J.A. Miralles, J.A. Font, J.A. Pons, 2004).
19. S. Matarrese, O. Pantano, D. Sáez, "General-relativistic approach to the nonlinear evolution of collisionless matter" *Phys. Rev. D***47**, 1311-1323 (1993).
20. A. Barnes, R.R. Rowlingson, "Irrotational perfect fluids with a purely electric Weyl tensor" *Class. Quantum Grav.* **6**, 949-960 (1989).
21. P. Szekeres, "Class of inhomogeneous cosmological models" *Commun. Math. Phys.* **41**, 55-64 (1975).
22. C.F. Sopuerta, "New study of silent universes" *Phys. Rev. D* **55**, 5936-5950 (1997).
23. M. Mars, "3 + 1 description of silent universes: a uniqueness result for the Petrov type I vacuum case" *Class. Quantum Grav.* **16**, 3245-3262 (1999).
24. N. Van den Bergh, L. Wylleman "Silent universes with a cosmological constant" *Class. Quantum Grav.* **21**, 2291-2299 (2004).
25. R. Emparan, H.S. Reall, "A Rotating Black Ring Solution in Five Dimensions" *Phys. Rev. Lett.* **88**, 101101 (2002).
26. R. Emparan, H.S. Reall, "Generalized Weyl solutions" *Phys. Rev. D* **65**, 084025 (2002).
27. K.P. Tod, "All metrics admitting supercovariantly constant spinors" *Phys. Lett. B* **121**, 241-244 (1983).
28. J.P. Gauntlett, J.B. Gutowski, C.M. Hull, S. Pakis, H.S. Reall, "All supersymmetric solutions of minimal supergravity in five dimensions" *Class. Quantum Grav.* **20**, 4587-4634 (2003).
29. R. Mann, C. Stelea, "Nuttier (A)dS Black Holes in Higher Dimensions" *Class. Quantum Grav.* **21**, 2937-2962 (2004).
30. F.R. Tangherlini, "Schwarzschild field in n dimensions and the dimensionality of space problem" *Nuovo Cimento* **27**, 636-651 (1963).
31. R.C. Myers, M.J. Perry, "Black-holes in higher dimensional space-times" *Annals of Physics* **172**, 304-347 (1986).
32. A.N. Aliev, V.P. Frolov, "Five-dimensional rotating black hole in a uniform magnetic field: The gyromagnetic ratio" *Phys. Rev. D* **69**, 084022 (2004).
33. K. Ait Moussa, G. Clément, C. Leygnac, "The black holes of topologically massive gravity" *Class. Quantum Grav.* **20**, L277-L283 (2003).

34. S. Deser, R. Jackiw, S. Templeton, "Three-Dimensional Massive Gauge Theories" *Phys. Rev. Lett.* **48**, 975-978 (1982).
35. I. Vuorio, "Topologically massive planar universe" *Phys. Lett. B* **163**, 91-95 (1985).
36. G. Clément, "Black hole mass and angular momentum in 2+1 gravity" *Phys. Rev. D* **68**, 024032 (2003).
37. P. Krtouš, J. Podolský, "Radiation from accelerated black holes in a de Sitter universe" *Phys. Rev. D* **68**, 024005 (2003).
38. J. Podolský, M. Ortaggio, P. Krtouš, "Radiation from accelerated black holes in an anti–de Sitter universe" *Phys. Rev. D* **68**, 124004 (2003).
39. P. Krtouš, J. Podolský, "Gravitational and electromagnetic fields near an anti–de Sitter-like infinity" *Phys. Rev. D* **69**, 084023 (2004).
40. P. Krtouš, `http://utf.mff.cuni.cz/~krtous/physics/CADS`.
41. P. Krtouš, J. Podolský, J. Bičák , "Gravitational and electromagnetic fields near a de Sitter–like infinity" *Phys. Rev. Lett.* **91**, 061101 (2003); P. Krtouš, J. Podolský, "Gravitational and electromagnetic fields near an anti-de Sitter-like infinity" *Phys. Rev. D* **69**, 084023 (2004); P. Krtouš, J. Podolský, "Asymptotic directional structure of radiative fields in spacetimes with a cosmological constant" submitted to *Class. Quantum Grav.*
42. M. Ortaggio, "Impulsive waves in the Nariai universe" *Phys. Rev. D* **65**, 084046 (2002); M. Ortaggio, J. Podolský, "Impulsive waves in electrovac direct product spacetimes with Λ" *Class. Quantum Grav.* **19**, 5221-5227 (2002).
43. M. Ortaggio, "Ultrarelativistic black hole in an external electromagnetic field and gravitational waves in the Melvin universe" *Phys. Rev. D* **69**, 064034 (2004).
44. R. Penrose, "Twistor quantization and curved space-time" *Int. J. Theor. Phys.* **1**, 61-69 (1968); R. Penrose, in *General Relativity: Papers in Honour of J. L. Synge*, Ed. L. O'Raifeartaigh (Oxford: Clarendon Press) (1972).
45. P.C. Aichelburg, R.U. Sexl, "On the gravitational field of a massless particle" *Gen. Rel. Grav.* **2**, 303-312 (1971).
46. M. Hotta, M. Tanaka, "Shock-wave geometry with nonvanishing cosmological constant" *Class. Quantum Grav.* **10**, 307-314 (1993); J. Podolský, J.B. Griffiths, "Impulsive gravitational waves generated by null particles in de Sitter and antide Sitter backgrounds" *Phys. Rev. D* **56**, 4756-4767 (1997).
47. J. Podolský, in *Gravitation: Following the Prague Inspiration*, Eds. O. Semerák, J. Podolský, M. Žofka (Singapore: World Scientific) (2002); M. Ortaggio, *On Some Properties of Impulsive Waves in General Relativity*, Ph.D. thesis, Università degli Studi di Trento (2002).
48. M.P. Machado Ramos, J.A.G. Vickers, "A spacetime calculus invariant under null rotations" *Proc. Roy. Soc. Lon. A* **1940**, 693-709 (1995).
49. M.P. Machado Ramos, J.A.G. Vickers, "A spacetime calculus based on a single null direction" *Class. Quantum Grav.* **13**, 1579-1587 (1996).
50. S.B. Edgar, J.A. Vickers, "Integration using invariant operators: conformally flat radiation metrics" *Class. Quantum Grav.* **16**, 589-604 (1999).
51. A. Karlhede, "A review of the geometrical equivalence of metrics in general relativity" *Gen. Rel. Grav.* **12**, 693-707 (1980).

52. S.B. Edgar, G. Ludwig, "A generalized Lie derivative and homothetic or Killing vectors in the Geroch-Held-Penrose formalism" *Class. Quantum Grav.* **17**, 1683-1705 (2000).
53. J.B. Griffiths, *Colliding Plane Waves in General Relativity* (Oxford: Oxford University Press, 1991).
54. J.B. Griffiths, M. Santano-Roco, "The characteristic initial value problem for colliding plane waves: The linear case" *Class. Quantum Grav.* **19**, 4273-4286 (2002).
55. J. Bičák, T. Ledvinka, B.G. Schmidt, M. Žofka, "Static fluid cylinders and their fields: global solutions" *Class. Quantum Grav.* **21**, 1583-1608 (2004).
56. J. Bičák, M. Žofka, "Notes on static cylindrical shells" *Class. Quantum Grav.* **19**, 3653-3664 (2002).
57. M. Žofka, "Comments on photonic shells" *Class. Quamtum Grav.* **21**, 465-470 (2004).
58. J. Geršl, P. Klepáč, J. Horský, "A charged rotating cylindrical shell" *Gen. Rel. Grav.* **36**, 1399-1413 (2004).
59. W. B. Bonnor, "Static magnetic fields in general relativity" *Proc. Phys. Soc. Lond. A* **67**, 225-232 (1954); M. A. Melvin, "Pure magnetic and electric geons" *Phys. Lett.* **8**, 65-68 (1964).
60. B.K. Datta, A.K. Raychaudhuri, "Stationary electromagnetic fields in general relativity" *J. Math. Phys.* **9**, 1715-1718 (1968).
61. G.T. Horowitz, D. Marolf, "Quantum probes of spacetime singularities" *Phys. Rev. D* **52**, 5670-5675 (1995).
62. D.A. Konkowski, T.M. Helliwell, C. Wieland, "Quantum singularity of Levi-Cività spacetimes" *Class. Quantum Grav.* **21**, 265-272 (2004).
63. W.B. Bonnor, "The static cylinder in general relativity" in *On Einstein's Path* ed. A. Harvey (New York: Springer, 1999).
64. M.F.A. Da Silva, A. Wang, F.M. Paiva, N.O. Santos, "Levi-Cività solutions with a cosmological constant" *Phys. Rev. D* **61**, 44003 (2000).
65. T.M. Helliwell, D.A. Konkowski, C.A. Reese, in preparation.
66. F.C. Mena, R. Tavakol, R. Vera, "Generalization of the Einstein-Straus model to anisotropic settings" *Phys. Rev. D* **66**, 044004 (2002).
67. J.M. Senovilla, R. Vera, "Impossibility of the cylindrically symmetric Einstein-Straus model" *Phys. Rev. Lett.* **78**, 2284-2287 (1997).
68. M. Mars, "On the uniqueness of the Einstein-Straus model" *Class. Quantum Grav.* **18** , 3645-3663 (2001).
69. P. Tod, F.C. Mena, "Matching of spatially homogeneous non-stationary space-times to vacuum in cylindrical symmetry" *Phys. Rev.* **D70** 104028 (2004).
70. G. Bergqvist, I. Eriksson, J.M.M. Senovilla, "New electromagnetic conservation laws" *Class. Quantum Grav.* **20**, 2663-2668 (2003).
71. M. Chevreton, "Sur le tenseur de superénergie du champ électromagnétique" *Nuovo Cimento* **34**, 901-913 (1964).
72. J.M.M. Senovilla, "Super-energy tensors" *Class. Quantum Grav.* **17**, 2799-2841 (2000).
73. J.M.M. Senovilla in "Symmetries and gravity in field theory", Eds. V. Aldaya

and J.M. Cerveró (Universidad de Salamanca 2004). gr-qc/0311033.

74. S. Deser, "A note on matter superenergy tensors" *Class. Quantum Grav.* **20**, L213-L215 (2003).
75. B. Edgar, "On the structure of the new electromagnetic conservation laws" *Class. Quantum Grav.* **21**, L21-L25 (2004).
76. J.D. Finley, J.K. McIver, "Non-Abelian infinite algebra of generalized symmetries for the $SDiff(2)$ Toda equation" *J. Phys. A* **37**, 5825-5847 (2004).
77. K. Takasaki, T. Takebe, "$SDiff(2)$ Toda Equation - Hierarchy, Tau Function and Symmetries -" *Lett. Math. Phys.* **23**, 205-214 (1991).
78. M.P. Machado Ramos, E.G.L.R. Vaz, J. Carot, "Double warped spacetimes" *J. Math. Phys.* **44**, 4839-4865 (2003).
79. H. J. Sussmann, "Orbits of families of vector fields and integrability of distributions" *Trans. Am. Math. Soc.* **180**, 171-188 (1973).
80. R. Hermann, *International Symposium on Non-Linear Differential Equations and Non-Lonear Mechanics*, Acad. Press, 325 (1963).
81. P. Stefan, "Integrability of systems of vector fields" *L. London Math. Soc.* **21**, 544-556 (1980).
82. G.S. Hall, B. Haddow, J.R. Pulham, "Weyl connections, curvature and generalised gauge changes" *Gravitation and Cosmology* **3**, 175-184 (1997); G.S. Hall, M.T. Patel, "On the theory of homothetic and affine orbits in spacetime" *Class. Quantum Grav.* **21**, 4731-4742 (2004).
83. G.S. Hall, "On the theory of Killing orbits in spacetime" *Class. Quantum Grav.* **20**, 4067-4084 (2003); G.S. Hall *Symmetries and Curvature Structure in General Relativity* World Scientific, (2004).
84. A. Garat, R. Price, "Nonexistence of conformally flat slices of the Kerr spacetime" *Phys Rev. D* **61**, 124011 (2000).
85. J.A. Valiente Kroon, "Nonexistence of Conformally Flat Slices in Kerr and Other Stationary Spacetimes" *Phys. Rev. Lett.* **92**, 041101 (2004).
86. J.A. Valiente Kroon, "Asymptotic expansions of the Cotton-York tensor on slices of stationary spacetimes" *Class. Quantum Grav.* **21**, 3237-3249 (2004).
87. L.P. Chimento, R. Lazkoz, "Constructing Phantom Cosmologies from Standard Scalar Field Universes" *Phys. Rev. Lett.* **91**, 211301 (2003).
88. R.R. Caldwell, "A Phantom Menace? Cosmological consequences of a dark energy component with super-negative equation of state" *Phys.Lett. B* **545**, 23-29 (2002).
89. J.M.M. Senovilla, "New class of inhomogeneous cosmological perfect-fluid solutions without big-bang singularity" *Phys. Rev. Lett.* **64**, 2219-2221 (1990).
90. J.B. Griffiths, J. Bičák, "Singularity-free stiff fluid cosmologies" *Class. Quantum Grav.* **12**, L81-L83 (1995).
91. M. Mars, "New nondiagonal singularity-free cosmological perfect-fluid solution" *Phys. Rev. D* **51**, R3989-R3991 (1995).
92. L. Fernández-Jambrina, L.M. González-Romero, "A wide family of singularity-free cosmological models" *Phys. Rev. D* **66**, 024027 (2002).
93. L. Fernández-Jambrina, L.M. González-Romero, "Exterior differential system for cosmological $G(2)$ perfect fluids and geodesic completeness" *Class. Quantum Grav.* **16**, 953-972 (1999).

A2: COMPLEX METHODS, TWISTORS, AND CONNECTION VARIABLES

JERZY LEWANDOWSKI

Instytut Fizyki Teoretycznej
Uniwersytet Warszawski, ul. Hoza 69
00-681 Warszawa, Poland
e-mail: lewand@fuw.edu.pl

1. Understanding the Type II Einstein-Maxwell Fields

Ezra Ted Newman

1. General comments: Given both a metric g (and a Maxwell field F) that satisfy the vacuum Einstein (or Einstein-Maxwell) equations how do we find the physical content of the solution. *An* approach is that one searches for the coordinate transformation from the given coordinates to (say) Bondi coordinates and one tries to give meaning to quantities or functions defined in the original metric g and field F.

2. We applied this approach to the class of metrics *and* fields known as the type II twisting Einstein-Maxwell fields, i.e., the Einstein-Maxwell fields with a single degenerate Weyl 'pnv' that coincides with the Maxwell 'pnv'. In particular we want to know the physical consequences/meaning of the existence of the twist.

The solutions (metric and Maxwell field) are found in a natural coordinate system (twisting) but which is far from a standard coordinate system. One then finds the coordinate transformation from these twisting coordinates to the Bondi coordinates and reads out the physical quantities; e.g., the mass, the charge, the electric and magnetic dipole moments and the spin angular momentum.

Both the spin and the magnetic moment are given directly in terms of the twist.

The leading term in the spherical harmonic decomposition of the twist involves a complex four-vector, "a complex displacement" , depending on the Bondi-time U. The spin angular momentum is given by the mass times the imaginary part of this displacement while the magnetic moment is given by the charge times the same imaginary displacement,

A surprising result is that the gyromagnetic ratio is that of the Dirac electron. This result generalizes the results found for the special case, the charged Kerr metric.

2. Curves of Higher Degree in Twistor Space

Sir Roger Penrose

In the standard (flat-space) twistor correspondence, projective lines (curves of order1) in CP^3 represent (complex) points in space-time. Witten has recently conjectured that curves of higher order (sometimes referred to as curves of higher "degree") have a specific role to play in the Feynman vertices for multiple-gluon processes. In this lecture I discussed the twistorial description of such curves and their role, generally, in twistor theory.

3. Perturbative Gauge Theory as a String Theory in Twistor Space

Peter Svrcek

In this talk I have reviewed the relation between twistor string theory and gauge theory scattering amplitudes that has been uncovered last year by E. Witten. The perturbative scattering amplitudes have many unexpected properties that are not apparent from the textbook recipes for computing them. Tree amplitudes in which the maximal number of gluons have the same helicity are described by a marvelously simple formula. The loop amplitudes also turn out to be unexpectedly simple.

Some of these properties can be explained by relating the N=4 super-Yang-Mills theory to the instanton expansion of a topological string theory in twistor space. This is the topological B-model and the target space is the supersymmetric version of twistor space. This representation of weakly coupled N=4 super-Yang-Mills theory as a string theory is an interesting counterpoint to the by now familiar description of the strongly coupled regime of the same theory via Type IIB superstring theory on AdS5xS5.

In twistor theory, is has been a longstanding problem to understand how to use twistors to describe perturbative field theory amplitudes. The twistor string theory leads to a proposal that relates the amplitudes to families of holomorphic curves in twistor space.

In collaboration with F. Cachazo and E. Witten, I have reconsidered the tree amplitudes in a way that is suggested by twistor transforms and by our study of differential equations obeyed by the scattering amplitudes and also by the use of MHV tree amplitudes in calculating loop amplitudes. We reinterpret the maximally helicity violating amplitude as representing, in some sense, a generalization of a local interaction vertex. We pick a specific off-shell continuation of the MHV amplitude and consider Feynman diagrams in which the vertices are tree level MHV amplitudes - with an arbitrary number of gluon lines - and the propagator is the standard Feynman propagator. We conjecture that this rules calculate scattering amplitudes to all orders and we obtain relatively short and simple expressions for certain amplitudes.

There are many aspects of the B-model on twistor space that remain unclear. One pressing question is the closed string sector. The closed strings seem to give some version of N=4 conformal supergravity, so the string theory introduced by Witten might be equivalent to super-Yang-Mills theory only in the planar limit.

4. Quasi-local holography of linear zero-rest-mass fields in Minkowski spacetime

László B Szabados

In the present contribution the connection between the following three concepts is discussed:

- 1. The quasi-local mass of the matter fields, associated with a closed spacelike 2-surface $\mathcal{S}$ in Minkowski spacetime,
- 2. The pure radiative solutions of specific matter field equations (Higgs, Yang–Mills, linear zero-rest-mass fields with any spin) on the globally hyperbolic domain $D(\Sigma)$ for which the common boundary of the Cauchy surfaces Σ is just $\mathcal{S}$,
- 3. The complete characterization of these radiative solutions in terms of data specified on the 2-surface $\mathcal{S}$ rather than the familiar Cauchy data on Σ ('classical quasi-local holography').

This problem is motivated by the result in general relativity that the vanishing of the Dougan–Mason quasi-local mass [1], associated with a topologically spherical 2-surface $\mathcal{S}$, is equivalent to the *pp*-wave geometry of the domain of dependence $D(\Sigma)$ with $\mathcal{S} = \partial\Sigma$ [2], and that the *pp*-wave metric on $D(\Sigma)$ is completely determined by the 2-surface data on $\mathcal{S}$ [3].

First it is shown that the quasi-local mass is zero precisely when the matter fields on $D(\Sigma)$ are of pure radiative type in the sense that $T_{ab}L^b = 0$ for some *constant* null vector field L^a on $D(\Sigma)$, provided T^{ab} satisfies the dominant energy condition. All the solutions of the Higgs and Yang–Mills field equations with compact semisimple gauge groups are determined for which the quasi-local mass is zero. These are all plane waves (provided the Higgs field is massless and linear) and appropriate generalizations of plane waves ('Yang–Mills *pp*-waves'), respectively. A totally symmetric, trace-free tensor field $T_{a_1 \dots a_{2s}}$ is found, which is a natural generalization of the energy-momentum tensor of the Maxwell field and of the Bel–Robinson tensor for the linearized gravitational field, and by means of which the *pp*-wave character of the solutions of the linear zero-rest-mass field eqautions with spin s on $D(\Sigma)$ can be characterized equivalently. All these solutions are represented by a constant spinor field O^A and a complex function $\phi(u, \zeta)$ on $D(\Sigma)$, where $\phi(u, \zeta)$ is C^2 in the coordinate u, labelling the wave-fronts, and complex analytic in ζ, the holomorphic coordinate in the intersection of the wave-fronts and Σ. Finally, provided a convexity condition for the 2-surface $\mathcal{S}$ is satisfied, it is shown that such a pair (O^A, ϕ) on $D(\Sigma)$ is completely determined by a constant spinor field O^A on the 2-surface $\mathcal{S}$ and two real valued functions $(f(u, w), g(u))$ of two and one variables, respectively, where w is a coordinate on the intersection of $\mathcal{S}$ and the wave-fronts. The data set $(O^A, f(u, w), g(u))$ on $\mathcal{S}$ is unconstrained, and the field ϕ at any point $p \in D(\Sigma)$ is expressed as an appropriate contour integral of the 2-surface data $(f(u, w), g(u))$.

Therefore, similarly to general relativity, a classical version of the holography for Yang–Mills and linear zero-rest-mass fields with any spin in Minkowski spacetime works in the sense that the zero-quasi-local-mass radiative configurations on a four dimensional globally hyperbolic domain can be characterized completely by data on the two dimensional 'edge' of the globally hyperbolic domain, a 'quasi-local screen' as well.

5. Quasi-localization of the Nester-Witten Integral

Roh-Suan Tung

The Nester-Witten integral [4] is important for the quasi-local expression of energy-momentum. When integrated over a finite spatial region, it defines a quasi-local energy-momentum for any choice of the spinor field on the boundary. Several quasi-local energy-momentum expressions of this type have been investigated [5,6]. For a truely quasi-local expression, one wants to determine the spinor field quasi-locally (i.e., it should depend on the fields only on the two-surface). An example of such constructions was given by Dougan and Mason [7]. Here we give an alternative which directly links to several non-spinorial constructions. Given a spatial two-surface, a Dirichlet symplectic vector is uniquely defined [8]. We show that, for a non-spacelike Dirichlet symplectic vector, if the dominant energy condition holds on a spanning three-surface and the two-surface is convex, then the Nester-Witten integral is non-negative (for spinors satisfying a Witten-type equation with the two-surface boundary condition) and vanishes in a flat spacetime. A special case (when the twist vanishes) reduces to a boost-invariant generalization of the Brown-York quasi-local mass [9, 10].

6. Multipole moments for isolated horizons

Jonathan Engle, Abhay Ashtekar,
Tomasz Pawlowski, Christian Van Den Broeck

Prior to this work [11], there were good notions of multipoles in the contexts of linearized and perturbative GR, field multipoles for stationary, asymptotically flat space-times in full GR, and source multipoles from matter in full GR. There was no good notion of "source" multipoles for a *black hole* in full GR. The present work partially fills this gap. Specifically we define multipoles for *type II isolated horizons.* Briefly put, a type II isolated horizon Δ is a black hole whose horizon geometry is axisymmetric and in equilibrium.

We first define "geometric multipoles"; for this we use a notion of "natural" coordinates on cross-sections of the horizon. For every axisymmetric metric q on a two-sphere, there exist coordinates (ζ, ϕ), unique up to rigid shifts in ϕ, such that q takes the form

$$q = f(\zeta)^{-1}\mathrm{d}\zeta^2 + f(\zeta)\mathrm{d}\phi^2. \tag{1}$$

Using a certain $U(1)$ connection V defined on the horizon, the geometric multipoles are then defined by

$$I_n + iL_n := -\int_S Y_n^0(\zeta)\mathrm{d}V \overset{in\ vac.}{=} -\int_S Y_n^0(\zeta)\Psi_2\, {}^2\epsilon \tag{2}$$

We present in our paper an algorithm for explicit construction of an isolated horizon geometry with a given set of multipoles and area. In addition, there is a uniqueness result: geometric multipoles and area together completely determine horizon geometry up to diffeomorphism.

Next, to motivate a definition of dimensionful mass and angular momentum multipoles, we consider an effective mass density and mass surface current density defined by

$$\tilde{\rho}_\Delta = \frac{1}{8\pi} M_\Delta \tilde{\mathcal{R}} \quad (\tilde{j}_\Delta)_a = \frac{1}{8\pi G} \tilde{\omega}_a. \tag{3}$$

where $\tilde{\mathcal{R}}$ is the scalar curvature of the cross-section, and $\tilde{\omega}_a$ is the pull-back of the rotation one-form to the cross-section. Expressions for the multipoles are obtained by importing expressions for electric and magnetic multipoles in magnetostatics using the structure available on the isolated horizon. The result is

$$M_n = \sqrt{\frac{4\pi}{2n+1}} \frac{M_\Delta R_\Delta^n}{2\pi} I_n \quad J_n = \sqrt{\frac{4\pi}{2n+1}} \frac{R_\Delta^{n+1}}{4\pi G} L_n. \tag{4}$$

Here are some arguments in favor of these mass and angular momentum multipoles. $J_0 = 0$, as physically expected. $M_1 = 0$ – we are automatically in the "center of mass frame". $M_0 = M_\Delta$ by construction and $J_1 = J_\Delta$ by inspection. If $\mathrm{Im}\Psi_2 = 0$ on Δ, then all J_n's vanish – this is the case in particular if Δ is a static Killing horizon. If the horizon geometry is spherically symmetric, all the multipoles vanish except M_0. Lastly, the J_n's and M_n's are constructed solely from the intrinsic geometry (q_{ab}, ω_a); in particular neither stationarity of space-time nor asymptotic flatness is required.

7. Ernst equation, Fay identities and variational formulas on hyperelliptic curves

Christian Klein, Dmitri Korotkin and Vasilisa Shramchenko. [12]
Stable wave packets, so called solitons, are well known solutions to integrable evolution equations. Solutions to such equations in terms of multidimensional theta functions defined on some Riemann surface describe periodic or quasi-periodic wave phenomena. An elegant approach to construct theta-functional solutions uses a powerful identity for theta functions due to Fay [13]. This identity generalizes a well-known identity of projective geometry on $\mathbb{C}P1$ to Riemann surfaces: Let $P_i \in \mathbb{C}P1$, $i = 1, 2, 3, 4$ be four arbitrary disjoint points. Then the cross ratio function λ_{1234} is subject to the identity

$$\lambda_{1234} + \lambda_{1324} = 1, \quad \lambda_{1234} := \frac{(P_1 - P_2)(P_3 - P_4)}{(P_1 - P_4)(P_3 - P_2)}. \tag{5}$$

Fay's identity is extremely useful since it holds for arbitrary position of the four points. By considering the limit of coinciding points one thus obtains identities for derivatives of theta functions which can be used to establish formulas for theta-functional solutions to integrable evolution equations.

If this program is to be carried out for the Ernst equation

$$\mathrm{Re}\mathcal{E}\Delta\mathcal{E} = (\nabla\mathcal{E})^2 \tag{6}$$

for some complex valued potential $\mathcal{E}$ depending only on the cylindrical coordinates ρ, ζ, one faces the problem that the corresponding algebro-geometric solutions due to Korotkin [14] are defined on a hyperelliptic curve whose branch points depend on the physical coordinates. Thus the curve itself is 'dynamical' in this sense. In this case the modular dependence of the theta functions is important, not the dependence on the argument which leads to a different behavior of the solutions: the solitons to the Ernst equation, which is equivalent to the stationary axisymmetric Einstein equations in vacuum, correspond to black holes, the theta-functional solitons are not periodic. To apply Fay's identity in this context, one has to use variational formulas for Riemann surfaces. Doing this one obtains formulas in terms of theta functions for the Ernst potential and its derivatives and the complete Weyl-Lewis-Papapetrou metric.

8. Holonomy theory and applications in General Relativity

Graham Hall

Let (M, g) be a space-time and let $p \in M$. Let $C^k(p)$ denote the set of all piecewise C^k $(1 \leq k \leq \infty)$ closed curves at p and for $c \in C^k(p)$ let f_c denote the obvious isomorphism $T_pM \to T_pM$ defined by parallel transport around c. The set of all such maps f_c arising from all members of $C^k(p)$ is a group under the usual operations. This group is, up to isomorphism, independent of p and k and is the *holonomy group* Φ of M. Φ is a Lie subgroup of the Lorentz group and if M is simply connected, Φ is connected and determined by its Lie algebra which is a Lie subalgebra of the Lie algebra A of the Lorentz group. Since a classification of subalgebras of A is known one has a classification of space-time holonomy groups into 15 types denoted by the symbols $R_1, \ldots, R_{15}$ with R_1 flat and R_{15} the most general. One can investigate completely the possible holonomy types for perfect fluids, vacuum and null and non-null Einstein-Maxwell space-times, the associated holonomy structure suggesting natural coordinate systems for the metric. Holonomy structure can also be used to completely determine the affine vector fields in space-times and is useful for a systematic investigation of projective vector fields (whose local flows are the most general geodesic preserving transformations). For the latter study it is convenient to introduce a classification of the Riemann tensor into *curvature types* similar to the Petrov types for the Weyl tensor and then to make a decomposition of M into regions of constant curvature type. It turns out that the situation for proper (non-affine) projective vector fields can be resolved in all but three holonomy types and examples can be found in these latter three cases.

9. A canonical analysis of the 2+2 Hamiltonian using connection variables

Ray d'Inverno and James Vickers

We describe some recent work [15] which considers a Hamiltonian analysis of a double null $2+2$ Hamiltonian description of General Relativity in terms of self-dual two forms and the associated $SO(3)$ connection variables. Given a $2+2$ double null description of spacetime into spacelike 2-surfaces together with two null hypersurfaces, we decompose the first order action given by [16] into variables lying in the spacelike 2-surfaces and variables in the two dimensional normal direction.

Two different formulations are given. In the first we follow the approach of [17] in which a mixture of frame variables and 2-form variables is used and which generalises the work of [18], while the second formulation is entirely in t erms of the 2-form variables. Both sets of variables form an overdetermined system but the double null condition results in the constraints taking on a simple form. One of the advantages of working with a null foliation, as pointed out by [18], is that the Hamiltonian constraint becomes second class. However if one chooses a null 3+1 decomposition there is no natural projection operator associated with the foliation. A further problem is that the transformations which preserve the foliation is the group of null rotations which is algebraically awkward to work with. Both these problems are avoided if one works with a 2+2 description in which the projection is well defined and the relevant group is that of spin and boost transformations.

In both formulations we obtain the first and second class constraints together with the full constraint algebra and are able to give a clear geometrical interpretation of the algebra.

10. Quaternions, twistors & the algebrodynamical program

Vladimir V. Kassandrov

In the framework of "neoPithagorean" philosophy the *algebrodymamical* field theory based only on intrinsic properties of exceptional mathematical structures has been developed in our works. Specifically, fields are there considered as differentiable functions of *biquaternion* (**B**) variable, and the only basic field equations are the generalized *Cauchy-Riemann constraints* (i.e., the conditions of **B**-differentiability). These turn to be Lorentz invariant and *nonlinear* (in result of noncommutativity of **B**): in particular, the *complexified eikonal equation* (CEE) $\eta^{\mu\nu}\partial_\mu S \partial_\nu S = 0$ replaces the linear Laplace equation of complex analysis being satisfied by any component $S \in \mathbf{C}$ of the primary **B**-field.

Recently, we obtained *general solution of CEE* which was found to consist of two and only two classes with respect to (ambi)twistor structure of the CEE. One of them is represented by the well-known *Kerr theorem* and geometrically relates to the structure of *null geodesic shear-free congruences.*

Each solution of this class defines a set of fundamental gauge fields which satisfy free (Maxwell and Yang-Mills) equations. Nonetheless, bounded singularities of these fields (poles or closed strings) may be treated as *particle-like formations* and manifest even some properties of elementary particles. In particular, *for any singular "source" its electric charge is "self-quantized"*, i.e. multiple to a minimal "elementary" charge arising in theory as the charge of fundamental static solution with ring singularity (analogue of the Kerr-Newman solution in GTR).

Field singularities may be considered as *caustics* of fundamental null congruence defined by the primodial twistor field (**B**-field). For the "World solution" this field is necessarily *multivalued*, and the congruence is locally a *superposition* of a great number of "subflows" – constituents of the unique "primodial light" flow and of the "time flow" itself.

Generically, the algebrodynamical program can be thought of as a revival of W. Hamilton's ideas about fundamental role of quaternions in nature. It is closely related also to R. Penrose's twistor program and to E. Newman's complex space-time representation.

References

1. A.J. Dougan, L.J. Mason, Phys. Rev. Lett. **67** 2119 (1991)
2. L.B. Szabados, Class. Quantum Grav. **10** 1899 (1993)
3. L.B. Szabados, Class. Quantum Grav. **13** 1661 (1996)
4. J. M. Nester, in "Asymptotic Behaviour of Mass and Space-Time Geometry", edited by F. J. Flaherty, Lecture Notes in Physics Vol. 202 (Springer, Berlin, 1983), pp.155-163.
5. L. B. Szabados , Living Reviews in Relativity 7, (2004) 4.
6. C. M. Chen, J. M. Nester and R. S. Tung, gr-qc/0209100.

7. A. J. Dougan and L. J. Mason, Phys. Rev. Lett. 67, (1991) 2119-2122.
8. S. C. Anco and R. S. Tung, J. Math. Phys. 43 (2002) 5531-5566; J. Math. Phys. 43 (2002) 3984-4019.
9. J. D. Brown, S. R. Lau and J. W. York, Annals of Physics 297, (2002) 175.
10. C.-C. M. Liu and S.-T. Yau, Phys. Rev. Lett. 90 (2003) 231102.
11. J Engle, A Ashtekar, T Pawlowski, C Van Den Broeck Class. Quant. Grav. **21** (2004) 2549-2570
12. C. Klein D. Korotkin, V. Shramchenko Math. Res. Lett., Vol. 9, 27-47 (2002)
13. Fay, John D., Theta-functions on Riemann surfaces, Lect. Notes in Math. **352**, Springer (1973).
14. Korotkin, D., Finite-gap solutions of the stationary axisymmetric Einstein equation, Theor. Math. Phys. **77** 1018-1031 (1989).
15. P.Lambert, Canonical Analysis of Double Null Relativistic Hamiltonian Dynamics, PhD Thesis, University of Southampton, (2004).
16. T. Jacobson and L. Smolin (1989). *Phys. Lett. B* **196** 39-42.
17. R.A. d'Inverno and J.A. Vickers. (1995). *Class. Quantum Grav.* **12** 75
18. J N Goldberg, D C Robinson and C Soteriou. (1992). *Class. Quantum Grav.* **9** 1309-1328.

A3: MATHEMATICAL STUDIES OF THE FIELD EQUATIONS

JAMES ISENBERG
Department of Mathematics and Institute for Theoretical Science
University of Oregon
Eugene, OR 97403, USA.
e-mail: jim@newton.uoregon.edu

Even in these days of remarkable advances on the experimental side of gravitational physics, general relativity remains very much a mathematical science. Not surprisingly, the use of serious mathematical ideas and tools permeated many of the talks and many of the workshops of GR17. What especially distinguishes the works presented in this workshop is the predominance of ideas and tools from partial differential equation theory, and the focus on their application to questions arising in classical general relativity.

Over 100 abstracts were submitted to the A3 workshop. Of these, 36 were chosen for oral presentation, and a number of the rest were chosen for the poster sessions. In this review, we discuss some of the more interesting results which those attending the Workshop on Mathematical Studies of the Field Equations learned about.

1. Cosmological Solutions with Accelerated Expansion

The recent (somewhat surprising) observational evidence that the expansion of our universe is accelerating has led theorists to look for solutions of Einstein's equations which exhibit accelerated expansion. Such behavior is not found in vacuum solutions; however it is found if one includes a cosmological constant Λ in the equations, or if one adds certain types of non stiff fluids or scalar fields.

Work done over twenty years ago predicts that solutions of the equations with positive Λ should exhibit fairly special asymptotic behavior in the expanding direction, with the geometry approaching that of a DeSitter spacetime. The work which Alan Rendall presented at the workshop gives strong support for this conjecture. He uses Fuchsian techniques to to show that large families of solutions with positive scalar Λ, have the conjectured asymptotics, and has done preliminary work in extending these results to the Einstein equations with fluids and nonlinear scalar fields present. He also finds interesting evidence for the appearance of Gowdy-like spikes.

Rendall's work focuses on the behavior of general solutions in the asymptotic (large expansion) region. In contrast, the work presented by Hayoung Lee involves the global properties of spatially homogeneous solutions. She proves global existence and geodesic completeness for spatially homogeneous solutions of the Einstein equations with Λ positive, and with Vlasov fields and nonlinear scalar fields present. She shows that the Vlasov fields have little influence on the properties of these spacetimes, and finds that asymptotically they resemble not-tilted solutions with dust.

Makoto Narita presented work which explores ways in which accelerating expansion may result from string theory compactification. He considers an effective action resulting from such compactification, and shows that the effective fields can lead to solutions with accelerating expansion. The particular family of solutions he examines has the symmetry of Gowdy spacetimes. Oldrich Stolin looks at spacetimes with the same sort of symmetry. He examines the effect of the presence of a cosmological constant on the dynamical reduction which this family of spacetimes admits in the vacuum case.

The work reported by Masayuki Tanimoto does not deal with solutions with accelerated expansion. It does, however, investigate the asymptotic behavior of fields in an expanding spacetime. In particular, he shows that if one examines the evolution of a linear field in a background (locally) spatially homogeneous spacetime evolving with one fixed direction E_s (along a circle) expanding slower than the other directions, then one finds that asymptotically the derivative terms along E_s dominate the derivative terms along the other directions. Thus one obtains asymptotically a simplification of the evolution equation for the linear field in such a background expanding spacetime.

Jiri Bicak and Jiri Podolsky also reported on work which examines radiation fields in a background spacetime; in their case, the background is DeSitter space. Relying on the conformal relationship between the DeSitter and Minkowski spacetimes, they find explicit formulae for the asymptotic behavior of the radiation, with an important role played by the algebraic structure of the fields at $\mathcal{I}$.

2. Cosmological Singularities

One of the biggest challenges in mathematical relativity is to determine the behavior of the gravitational field in the neighborhood of the singularity in a generic cosmological solution of Einstein's equations. The Hawking-Penrose theorems predict that a singularity (in the sense of geodesic incompleteness) generically occurs, but they don't tell us whether the singularity is marked by unbounded curvature, by the formation of a Cauchy horizon, or by something else. Penrose's Strong Cosmic Censorship (SCC) conjecture suggests that generically, a cosmological singularity involves unbounded curvature.

Hans Ringstrom presented results at the workshop which constitute a significant step toward understanding whether SCC holds or not. He proves that SCC is true for the Gowdy spacetimes on the torus. The Gowdy solutions are of course very specialized, with each of them admitting a two dimensional isometry group. And there is a large body of numerical and asymptotic work suggesting that SCC holds for the Gowdy solutions. However, a rigorous proof of this had been very elusive, up until Ringstrom's just completed work. Like much of the earlier numerical and asymptotic work, Ringstrom's analysis focuses on first showing that the Gowdy solutions are asymptotically velocity term dominated (AVTD) in a neighborhood of the singularity. One might hope that some of the techniques used by Ringstrom could be useful for proving that more general families of solutions, such as the polarized $U(1)$ spacetimes, are also AVTD, and also satisfy SCC.

For less restricted families of spacetimes, numerical studies show that velocity dominated behavior is not to be found. Oscillatory, or possibly even more chaotic behavior is expected. At the workshop, Claes Uggla discussed his work on a "renormalized" formalism which he suggests should be very useful in studying the gravitational fields in a neighborhood of a cosmological singularity. In his formalism, the gravitational field is described by a set of gauge invariant variables for which the expansion has been quotiented out. He argues that these variables allow one to regularize the analysis near the singularity, and therefore make it accessible to the tools of dynamical systems analysis.

3. Singularities and Regularity

One approach to the study of spacetime singularities focusses on the regularity of the gravitational fields. This approach addresses the possibility that the formation of a singularity may signal the breakdown of differentiability rather than a physical effect like curvature blowup. James Vickers' workshop talk discussed his recent work with collaborators which develops a "nonlinear distribution geometry" that can be adapted to spacetimes and used to study generalized (low regularity) geodesics. There are existence and uniqueness results for these geodesics, along with a theory of their Jacobi fields. He described recent attempts to prove singularity theorems based on these generalized geodesics.

Istvan Racz presented a brief survey of spacetime extensions of the sort one uses to study the formation of spacetime singularities. Noting the crucial role of regularity in exploring extensions, he discussed results which show that if one focusses on C^{1-} extensions of globally hyperbolic spacetimes which are causal geodesically incomplete, the only obstruction to extension is unbounded curvature.

4. Gravitational Collapse

The issue of Strong Cosmic Censorship arises not just in the study of cosmological solutions: one encounters it as well in investigating the nature of gravitational collapse, since Cauchy horizons can develop in black hole interiors. Mihalis Dafermos reported at the workshop on his extensive recent studies of black hole solutions of the Einstein-Maxwell-Klein-Gordon (EMKG) equations, in which the question of SCC is addressed. His work shows the stability of the existence of a Cauchy horizon in families of spherically symmetric solutions of the EMKG equation with certain specified regularity conditions, thus showing that a particular (restrictive) model formulation of SCC is false. Interestingly, he finds that this stability of the Cauchy horizon is lost if one chooses a different order of regularity. Dafermos's work (some in collaboration with Igor Rodnianski) also shows that, for the families of collapsing solutions which he studies, both Price's Law of radiation decay and the Israel-Poisson formula for mass-inflation hold.

One of the more interesting features of our models of gravitational collapse is the critical collapse phenomenon, which was discovered numerically by Matt Choptuik over ten years ago and has been extensively studied numerically since then by a number of workers. A key feature of critical collapse is the role played by self-similar solutions. Although the numerical evidence for critical collapse behavior is very strong, mathematical verification has been elusive. The workshop contributions of Gerard Clement and Louise Nolan both involve mathematical studies of collapse and the role of self-similar solutions. Clement examines spherically symmetric collapse of scalar fields in $2+1$ dimensions, and the effect of the addition of a negative cosmological constant; while Nolan looks at cylindrically symmetric, homothetically self-similar, vacuum and dust solutions and their global properties.

5. Dynamical Horizons

A black hole is defined by its event horizon, which is only fully determined after a collapsing astrophysical object has settled into an equilibrium state. One would like to be able to study the physics of a black hole long before this, during its period of formation. Abhay Ashtekar and his collaborators have developed the notion of a dynamical horizon as a way to do this. A dynamical horizon in an evolving spacetime is a three dimensional spacelike hypersurface which is foliated by trapped surfaces. Its structure allows one to define fluxes of energy and angular momentum through a surface; consequently, they permit one to monitor radiation propagation in a spacetime containing a gravitationally collapsing object.

Three talks at the workshop discussed work on dynamical horizons. Badri Krishnan's talk provided an overview: He defined them, described how to measure the energy and angular momentum fluxes through surfaces, related these fluxes to surface area changes, and discussed the dynamical horizon version of the first and second laws of black hole mechanics. He also illustrated the relationship between a dynamical horizon and a stationary horizon in a spacetime which models a collapse.

Stephen Fairhurst's talk at the workshop described his efforts (with Ivan Booth) to develop an action principle and a Hamiltonian consistent with the structure of a dynamical horizon, for a spacetime which contains one. Use of the Hamiltonian gives him an alternative way to define energy and energy fluxes.

There is a close relationship between dynamical horizons and the future outer trapping horizons, which were defined and studied by Heyward. Ivan Booth's talk at the workshop explored this relationship, and related both to Ashtekar's idea of an isolated horizon. In developing these comparisons, Booth provides an instructive way to see where the zeroth, first, and second laws of dynamical horizon mechanics come from.

6. The Constraint Equations

Researchers have learned during the past few years how to apply the technique of "gluing" to solutions of the Einstein constraint equations, and it has turned out to be a very effective tool. Versions of gluing have been used to show that for rather general sets of asymptotically Euclidean data, one can scoop out any compact region and find another set of initial data which contains that identical region and is exactly Kerr outside a larger compact region. Other versions ("connected sum gluing") have been used to produce multi-black hole initial data sets, cosmological solutions with black holes, and initial data sets with arbitrary numbers of wormholes. Daniel Pollack reported at the workshop on his recent work (with Piotr Chruściel and the author) which produces what is likely the strongest connected sum gluing theorem obtainable: it shows that connected sum gluing can be carried out for generic solutions of the constraint equations, and it shows that the changes in the data resulting form the gluing can be localized.

One of the more important tasks for those studying the Einstein constraint equations is that of finding solutions which accurately model physical systems. One of the difficulties in doing this is identifying quantities which measure coordinate-invariant physical information. Sergio Dain's presentation at the workshop introduced a promising new quantity of this sort. Defined in terms of approximate symmetries and the kernel of the linearized constraint operator, Dain's quantity measures in some sense the departure of a set of asymptotically Euclidean data from a set of data for a stationary solution. One can argue that it effectively measures the gravitational radiation generated in the spacetime.

7. Gauge Choices

In most constructions of solutions of Einstein's equations via the Cauchy formulation, either maximal or constant mean curvature (CMC) foliations are used to specify the choice of time. This leads one to ask if every globally hyperbolic maximally-developed solution admits such a foliation. A number of years ago, Robert Bartnik showed that there are Einstein-matter solutions which admit neither a CMC nor a maximal slice; however, the vacuum case remained unresolved. At the workshop, the author of this review reported on his recent work (with Chruściel and Pollack) which shows that there do exist vacuum spatially compact globally hyperbolic maximally developed spacetimes which admit no CMC or maximal slice. The proof relies on the connected sum gluing technique mentioned above.

One of the ways to control the gauge choice in an evolving solution is via the lapse function, N. Recent studies by James York and others suggest that there are a number of advantages to using the "densitized lapse" $\alpha = \frac{N}{\sqrt{\gamma}}$ (where $\sqrt{\gamma}$ is the local volume element) instead of N. Robert

Jantzen's report carries this idea further, showing how it leads naturally to the "Taub time" gauge in spatially homogenous solutions, and describing the natural links of this approach to the minimal distortion shift gauge choice.

Since the inception of general relativity, there has always been interest in discovering the gauge-free constraint-free "degrees of freedom" of the general relativistic gravitational field. Bryan Kelleher spoke at the workshop about his program for doing this. It is based on the Lichnerowicz-Choquet-Bruhat-York conformal treatment of the constraints, together with a choice of a CMC foliation and certain other gauge-fixing conditions. He works out a canonical formalism for incorporating these features, and examines the resulting geometrodynamical theory.

8. Mass

The proof of the Positive Mass Theorem (PMT), completed over 25 years ago, is viewed as one of the triumphs of mathematical relativity. Both approaches to carrying out the proof–that of Richard Schoen and Shing-tung Yau and that of Edward Witten–are quite intricate, and both have led to a number of other significant developments in the field. At the workshop, Greg Galloway presented a new proof of a version of the PMT. Although this version of the PMT is a bit weaker than the Schoen-Yau and the Witten versions (it requires the spacetimes to satisfy a fairly rigid condition which Galloway and his collaborator Chruściel call "asymptotically Schwarzschildian"), the compensation is that the proof is significantly simpler. The proof relies on techniques similar to those used in the proof of the Hawking-Penrose singularity theorems.

The Penrose inequality, which conjectures that the mass is not just positive, but is bounded below by a quantity related to the area of any apparent horizons which are present, is a natural generalization of the PMT. The recent work of Gerhard Huisken and Thomas Ilmanen, and of Hugh Bray, shows that a version of the Penrose inequality holds at least in the case of time symmetric vacuum data. Gilbert Weinstein's workshop talk shows that one natural generalization of the Penrose inequality to Einstein-Maxwell data with charge present does not hold. He uses gluing techniques to construct counterexamples to the conjectured inequality.

It is well known that mass in general relativity is a global quantity, which is not readily localized. Over the years, there have been a number of attempts to define a quasi-local measure of mass in a bounded region, but most of these attempts have had problems. At the workshop, Lauloi So and Jim Nester described a new version of quasi-local mass. This version is based a covariant Hamiltonian formulation of general relativity and its application to regions in a given spacetime with finite boundary. One of its virtues is a close tie to the Bel-Robinson energy density.

9. Relativistic Elasticity

There are at least three motivations for developing a relativistic theory of elasticity compatible with Einstein's theory of gravitation: a) the desire to understand the interaction of certain "elastic" astrophysical bodies with gravitational radiation; b) the attempt to formulate a rigorous theoretical picture of how elastic "Weber bars" respond to and detect incoming gravitational radiation; and c) the drive to meld the two inherently geometric field theories of general relativity (curvature of the spacetime manifold) and elasticity (dynamics of mappings between the body manifold and the ambient space manifold).

There have been a number of attempts in the past to develop a theory coupling elasticity to general relativity, most notably that of Brandon Carter and Carlos Quintana. Robert Beig, Berndt Schmidt, and their collaborators have begun a new program aiming to obtain such a coupled theory. A goal of the program is to show that the resulting field equations constitute a well-posed system. Beig's talk at the workshop reported on some of their progress, focussing on the existence and uniqueness of solutions of the elasticity equations on a fixed but arbitrarily curved background spacetime, and also on the nature of solutions for the equilibrium system in an appropriate Newtonian limit.

Boundary conditions are of key importance if one wishes to model the dynamics of an elastic body. The inclusion of boundary conditions in the theory was the focus of the work which Michael Wernig-Pichler presented. Working from a Lagrangian formulation, he derives the initial value boundary value PDE system for the Einstein-elasticity field theory. He then proceeds to prove local existence for small data for this system, assuming that the elastic body is homogeneous, isotropic, and hyperelastic.

10. Asymptotic Structure of Spacetimes

The definition of the asymptotic structure of asymptotically flat solutions of Einstein's equations via conformal completion has proven to be a very useful tool for the study of radiation in astrophysical spacetimes. The recent proof based on the work of Justin Corvino, Helmut Friedrich, Piotr Chruściel and Erwan Delay that there are large families of solutions which contain a complete future null infinity $\mathcal{I}^+$ emphasizes the utility of this approach.

Not yet understood is what conditions on a set of asymptotically Euclidean data guarantee that the development of that data will admit a conformal completion, with a resulting smooth $\mathcal{I}$. Juan Valeinte Kroon presented work at the workshop which addresses this question. Using the cylindrical picture of spatial infinity developed by Friedrich, he is able to identify features of a set of data which signal the development of troublesome logarithmic terms at $\mathcal{I}$. He is then able to formulate a conjecture regarding the sorts of time symmetric data which will lead to smooth asymptotic structure.

Two other talks at the workshop dealt with generalizations of the conformal completion picture. Naoyo Okuyma tries to generalize this picture to theories of gravity which contain higher order derivatives, and which involve dimensions higher than $3 + 1$. Okuyama obtains this generalization, and uses it to define and calculate certain conserved quantities for the gravitational field of his higher order theories.

Claudio Dappiaggi discussed his ideas for doing a form of holography on asymtoptically Euclidean solutions, rather than on the usual asymptotically Anti DeSitter spacetimes. He works on $\mathcal{I}$, and attempts to set up a field theory there which is invariant under the action of the BMS group. He is able to obtain a full particle spectrum on $\mathcal{I}$, and is working to complete his program.

11. Other Talks

The workshop talks which I have not yet mentioned fit less conveniently into thematic topics for the workshop. I discuss each of them briefly here.

Thierry Daudé presented a fairly complete picture of the time-dependent scattering theory for charged, massive Dirac fields in a Kerr-Newman background. Using Newman-Penrose formalism, he examines scattering from the point of view of an observer at infinity, with the black hole horizon serving as a second asymptotic region. He shows that the energy of the Dirac field does not get trapped in compact regions.

Christian Klein's talk described his recent work on a coalescing binary system which is presumed to have the symmetry of a helical Killing field. Quotienting out the symmetry, he obtains a field theory which is equivalent to $2 + 1$ gravity coupled to a non-linear sigma model (or wave map). He discusses the change of equation type from hyperbolic to elliptic which occurs in passing from the exterior to the interior of the light cylinder, and he examines some of the asymptotic behavior of the fields as one approaches the horizon and the light cylinder.

The work presented by Todd Olynyk concerns monopole-like solutions of the Einstein-Yang-Mills-Higgs (EYMH) equations. He proves the existence of static spherically symmetric EYMH solutions of this type. In doing so, he avoids using the "shooting method", since while this is useful and efficient for ODEs (i.e., for spherically symmetric solutions) it is not useful for studies of more general, nonspherically symmetric, solutions.

Itai Seggev's workshop talk addressed the issue of the well-posedness of a scalar wave equation in a non globally hyperbolic (but causal) background spacetime. He extends some of his earlier results on static spacetimes to stationary spacetimes. These results guarantee existence of smooth solutions, although well-posedness has not yet been shown. His proof relies on a Hilbert space formulation.

It has been argued that if a wormhole in spacetime is to last long enough to be traversable, it must contain matter fields which violate the averaged null energy condition (ANEC). One may then ask what the consequences of the presence of such matter might be. Naresh Dadhich's presentation addressed this question, and also considered "how much" ANEC violation is needed to produce the desired wormholes. Working with simple models, he finds that perhaps the amount of necessary violation is small, and therefore may leave the positive mass theorem and other such desirable features intact.

The standard approach to the study of equilibrium states of self-gravitating fluid bodies is based on the analysis of solutions to the Tolman-Oppenheimer-Volkoff equation. While this approach is very effective for certain classes of equations of state, it leads to significant analytical difficulties if one examines more general matter models. Mark Heinzle's workshop presentation described an alternative formulation of the problem which appears to be able to handle some of these more general equations of states. Using normalized field variables, it casts the system into the form of a regular autonomous three-dimensional dynamical system. Heinzle shows how this formulation can be used to obtain the mass-radius diagram for static stars of various equations of state.

Sigbjorn Hervik's workshop talk also discussed the casting of a gravitational physics problem into the form of a dynamical system. He studies the future asymptotic behavior of spatially homogeneous cosmological models containing non vanishing perfect fluids whose motion is "tilted" in the sense that the fluid velocity is not orthogonal to the isometry group orbits. Focussing on Bianchi Class B cosmologies, he finds fairly rich and complicated dynamics in various families of solutions, such as those of Bianchi Type VII_h. He also shows that the asymptotic behavior of certain vacuum plane wave solutions is largely unaffected by the addition of the titled fluids.

A4: ANALYTIC APPROXIMATION AND PERTURBATION METHODS

BALA R IYER
Raman Research Institute
Bangalore 560080, INDIA
e-mail: bri@rri.res.in

The oral presentations of the Workshop A4 on Analytic Approximation and Perturbation Methods are summarised. Topics covered include post-Newtonian approximations, radiation reaction in general relativity, post-Minkowskian approximations, rotational perturbations, black hole perturbations and quasi-normal modes, Efroimsky formalism, background independent gravitational waves and cosmological perturbations.

1. Introduction

The topic of analytical approximation and perturbation methods has seen renewed activity over the last fifteen years due to problems arising from computations of gravitational waves in astrophysics and growth of perturbations in cosmology. This has led to systematic work on post-Newtonian approximations, post-Minkowskian approximations, perturbations around black hole backgrounds, radiation reaction in general relativity, high-frequency approximations, quasi-normal modes, non-radial oscillations, and rotational perturbations of stars and black holes on the one hand and cosmological perturbations on the other. All these topics find representation in the contributions to this workshop and the recent progress in these areas was summarised in the oral presentations of workshop A4.

2. Post-Newtonian Approximations

The workshop commenced with the presentation of the latest results on dynamics and gravitational radiation of compact binaries by Gilles Esposito-Farèse. The equations of motion [1] and the gravitational radiation [2] of two point masses in harmonic coordinates, have been derived using dimensional continuation to regularize the singularities occurring at the location of the particles. At the third post-Newtonian order, the singularities manifest as poles proportional to $1/(d-3)$, where d is the spatial dimension. It is shown that these can be *renormalized* away by suitable shifts of the two worldlines, which correspond physically to dressing each point particle with an infinite dipole. The ensuing, finite results do not involve any undetermined coefficient, and thereby fix the values of the four "ambiguity" parameters found within the Hadamard regularization scheme: $\lambda = -1987/3080$ ($\Leftrightarrow \omega_s = 0$), $\xi = -9871/9240$, $\kappa = 0$, and $\zeta = -7/33$. The small numerical value of the combination $\hat{\theta} \equiv \xi + 2\kappa + \zeta - \frac{7}{3}\lambda \approx 0.22$ relevant for circular orbits indicates that the post-Newtonian series converges well. This work completes the gravitational wave predictions needed for the data analysis of the LIGO and VIRGO interferometers by providing post-Newtonian corrections to the gravitational wave phase of inspiralling compact binaries to order 3.5PN beyond the leading quadrupole formula for phasing.

What can one say about the post-Newtonian corrections to the gravitational wave amplitude beyond those used at present in gravitational wave data analysis [3]? The computation of the 2.5PN gravitational wave polarisations from inspiralling compact binaries in circular orbits was summarised by Bala Iyer. In this work [4], "ready-to-use" gravitational waveforms up to 2.5PN order (the plus and cross polarisations of the gravitational waves) for inspiralling compact binaries in circular orbits are explicitly computed using the multipolar post-Minkowskian and matching formalism. Required inputs include mass-type quadrupole at 2.5PN order, mass octupole and current quadrupole at 2PN. The nonlinear hereditary contributions to the 2.5PN polarizations are fully accounted. The hereditary terms include the 'tails' and the nonlinear memory effect. The tails are due to the scattering of the linear waves by the total mass energy (ADM mass M). The memory effect can be viewed as due to re-radiation by linear waves. Such templates including all amplitude corrections and harmonics may be useful for improved parameter estimation of chirp signals in LISA.

Though for long lived compact binaries the quasi-circular approximation is adequate, in recent times scenarios involving compact eccentric binaries are also being suggested as potential gravitational wave sources. In this context, Achamveedu Gopakumar discussed a method for analytically constructing high-accuracy templates for the gravitational wave signals emitted by compact binaries moving in *inspiralling eccentric* orbits [5]. By using an improved 'method of variation of constants' the three time scales present in the system, namely those associated with the orbital motion, periastron precession and radiation-reaction, are combined without making the usual approximation of treating the radiative time scale in an adiabatic manner. The method is explicitly implemented to obtain $h_{+,\times}$, whose amplitude is Newtonian, but with the time evolution fully 2.5PN accurate.

To generalise the above work to 3PN order, one requires the third post-Newtonian accurate generalized quasi-Keplerian parametrization for compact binaries in eccentric orbits. Raoul-Martin Memmesheimer [6] presented such a parametrization for two non-spinning compact objects moving in an eccentric orbit. The orbital elements of the parametrization are explicitly given in terms of the 3PN accurate conserved orbital energy and angular momentum in both the Arnowitt, Deser, and Misner-type and harmonic coordinates. Due to the presence of certain 3PN accurate gauge invariant orbital elements, the parametrization should be useful to analyze the compatibility of general relativistic numerical simulations involving compact binaries with the corresponding post-Newtonian descriptions. If required, the present parametrization can be used to compute post-Newtonian corrections to the currently employed 'timing formula' for the radio observations of relativistic binary pulsars.

Can one generalise the above works on eccentric, non-spinning binaries to spinning compact binaries moving in eccentric orbits? Christian Koenigsdoerffer discussed this in his presentation [7] by considering PN accurate dynamics consisting of 3PN accurate conservative orbital dynamics, associated with non–spinning compact objects, influenced by the leading order spin–orbit interactions. The orbital elements of the representation were explicitly obtained in terms of the conserved orbital energy, angular momentum and a quantity that characterizes the leading order spin–orbit interactions in ADM–type coordinates. Expressions are derived for gravitational wave polarizations suitable to describe gravitational radiation from spinning compact binaries of arbitrary mass ratio moving in eccentric orbits.

Laszlo A. Gergely emphasized the need to go beyond the point-particle circular orbit approximation by pointing out that black holes have significant spin, and neutron stars can also have mass quadrupole and magnetic dipole moments. To analyse these scenarios, in [8, 9] they have obtained a closed system of first order differential equations describing the orbital dynamics of such a system, including the spin-orbit [10], spin-spin and self-interaction spin [11, 12], quadrupole-monopole [13] and magnetic dipole - magnetic dipole [14] contributions.

What are the post-Newtonian equations of motion for general systems of arbitrarily structured bodies? Étienne Racine discussed recent work

[15, 16] on 1PN translational equations of motion for systems of arbitrarily structured, strongly self-gravitating bodies, including for the first time the couplings to all multipole moments. The steps to the equations of motion include (i) defining for each body mass and current multipole moments, gravitoelectric and gravitomagnetic tidal moments, and a center-of-mass worldline in terms of the metric in a weak-field region surrounding that body; (ii) deriving for each body, using a surface integral method, relations between these quantities previously derived for the special case of weakly self-gravitating bodies by Damour, Soffel and Xu [17, 18]; (iii) eliminating the tidal moments to obtain the coupled equations of motion.

Finally, Chongming Xu talked about the post-Newtonian quasi-rigid body and its Dynamical Equation [19–21]. At 1PN approximation, a complete model of a quasi-rigid body is constructed by means of a new constraint on the mass current density and mass density. In this quasi-rigid body model most of the relations, such as a spin vector proportional to the angular velocity, the definition on the moment of inertia tensor, the key relation between the mass quadrupole moment and the moment of inertia tensor, the rigid rotating formulae for the mass quadrupole moment and the moment of inertia tensor, are just extensions of the main relations in the Newtonian rigid body model. The key relation obtained in this paper might be useful in future applications to problems in geodynamics and astronomy. The dynamical equation of 1PN quasi-rigid body is deduced from the general rotational laws of motion at the 1PN approximation. The torque-free motion of a quasi-rigid body, especially PN Euler-Poinsort's construction is obtained. The result is similar to the Newtonian one except for the additional 1PN correction.

Work on radiation reaction within the post-Newtonian approach was also discussed but included in the next section.

3. Radiation Reaction

The presentation on radiation reaction started off with Samaya Nissanke [22]. It was related to the computation of the 3.5PN radiation reaction effect in the local equations of motion of a compact binary (in a general frame) in harmonic coordinates. The derivation is based on a direct post–Newtonian iterated expansion of the near–zone metric and the equations of motion for an extended isolated system in harmonic coordinates at the 3.5 PN approximation, using appropriate 'asymptotically matched' retarded flat space-time potentials [23, 24]. The formalism is subsequently applied to binary systems of point particles, using the standard Hadamard self-field regularisation. The 3.5PN radiation-reaction term in the acceleration of the particles in a general frame is obtained. Restricting the expression to the centre-of-mass frame, the 3.5PN term in the equations of motion is found to be in perfect agreement with the general expression of the 1PN radiation reaction force, derived by energy and angular–momentum balance equations [25,26] and 3.5PN results obtained via direct integration of relaxed Einstein equations [27].

Tadayuki Ohta [28] discussed radiative and non-radiative forces acting on test bodies in a far-zone of the radiating sources. In the ADM formalism under one globally defined coordinate systems the Newtonian-type equations for the deviation between two test bodies are derived. The leading terms of r^{-1} consist of the force due to quadrupole radiation and the non-radiative (instantaneous) force coming from h_{00} component of the metric. Both forces are proportional to the relative velocity of the two bodies.

Wataru Hikida [29] presented the formulation of a new analytical method for regularizing self-force acting on a particle of small mass μ orbiting a black hole of mass M, where $\mu \ll M$, in the black hole perturbation approach. Using this formalism, the regularized self-force on the scalar particle in circular orbit around a Schwarzschild black hole is analytically evaluated up to 18PN order to assess the effectiveness of this new scheme. The convergence of the post-Newtonian expansion for the regularized self-force is finally investigated [30].

Hiroyuki Nakano [31] discussed the gravitational self-force on a particle in a black hole space-time [32–34]. For a point particle, the full (bare) self-force diverges. In order to obtain the regularized self-force, one considers the "mode-sum/decomposition regularization", and introduces the concept of a "finite gauge transformation". In the Regge-Wheeler gauge, the first post-Newtonian order self-force for a particle in circular orbit around a Schwarzschild black hole has been obtained.

Finally, C. Lousto summarised his work [35] on calculation of the self-force on a particle in circular and head-on orbit around a Schwarzschild black hole. A version of the mode-sum scheme adapted to the Regge-Wheeler gauge is used to regularise the force that is computed numerically by integration of the Zerelli and Regge-Wheeler equations.

4. Post-Minkowskian expansions of the world function and time transfer

The Post-Minkowskian expansions have also been used to elegantly treat the problem of time transfer. This was summarised by Christophe Le Poncin-Lafitte [36]. In suitably chosen domains of spacetime, the world function is a powerful tool for modelling the deflection of light and the time/frequency transfer. A recursive procedure for expanding the world function into a perturbative series of ascending powers of the Newtonian gravitational constant G is worked out. It is shown rigorously that each perturbation term is given by a line integral taken along the unperturbed geodesic between two points. Once the world function is known, it becomes possible to determine the time transfer functions giving the propagation time of a photon between its emission and its reception. It is established that the direction of a light ray as measured in the 3-space relative to an observer can be derived from these time transfer functions, even if the metric is not stationary. It is shown how to derive these functions up to any given order in G from the perturbative expansion of the world function.

5. Collapse

Unlike for gravitational waves from inspiralling systems, there is not much analytical work on gravitational radiation from collapse. Ken-ichi Nakao [37] talked of extremely high speed collapse of cylindrical perfect fluid. By the use of a recently proposed approximation scheme, the gravitational collapse of cylindrically distributed dust fluid is studied and the generation of gravitational waves at the naked singularity formed by this gravitational collapse is investigated. The boundary condition at the naked singularity is imposed so that all the metric variables are finite there. The energy carried by the gravitational radiation from the naked singularity is found to be an increasing function of the energy per unit length of the cylindrical dust and diverges at some critical value of it.

6. Rotational Perturbations

The issues related to rotation, rotational perturbations and non-radial perturbations of stars in general relativity is a difficult but interesting area of research. In this regard, Herbert Pfister talked about a model for linear dragging. A Reissner-Nordström shell (mass M, charge q, radius R) is linearly accelerated by a small, dipolar, and momentarily static charge distribution in its exterior. The Einstein-Maxwell equations for this perturbation of Reissner-Nordström are solved, and the dragging d of interior neutral test particles (linear acceleration of these particles, scaled by the acceleration of the shell) is calculated. In the weak field case ($M/R << 1$, $q/R << 1$), and for a charge distribution proportional to r^{-5}, one gets $d = 4M/3R$, coinciding with Thirring's value for a rotating shell. This represents a partial proof of the hypothesis of a quasi-global equivalence principle in general relativity [38].

Raul Vera discussed a consistent theory of isolated rotating stationary and axisymmetric finite objects to first and second order in perturbation theory around static configurations [39]. The functions that determine the asymptotically flat vacuum exterior region satisfy an elliptic system of partial differential equations. The boundary conditions at the surface of the body for the asymptotically flat exterior problem, which arise from the matching procedure, are of Cauchy type and thereby constitute an overdetermined data set. The necessary and sufficient compatibility conditions on the Cauchy data for the perturbed exterior problem from a static configuration up to the second order have been found. These conditions therefore constitute necessary and sufficient conditions that any second order stationary and axially symmetric perturbation of a static interior metric should satisfy in order to define true perturbations of an isolated finite body.

In the period since the discovery of the Kerr metric, there have been many attempts to find closed interior solutions which match the exterior smoothly. This has contributed to a lack in the development in the theory of rotating fluid bodies in general relativity. Paul Messenger presented work on spherically symmetric fluid spheres endowed with slow rotation matching to Kerr empty space-time [40]. A perturbation technique was adopted, up to and including quadratic terms in angular velocity parameter. The rotating second order source metric was matched the second order Kerr metric using the Darmois matching conditions. It was shown that it was possible to match the source with the Kerr metric, while satisfying the required Einstein's equations. The approach was illustrated by using the Schwarzschild metric as an example, with the equation of the fluid boundary presented, together with expressions for pressure and density, with pressure being zero at the boundary.

Nils Andersson discussed progress in developing a Lagrangian perturbation framework for rotating non-relativistic superfluid neutron stars [41]. He described the first generalization of classic work on the stability properties of rotating stars to models which account for the presence of potentially weakly coupled superfluid components. The analysis is relevant for the conditions that prevail in the outer core of mature neutron stars, and has implication for dynamical and secular instabilities of such stars.

Linear perturbation theory is an appropriate way of describing small oscillations of stars, while a mild non-linearity is still tractable perturbatively but requires to consider mode coupling. Andrea Passamonti talked of gauge invariant formalism for coupling of radial and non-radial oscillations of relativistic stars. A relativistic formalism has been introduced to study, in the time domain, the coupling of radial and non-radial first order perturbations of a static spherical star [42]. This is based on 2-parameter perturbation theory [43], where the two sets of linear perturbations are treated separately. In this framework, one can introduce gauge invariant variables (with a fixed gauge for the radial perturbations) describing the coupling; the corresponding equations have the same differential structure as those for non-radial linear perturbations, with additional source terms from the product of first order radial and non-radial perturbations. Possible applications include investigations of the gravitational wave signal and the possible existence of resonances, parameter amplification, and/or changes in the damping time of non-radial perturbations. The evolution and constraint equations are presented and the implementation of numerical computations discussed, with special attention to the surface boundary conditions for the non-linear perturbations describing the coupling [42].

7. Black Holes

The subject of black hole perturbations is a mature one with a history of more than 40 years. Yet interesting work related to its extensions in other theories of gravity continue. Hideo Kodama talked of perturbative uniqueness of black holes in four and higher dimensions near the static maximally symmetric limit. By utilising the gauge-invariant formulation [44,45], it is shown that there exists no bounded stationary regular perturbation (that falls off at infinity in the case $\Lambda < 0$) for a static black hole with constant curvature horizon in any spacetime dimensions, except for those corresponding to variations of moduli parameters and rotational perturbations characterised by angular momenta [46]. This result indicates that the Gibbons-Lü-Page-Pope solutions [47] are the only stationary regular black hole solutions at least near the static and spherically symmetric limit in all dimensions and that a similar uniqueness theorem holds for topological black holes as well.

Ilya Mandel discussed the validity of the quasi-stationary approximation to the binary black hole problem for non-tidally locked black holes. A spherically symmetric scalar field serves as a model for gravitational waves. The standing-wave scalar field leads to a naked singularity in place of a Schwarzschild black hole. The downgoing scalar field solution can be obtained by the addition of a perturbative radiation reaction field to the standing-wave solution. For field frequency and amplitude that correspond to a binary black hole separation of less than $5M$, the downgoing field can be recovered to an accuracy of a few percent all the way to the region just outside the black hole horizon ($\approx 2.001M$) [48].

8. Efroimsky Formalism versus Isaacson's Formalism

The relation between different perturbation schemes needs to be understood and in this talk Otaker Svitek [49, 50] discussed the Efroimsky perturbation scheme for consistent treatment of gravitational waves. Their influence on the background was summarized and compared with the classical high-frequency approach of Isaacson [51,52]. It was demonstrated that the Efroimsky method is not consistent if high-frequency gravitational waves are considered. A possible modification of the Efroimsky formalism was proposed which could resolve this drawback. The gauge invariance of this modification has been investigated in detail [53].

9. Quasi-normal modes

Suneeta Varadarajan [54] talked on the problem of identifying signatures of incompleteness of quasi-normal modes. A model for studying criteria for (in)completeness of quasi-normal modes was presented. This is a model for perturbations of the Schwarzschild-de Sitter black hole. The time decay of fields in this black hole shows some striking features due to the presence of two horizons, the black hole and cosmological horizon - in particular, intermediate time power law tails. The model is an analytical approximation for studying the time decay of scalar fields in this black hole background. It mirrors all the striking qualitative features of the time decay. Further, it is exactly solvable and could be used to find signatures for intermediate time power law tails in the analytical behaviour of the Green's function for the problem.

10. Background-independent gravitational waves

Luca Lusanna posed the question about background independent gravitational waves and proposed an answer to it. In the rest-frame instant form of tetrad gravity [55] in non-compact, globally hyperbolic, asymptotically flat at spatial infinity space-times of the Christodoulou-Klainermann type, the admissible 3+1 splittings of space-time have the space-like leaves (both Cauchy and simultaneity 3-surfaces [56]) tending in a direction-independent way to space-like hyper-planes orthogonal to the ADM 4-momentum at spatial infinity, where there is an asymptotic Minkowski metric. A Shanmugadhasan canonical transformation separates the Dirac observables (tidal effects) from the gauge variables (inertial effects [57]). The weak ADM energy is the canonical Hamiltonian. In a completely fixed Hamiltonian 3-orthogonal (non-harmonic) gauge it is possible to make a Hamiltonian linearization without introducing a background [58] and to solve all the constraints and the Hamilton equations, identifying post-Minkowskian background- independent gravitational waves. In presence of a perfect fluid this will allow to get a post-Minkowskian version of the quadrupole emission formula. Application to Hamiltonian numerical gravity is under investigation.

11. Cosmological Perturbations

In addition to perturbations of stars and black holes the other vigorous area of research relates to cosmological perturbations. Most papers related to this will be found in the workshop B-2 with applications to issues in cosmology. Hence only a couple of papers also of general interest to the topic of analytical approximations and perturbations were discussed. In a comprehensive paper under way on linear perturbations of Friedmann-Robertson-Walker universes Jiri Bicak has studied perturbations in a general gauge and then formulated suitable "instantaneous" gauge conditions which increase the number of elliptic equations for perturbations. This formalism was used recently to analyze toroidal motions of the cosmic fluid and their influence on dragging of local inertial frames [59]. The dragging effects are instantaneous and depend on whether angular velocities or angular momenta of the fluid are prescribed. The perturbation equations are solved for all three types of FRW universes in terms of vector spherical harmonics and Legendre functions. In the case of the closed universes the Legendre-type equation requires a special mathematical treatment: the solutions are given in terms of the derivatives of the Legendre functions with respect to the order.

Finally, Christoph Schmid talked on the cosmological origin of inertia and the Mach's principle [60]. The axes of gyroscopes experimentally define local non-rotating frames. But what physical cause governs the time-evolution of gyroscope axes? The author considers cosmological vorticity perturbations (i.e. vector perturbations) of a spatially flat FRW universe. His results (valid for any equation of state for matter) are: 1) The dragging of a gyroscope axis by rotational perturbations beyond the H-dot radius (H = Hubble constant) is exponentially suppressed.- 2) If the perturbation is a homogeneous rotation inside a radius significantly larger than the H-dot radius, then the dragging of the gyroscope axis by the rotational perturbation becomes exact.- 3) The time-evolution of a gyroscope axis exactly follows a specific average of the matter inside the H-dot radius.- In this precise sense Mach's Principle (that axes of local non-rotating frames precisely follow some average of the motion of cosmic matter) is a consequence of cosmology with Einstein gravity.

12. Poster session

In addition to the oral presentations summarised above there were the following poster presentations.

(1) Xuejun Wu, Chongming Xu and Michael Soffel
General Relativistic Theory of Elastic Astronomical Body
(2) Gerold Betschart and Chris Clarkson
Scalar and electromagnetic perturbations on LRS class II spacetimes
(3) Eiji Mitsuda
Electromagnetic burst from naked singularity
(4) Kouji Nakamura
General flamework of second order gauge invariant perturbation theory
(5) A.M.Baranov and S.F.Tegai
An approximate radiating star model
(6) Yoshino Hirotaka
Anti-de Sitter black-hole spacetimes with deformed boundaries"
(7) Nikodem Szpak
Quasinormal mode expansion for solutions of the wave equations
(8) Thomas Waters
Perturbations of self-similar Vaidya spacetime.
(9) Krzysztof Roszkowski

The propagation of waves on the Schwarzschild background

(10) Michael Brady and Eriksson Daniel
Rotating Cosmological Models of Bianchi Type V

(11) Hironobu Furuhashi
The Evolution of the Scalar Field on the Black Hole Model

(12) H. Arcos Velasco
Harmonic deformations of a Kerr-Newman spacetime source.

References

1. L. Blanchet, T. Damour and G. Esposito-Farèse, *Phys. Rev. D* **69**, 124007 (2004).
2. L. Blanchet, T. Damour, G. Esposito-Farèse and B.R. Iyer, *Phys. Rev. Lett.* **93**, 091101 (2004).
3. L. Blanchet, B. R. Iyer, C. M. Will and A. G. Wiseman, *Class. Quant. Grav.* **13**, 575 (1996).
4. K. G. Arun, L. Blanchet, B. R. Iyer and M. S. S. Qusailah, *Class. Quant Grav.* **21**, 3771 (2004).
5. T. Damour, A. Gopakumar and B. R. Iyer, *Phys. Rev. D* **70**, 064028 (2004); `gr-qc/0404128`.
6. R.-M. Memmesheimer, A. Gopakumar, and G. Schäfer, *Phys. Rev. D* **70** 104011 (2004).
7. C. Koenigsdoerffer and A. Gopakumar, *Phys. Rev. D* **71** 024039 (2005).
8. B. Mikoczi and L. A. Gergely, Orbital evolution of radiating compact binaries I. All linear effects, `(In preparation)`.
9. B. Mikoczi, M. Vasuth and L. A. Gergely, Orbital phasing in inspiraling compact binaries, `(In preparation)`.
10. L. A. Gergely, Z. Perjes and M. Vasuth *Phys. Rev. D* **58**, 124001 (1998).
11. L. A. Gergely, *Phys. Rev. D* **61**, 024035 (2000).
12. L. A. Gergely *Phys. Rev. D* **62**, 024007 (2000).
13. L. A. Gergely and Z. Keresztes, *Phys. Rev. D* **67**, 024020 (2003).
14. M. Vasuth, A. Mihaly, Z. Keresztes and L. A. Gergely, *Phys. Rev. D* **68**, 104006 (2003).
15. E. Racine and E. E. Flanagan, *Phys. Rev. D* **71** 044010 (2005).
16. E. Racine, (2004); `gr-qc/0405058`.
17. T. Damour, M. Soffel and C. Xu, *Phys. Rev. D* **45**, 1017 (1992).
18. T. Damour, M. Soffel and C. Xu, *Phys. Rev. D* **47**, 3124 (1993).
19. C. Xu and X. Wu, *Phys. Rev. D* **63**, 064001 (2001).
20. C. Xu, J. Tao and X. Wu, *Phys. Rev. D* **69**, 024003 (2004).
21. C. Xu, J. Tao, T. Huang and X. Wu, *Chin. Phys. Lett.* **21**, 1884 (2004).
22. S. Nissanke and L. Blanchet, *Class. Quantum Grav.* **22** 1007-1032 (2005).
23. L. Blanchet, G. Faye and B. Ponsot. *Phys. Rev. D* **58**, 124002 (1998).
24. L. Blanchet and G. Faye. *Phys. Rev. D* **63**, 062005 (2001).
25. B. R. Iyer and C.M. Will. *Phys. Rev. Lett.* **70**, 113 (1993).

26. B. R. Iyer and C.M. Will. *Phys. Rev. D* **52**, 6882 (1995).
27. M. Pati and C. Will, *Phys. Rev. D* **65**, 104008 (2002).
28. T. Ohta, Radiative and non-radiative forces acting on test bodies in a far-zone of the radiating source, `(In preparation)`.
29. W. Hikida et al. *Prog. Theor. Phys.* **111** 821 (2004).
30. W. Hikida et al. *Prog. Theor. Phys.* **113** 283-303 (2005).
31. H. Nakano, N. Sago and M. Sasaki, *Phys. Rev. D* **68**, 124003 (2003).
32. L. Barack, Y. Mino, H. Nakano, A. Ori and M. Sasaki, *Phys. Rev. Lett.* **88**, 091101 (2002).
33. N. Sago, H. Nakano and M. Sasaki, *Phys. Rev. D* **67**, 104017 (2003).
34. S. Detweiler and E. Poisson, *Phys. Rev. D* **69**, 084019 (2004).
35. C. Lousto, Self-force evolution of circular orbits around a Schwarzschild black hole, `(In preparation)`.
36. C. Le Poncin-Lafitte, B. Linet and P. Teyssandier, *Class. Quant. Grav.*, **21**, 4463 (2004).
37. K. Nakao, *Class. Quant. Grav.* **21**, 2101 (2004).
38. H. Pfister and K. H. Braun, *Class. Quant. Grav.* **2**, 909 (1985).
39. M. A. H. MacCallum, M. Mars and R. Vera, First and second order metrics of stationary and axially symmetric slowly rotating stars, `(In preparation)`.
40. R. J. Wiltshire and P. Messenger, Spherically symmetric fluid spheres endowed with slow rotation matching Kerr spacetime, `(In preparation)`.
41. N. Andersson and K. Grosart, Lagrangian perturbation theory of nonrelativistic rotating superfluid stars, `(In preparation)`.
42. A. Passamonti, M. Bruni, L. Gualtieri, and C. F. Sopuerta, **Phys. Rev. D 71** 024022 (2004).
43. C. F. Sopuerta, M. Bruni, and L. Gualtieri, *Phys. Rev. D* **70** 064002 (2004).
44. H. Kodama, A. Ishibashi and O. Seto, *Phys. Rev. D* **62**, 064022 (2000).
45. H. Kodama and A. Ishibashi, *Prog. Theor. Phys.* **110**, 701 (2003).
46. H. Kodama, *Prog. Theor. Phys.* **112**, 249-274 (2004).
47. G.W. Gibbons, H. Lü, D.N. Page and C.N. Pope, *J. Geom. Phys.* **53** 49-73 (2005).
48. I. Mandel, Towards a solution of the binary black hole problem with non-tidally-locked horizons. gr-qc/0505149 (2005).
49. M. Efroimsky, *Class. Quantum Grav.* **9**, 2601 (1992).
50. M. Efroimsky, *Phys. Rev. D* **49**, 6512 (1994).
51. R. A. Isaacson, *Phys. Rev.* **166** 1263 (1968).
52. P. R. Anderson, *Phys. Rev. D* **55**, 3440 (1997).
53. O. Svitek and J. Podolsky, *Class. Quantum Grav.* **21**, 3579 (2004).
54. S. Vardarajan, A model for studying signatures of incompleteness of quasi-normal modes, *(In preparation)*.
55. L. Lusanna, *Gen.Rel.Grav.* **33**, 1579 (2001); L. Lusanna and S. Russo, *Gen.Rel.Grav.* **34**, 189 (2002); R. De Pietri, L. Lusanna, L. Martucci and S. Russo, *Gen.Rel.Grav.* **34**, 877 (2002).
56. D. Alba and L. Lusanna, `gr-qc/0311058`.
57. L. Lusanna, in *Progress in General Relativity and Quantum Cosmology (Nova Science, NY, 2004)*.

58. L. Lusanna and M. Pauri, gr-qc/0403081; gr-qc/0407007.
59. J. Bicak, D. Lynden-Bell and J. Katz, *Phys. Rev. D* **69**, 064011 and 064012 (2004).
60. C. Schmid, gr-qc/0409026.

A5: NUMERICAL RELATIVITY AND ALGEBRAIC COMPUTING

BERND BRÜGMANN
Institute for Gravitational Physics and Geometry
Penn State University
University Park, PA 16802, USA
e-mail: bruegman@gravity.psu.edu

1. Introduction

As chairman of the session on Numerical Relativity and Algebraic Computing I received 65 abstracts. Only three of those focussed on algebraic computing, although I would estimate that about one half of all research in numerical relativity relies on algebraic computing in more than a minimal way. According to Takashi Nakamura in the proceedings of GR15, there were 34 abstracts submitted for numerical relativity for GR15 and 17 abstracts for GR13. Both the growing number of abstracts and the high quality of the submitted work is an indication that numerical relativity is a rapidly growing field with a very lively, young community.

I was very happy with the overall quality of the submitted abstracts, scheduling 39 short talks (8 + 2 minutes) and 22 posters. What worked very well, based on feedback from the participants, were the short 1 minute poster advertisements that introduced the researcher to the audience and added encouragement to attend the poster session.

In the following I briefly summarize the oral presentations by giving a condensed version of the submitted abstracts. Since references to publications were not required for the submission of abstracts, they were in general missing, but the interested reader should have no problems finding the corresponding work online given the descriptions below.

2. Summary of oral presentations

Peter Aichelburg, on work with Bizoń and Tabor: *Fine structure phenomena in critical collapse.* A report was given on new numerical studies of critical phenomena in black hole formation for the spherically symmetric SU(2)-sigma model (wave maps) coupled to gravity. In order to understand better the character of the continuously/discretely self-similar transition, high precision computations near the bifurcation point were performed and an unexpected fine structure was found.

Miguel Alcubierre, on work with Salgado: *Black holes, solitons and scalar hair.* A numerical analysis shows that the Einstein field equations allow static and spherically symmetric black hole solutions with scalar-field hair in asymptotically flat spacetimes. When the horizon radius of the black hole tends to zero, regular scalar solitons are found. The configurations, although unstable under spherically symmetric linear perturbations, are regular and thus can serve as counterexamples to the no-scalar-hair conjecture. A non-linear evolution of the soliton configuration was performed, showing a collapse into a black hole or an expanding domain wall, depending on the kind of perturbation.

Beverly Berger: *Why Enforce the Hamiltonian Constraint in Numerical Relativity?* The indefinite sign of the Hamiltonian constraint means that solutions to Einstein's equations must achieve a delicate balance — often among almost canceling large terms. If numerical errors cause violation of the Hamiltonian constraint, the failure of the delicate balance can lead to qualitatively wrong behavior rather than just a lesser accuracy. This issue is different from instabilities caused by constraint-violating modes. Examples of numerical simulations of collapsing cosmological spacetimes with and without Hamiltonian constraint enforcement were presented.

Nigel Bishop: *Computing the evolution of self-gravitating objects orbiting a black hole.* The ultimate aim of this project is to compute the evolution of a system containing a self-gravitating object in close orbit around a black hole. Recent results were reported, including the lack of sensitivity of the evolution on the initial gravitational data, as well as the requirements for a successful computation of the emitted gravitational radiation.

Gioel Calabrese: *Spherical excision for moving black holes.* This talk presented a new excision method for moving black holes using multiple overlapping coordinate patches, such that each boundary is fixed with respect to at least one coordinate system. The method was demonstrated by evolving a Klein-Gordon scalar field around a boosted Kerr black hole in axisymmetry.

Manuela Campanelli: *The coalescence remnant of a spinning binary black hole.* Using a combination of numerical and perturbative approximation methods (the Lazarus approach), it is found that for moderately spinning holes the remnant Kerr black hole formed at the end of an inspiral process has a rotation parameter $a/M \approx 0.72 + 0.32(s/m_H)$, suggesting it is difficult (though not excluded) to end up with near maximally rotating holes from such scenarios.

John Carminati: *The Maple packages DESOLV, PETROV, and GHPII.* In this presentation, several aspects of the Maple packages DESOLV, PETROV, and GHPII were discussed. DESOLV finds Lie point symmetries and similarity solutions of partial differential equations. PETROV classifies the Weyl tensor using a new algorithm, and GHPII is an implementation of the Geroch-Held-Penrose formalism.

Dale Choi: *Collision dynamics of boson stars.* The results from a numerical investigation of dynamics in the head-on collision of Boson stars were presented. The Einstein-Klein-Gordon equations were solved in axisymmetry with the Boson stars being represented by the massive complex scalar field. This study indicated the existence of two regimes: a "solitonic" regime where the head-on collision exhibits the solitonic nature of the boson stars and a "merging" regime where the collision of the two Boson stars results in a merged single Boson star.

Peter Diener, on work with Lehner, Neilsen, Tiglio, and Ventrella: *Evolving Brill wave spacetimes using Carpet.* Brill wave spacetimes have been shown in 2D to exhibit critical behaviour. The resolution requirements to show this are so severe that it has not been possible to reproduce these results in 3D. In this talk recent results were presented for the evolution of axi-symmetric Brill wave spacetimes using 3D codes with Fixed Mesh Refinement (FMR). In addition to FMR the cartoon technique was used to further extend the range of achievable resolutions.

Harald Dimmelmaier: *"Mariage des maillages": A new 3D general relativistic hydro code. II. Applications to astrophysical phenomena and outlook.* This talk presented first applications of a new three-dimensional general relativistic hydrodynamics code to highly relativistic stellar configurations. Among these were simulations of rotational stellar core collapse to a neutron star and long-term evolutions of highly perturbed rotating single neutron stars without the assumption of symmetries.

John Friedman: *Helically symmetric models of binary systems.* When a truncated set of Einstein equations is solved to construct initial data, the relation between radius and angular velocity for a circular orbit is not enforced, and the initial data yield elliptical orbits. Work by Friedman and Uryu seeks to address this problem by modeling compact binary systems in general relativity as vacuum or perfect-fluid spacetimes with a helical

Killing vector, heuristically, the generator of time-translations in a corotating frame. A code has been constructed to solve the full Einstein equation for a binary neutron-star system with helical symmetry.

David Garfinkle: *Numerical simulations of generic singularities.* Numerical simulations of the approach to the singularity in vacuum spacetimes were presented. The spacetimes examined have no symmetries and can be regarded as representing the general behavior of singularities. It is found that the singularity is spacelike and that as it is approached, the spacetime dynamics becomes local and oscillatory.

Adrian Gentle: *The constraints as evolution equations for numerical relativity.* A new approach was described to rewriting the constraints as first-order evolution equations, thereby guaranteeing that they are satisfied to within the accuracy of the discretisation scheme. This introduces a set of four subsidiary constraints which are far simpler than the standard constraint equations. Preliminary numerical results were described.

Achamveedu Gopakumar, on work with Schäfer: *A minimal no-radiation approximation to Einstein's field equations.* An approximation scheme to Einstein's field equations in the Arnowitt-Deser-Misner canonical formalism was presented which corresponds to the magneto-hydrodynamics approximation in electrodynamics. The proposed scheme, where stationary spacetimes are exactly reproduced, should be useful to construct realistic initial data for general relativistic simulations as the Hamiltonian, in this approximation, is identical with the non-dissipative part of the Einsteinian one through the third post-Newtonian order.

Eric Gourgoulhon, on work with Bonazzola, Grandclément, and Novak: *Time Evolution of 3+1 Einstein equations via a constrained scheme.* The first results from a new numerical code for evolving 3+1 spacetimes were presented. This code implements a fully constrained scheme, based on maximal slicing and the Dirac gauge. These coordinate conditions result in an elliptic-hyperbolic system, which is solved by means of multi-domain spectral methods. Thanks to the use of spherical coordinates and the Dirac gauge, the hyperbolic part is reduced to two scalar wave equations which represent the two degrees of freedom of the gravitational field.

Carsten Gundlach: *Symmetric hyperbolicity and consistent boundary conditions for second-order Einstein equations.* There has been much interest recently in first-order symmetric hyperbolic formulations of the Einstein equations. Even well-posed systems suffer from instabilities, but there is some evidence that this problem is less severe in second-order formulations (such as BSSN) than in first-order formulations. This has led to the definition of symmetric hyperbolicity for second-order systems of equations and to a proof for two BSSN-like families of formulations of the Einstein equations.

Francisco Guzman: *Shift Conditions for Orbiting Black Holes in Numerical Relativity.* A simple scheme was presented to create a co-rotating coordinate system for orbiting binary systems in numerical relativity. The scheme is able to find a shift vector that brings orbiting binary systems into a co-rotating frame with respect to the coordinate system used to do the computations. Furthermore, it is able to track the orbiting system, adjusting automatically to account for changes in orbital velocity, as will be encountered during binary inspirals.

Sascha Husa: *Towards standard testbeds for numerical relativity: methodology and first results.* In recent years, many different numerical evolution schemes for Einstein's equations have been proposed to address stability and accuracy problems that have plagued the numerical relativity community for decades. Some of these approaches have been tested on different spacetimes, and conclusions have been drawn based on these tests. However, differences in results originate from many sources, including not only formulations of the equations, but also gauges, boundary conditions, numerical methods, etc. This talk was a status report on a program to build up a suite of standardized testbeds for comparing approaches to the numerical evolution of the Einstein equations.

Jose Jaramillo: *Excised black holes with isolated horizon boundary conditions.* In the setting of the numerical construction of black holes, Jaramillo took the Isolated Horizon framework developed by Ashtekar et al. as a guide-line for the explicit construction of a horizon in the appropriate physical regime. This permitted the derivation of definite prescriptions for the values of the lapse, shift and angular components of the three metric on the horizon, to be used as boundary conditions for the corresponding elliptical equations. Numerical results were presented mainly focused on the construction of initial data for black holes by using a conformal thin sandwich approach.

Bernard Kelly: *Black Hole Head-On Collisions Revisited.* Kelly and collaborators re-visited the binary black-hole head-on collision problem using the Penn State 'Maya' code. Evolutions are in full 3D, using the Kerr-Schild-like binary initial data, the BSSN formulation of Einstein's equations, and dynamic singularity excision. The resulting evolutions demonstrate numerical stablity, and are a natural starting point to more realistic and physical grazing collisions and radiation extraction in 3D.

Gabor Kunstatter: *Critical behaviour in Black Hole Formation: from 2d dilaton gravity to infinity.* A recently devised algorithm is used to examine the dependence on spacetime dimension, d, of the critical exponent in black hole formation via the spherically symmetric collapse of a scalar field. It is found that the value of the exponent reaches a maximum at around $d = 11$ spacetime dimensions, and approaches the value 2 asymptotical for large d. The large d limit of the theory is closely connected to CGHS dilaton gravity.

Pablo Laguna: *Numerical Simulations of Extreme-mass-ratio Binaries.* The final inspiral and merger of extreme-mass-ratio binaries lie within the sweet spot of LISA. These binary systems provide a unique opportunity of constructing a map of the space-time of the supermassive black hole and testing the black-hole uniqueness theorems. Results were presented from simulations based on modern formulations of numerical relativity using highly adaptive and parallel finite element methods.

Christiane Lechner: *Towards a numerical implementation of the Friedrich-Nagy system.* Friedrich and Nagy gave a formulation of the Einstein equations that admits a well posed initial boundary value problem for the half-plane. The basic unknowns in this formulation are the components of a tetrad, the connection coefficients and the Weyl tensor. A status report was given for a numerical implementation and testing of this system.

Carlos Lousto: *Radiation content of conformally flat initial data.* This talk discussed the radiation of energy and linear momentum emitted to infinity by the headon collision of binary black holes, starting from rest at a finite initial separation, in the extreme mass ratio limit. For these configurations the radiation produced by the initially conformally flat choice of the three geometry was identified. This identification suggests that the radiated energy and momentum of headon collisions will not be dominated by the details of the initial data for evolution of holes from initial proper separations $L_0 \geq 7M$.

Robert Mann, on work with Burnell, Malecki, Ohta: *Chaos in a Relativistic 3-body Self-Gravitating System* This work considers the problem of three body motion for a relativistic one-dimensional self-gravitating system. Despite the high degree of non-linearity in the relativistic system, it was found that the global structure of its phase space remains qualitatively the same as its non-relativisitic counterpart for all values of η that one was able to study. However the relativistic system has a weaker symmetry and so its Poincare section develops an asymmetric distortion that increases with increasing η. For the post-Newtonian system it was found that it experiences a KAM breakdown for $\eta \simeq 0.26$, above which the near integrable regions degenerate into chaos.

Reinhard Meinel, on work with Ansorg, Fischer, Kleinwächter, Petroff and Schöbel: *Equilibrium Configurations of Homogeneous Fluids in General Relativity.* By means of a highly accurate multi-domain pseudo-spectral method, classes of uniformly rotating, homogeneous and axisymmetric relativistic fluid bodies were investigated. In an appropriate parameter diagram, the well-known Maclaurin sequence of rotating fluid spheroids can be divided into segments, each of which is contained within one of the above mentioned classes as one of its Newtonian limits. One of the classes studied, which contains the recently studied relativistic Dyson rings, permits an extreme Kerr Black Hole limit.

Warner Miller: *Regge Calculus: A Discrete representation of General Relativity.* In Regge calculus, the principles of General Relativity are applied directly to a lattice spacetime geometry. This yields a finite representation of the theory based on the underlying physical principles. This geometric foundation may provide insight into how one can best discretize General Relativity for both quantum and classical calculations. Recent progress on the foundations and numerical applications of Regge Calculus was outlined.

Jerome Novak: *"Mariage des maillages": A new 3D general relativistic hydro code. I. Introduction and theoretical issues* A new code based on spherical polar coordinates was presented, which can be applied to study stellar core collapse and the resulting gravitational radiation, as well as phenomena like oscillations and triaxial instabilities of rotating neutron stars. The focus was on a novel approach of combining finite difference and spectral methods, which we call "Mariage des Maillages" (grid wedding). It results in high accuracy of the metric solver and provides a stable 3D evolution with reduced computational costs.

Roland Oechslin: *Conformally flat smoothed particle hydrodynamics: Application to neutron star mergers.* The influence of numerical and physical uncertainties on binary neutron star merger simulations was considered. Using the conformal flat approximation and working with an SPH hydro code coupled to an elliptic metric solver on an overlaid grid, the following quantities were varied: the numerical resolution, the position of the outer boundaries, the artificial viscosity scheme, the amount of radiation reaction added and the formulation of the hydrodynamics equations.

Harald Pfeiffer: *Optimal constraint projection for hyperbolic evolution systems.* Evolutions of Einstein's equations often show exponentially growing constraint violating modes. One proposed solution of this problem is to re-solve the constraints periodically, thus projecting the numerical solution back into the constraint surface. Here, a projection method was introduced with the important property that it projects to the closest constraint-satisfying solution.

Denis Pollney, on work with Koppitz, Schnetter, and Seidel: *Numerical studies of the merger of black hole binaries from close quasi-circular orbits.* Presented were the results of first simulations of equal-mass binary black hole mergers from initial data produced via the thin-sandwich helical Killing vector method. The runs are followed past merger, as indicated by the appearance of a common apparent horizon. A special feature of these simulations is the construction of a dynamically adjusted co-rotating frame.

Frans Pretorius: *Numerical Relativity Using a Harmonic Decomposition.* Results on experiments using a generalized harmonic decomposition for 3D numerical simulations in asymptotically flat spacetimes were discussed. Topics included specifying a gauge using source functions, compactifying to spatial infinity, a fully second order discretization scheme that does not introduce any additional variables, and black hole excision within this scheme.

Eric Schnetter: *Analysing Horizons in Numerical Relativity.* Methods were presented to analyse the world tubes of apparent horizons in numerical simulations. These are based on the Isolated and Dynamical Horizon formalism introduced by Ashtekar et al. Schnetter calculated spin and total mass for horizons with axial Killing vector fields and gave an expansion of spin and mass in terms of invariant multipole moments. Results were presented from 3+1-dimensional simulations for collapsing stellar cores and coalescing black holes.

Deirdre Shoemaker: *Dynamics of a scalar field on the "frozen" spacetime of a binary black hole system.* Numerical results of the dynamics of a scalar field on the frozen space-time of a binary black hole system in quasi-circular orbit were presented. The goal of this study is to map the emergence of a ringing potential as the binary system coalesces. The moment in the lifetime of a binary system when such potential forms could have dramatic effects on the signatures of the emitted gravitational radiation.

Martin Snajdr: *Accretion of a perfect fluid onto a moving black hole.* This talk discussed the status and preliminary results of numerical simulations of general relativistic accretion of a perfect fluid onto a moving black hole. The code is based on high resolution shock capturing methods. The objective of the work is to develop a fully adaptive 2D/3D parallel code capable of simulating a wide variety of interesting astrophysical scenarios.

Ulrich Sperhake: *3-d evolutions of single and binary black holes using a modified BSSN-formulation.* 3-dimensional evolutions of single and binary black hole spacetimes were presented using the BSSN-formulation with appropriate modifications. Particular emphasis was on the excision techniques used for simulating dynamic black hole spacetimes performed in the framework of fixed mesh refinement. It was demonstrated how these techniques are used to obtain long term stable evolutions for various types of dynamic space times involving single or binary black holes.

Ryoji Takahashi: *Gravitational Radiation Recoil.* Gravitational recoil is expected to occur in supermassive black hole binary collisions, which is of relevance for LISA. In order to understand the dynamics of recoiling black holes, as a first step head-on collisions were studied (with mass ratios up to $m_1/m_2 \approx 0.1$) as well as highly distorted black holes via fully 3D general relativistic codes with fix mesh refinement.

Jonathan Thornburg: *Black Hole Excision with Multiple Grid Patches.* In this presentation a finite differencing scheme was described which uses multiple $\{r \times \text{(angular coordinates)}\}$ grid patches, each patch using a different (nonsingular) choice of angular coordinates (e.g. the "inflated-cube"). This allows excision on a smooth $r =$ constant 2-sphere. Numerical results were presented for the time evolution of the asymptotically flat spacetime around a single excised black hole. The lifetime of these evolutions appears to be limited only by instabilities at the outer boundary, and not by any instabilities due to the multiple-patch scheme.

Cliff Will, on work with Berti, Iyer, and Mora: *A Post-Newtonian diagnostic of quasi-equilibrium binary configurations of compact objects obtained in numerical relativity.* Using equations of motion accurate to the third post-Newtonian (3PN) order ($O(v/c)^6$ beyond Newtonian gravity), expressions have been derived for the total energy E and angular momentum J of the orbits of compact binary systems (black holes or neutron stars) for arbitrary orbital eccentricity, incorporating certain finite-size contributions. It was described how these formulae may be used as an accurate diagnostic of the physical content of quasi-equilibrium configurations of compact binary systems of black holes and neutron stars generated using numerical relativity.

Acknowledgments

It is a pleasure to thank all participants for their effort to make this a successful session. This work was partially supported by NSF grant PHY-02-18750.

A6: ALTERNATIVE THEORIES OF GRAVITY

CLIFFORD M. WILL
Department of Physics, McDonnell Center for the Space Sciences
Washington University, St. Louis MO 63130 USA
e-mail: cmw@wuphys.wustl.edu

The extraordinary success of general relativity empirically, both in the solar system and in binary pulsars, has made it the "standard model" for gravity. Very little research is taken seriously any more that attempts to replace general relativity with some radically different approach, at least at the classical level. On the other hand, there is a significant thread of serious research on possible "extensions" of general relativity, motivated by progress in string theory, braneworlds and quantum gravity, and by cosmological puzzles such as the nature of dark matter and the cause of the acceleration of the universe. The session A6 on Alternative Theories of Gravity featured several interesting contributions in this spirit. A (necessarily subjective) selection of highlights of the session follows.

Éanna Flanagan (Cornell University) spoke on gravitation theories obtained from the Palatini variational principle; this talk was a kind of appetizer for Flanagan's longer Xanthopoulos Prize Lecture, given during the GR Society General Assembly. In an attempt to account for the acceleration of the universe purely within a metric gravitational theory (in other words, without resorting to dark energy), some authors have proposed actions for gravity that include nonlinear functions of the Ricci scalar. Working in the Palatini variational formalism, where the connection and metric are independent, he pointed out that such theories are equivalent to scalar-tensor theories in which the scalar field kinetic energy term is absent from the action. By integrating out the scalar field, the theory can be recast as general relativity coupled to a modified matter action. He demonstrated how this rules out some models that have been proposed of the Universe's recent acceleration. Loop corrections make the situation with these theories even worse, unless fine tuning is invoked. A more general class of theories, stable under loop corrections, can be formulated by taking the Lagrangian to be some function of (i) the Ricci scalar computed from the metric, and (ii) a second Ricci scalar computed from the connection. Flanagan showed that such theories can be recast as tensor-biscalar theories, and that some of them can be made to be compatible with solar system experiments. [1, 2]

In a similar vein, Sante Carloni, working with Peter Dunsby (Cape Town) described a detailed analysis of the dynamics of cosmological models based on R^n-gravity. [3] Writing the cosmological equations as an autonomous system of first order equations and using standard results of dynamical system theory, he studied families of exact solutions and analyzed their stability, both for vacuum and for non-minimally coupled matter, as a function of the power parameter n.

Richard Woodard (Florida) reviewed attempts to formulate viable theories of gravity that would embody the non-Newtonian dynamics of the set of ideas known as MOND, for MOdified Newtonian Dynamics, that have been proposed as a way to explain rotation curves of galaxies and other large scale phenomena in the universe, without having to invoke dark matter. By considering metric theories of gravity in the weak-field limit, with a metric approximated by $g_{\mu\nu} = \eta_{\mu\nu} + h_{\mu\nu}$, and a set of field equations given by $E_{\mu\nu} = 8\pi T_{\mu\nu}$, he concluded that in order to get MOND behavior, then at least one component of the function $E_{\mu\nu}$ has to have its leading order in a power series in h to be quadratic. Finding no way to accomplish this, he concluded that there exists no purely metric, covariant, minimally coupled theory that gives rise to correct MOND behavior. [4] He noted recent work by Bekenstein that gets around this by adding scalar and vector fields. [5]

Braneworlds have been the focus of much recent attention as a means to extend general relativity into extra dimensions, while keeping the non-gravitational interactions confined to a four dimensional "brane" embedded in the higher dimensional space (for a review of aspects of this topic, see Ref. [6]). Mariam Bouhmadi-López working with David Wands (Portsmouth) investigated the non-minimal coupling to the Ricci curvature of a scalar field on a four-dimensional brane and the effect this may have on the scalar field dynamics and cosmological evolution. She deduced the gravitational equations of a 4D brane with an induced gravity term coupled to a scalar field, in a 5D bulk with a cosmological constant. She then discussed the form of the equations in the 4-dimensional or 5-dimensional Einstein frames. [7]

Also on the topic of extra dimension theories, Lorenzo Fatibene and colleagues (Torino) discussed Chern-Simons theories and Lovelock gravity as a particular case of the general framework of gauge natural theories.

Higher curvature interaction terms arise in the study of back-reactions of quantum field energy as well as in the low-energy limit of string theory. Starting from the SO(2,2n) Chern-Simons form in (2n+1) dimensions they calculated the variation of conserved quantities in Lovelock gravity and Lovelock-Maxwell gravity in a fully covariant formalism. The definition of the variation of charges, following from the covariance of the theory, reproduces the formal expression for the entropy of Lovelock black holes. Moreover, it applies to the definition of the conserved quantities and to the definition of a first principle of black-hole thermodynamics in Gauss-Bonnet Gravity [8].

Extending both special and general relativity in ways that violate Local Lorentz Invariance has been a particularly active subject recently, motivated in part by quantum gravity, and also as a way to provide "test theories" to underpin experimental investigations. Eduardo J. S. Villaseñor with J. Fernando Barbero G. (U. Carlos III, Madrid) discussed a class of gravitational theories that introduce dynamical vector fields as a way to produce effective violations of LLI in gravitational physics. Specifically they showed that the families of effective actions considered by Jacobson et al. [9, 10] to study Lorentz invariance violations contain a class of models that represent pure general relativity with Euclidean signature. [11] They also pointed out that some members of this family of actions preserve Lorentz invariance in a generalized sense.

Two talks dealt with the subject of teleparallel gravity, gauge theories of gravity in which the fundamental tetrads have vanishing covariant derivative. Anna Barbosa, with V. C. Andrade and J. G. Pereira (São Paulo) reported on efforts to derive equations of motion in non-Abelian teleparallel theories using a duality prescription. Yuri Obukhov (Moscow and Cologne) discussed plane wave solutions in related theories of metric-affine gravity.

Two talks were concerned with observational issues. Sergei Kopeikin (Missouri) discussed efforts to extend the post-Newtonian theory of astronomical reference frames in general relativity (adopted officially by the 2000 General Assembly of the International Astronomical Union [12]) to the more general class of scalar-tensor theories. [13] Ian Jones discussed how measurement of the relative phases of arrival of gravitational wave harmonics emitted by eccentric binaries involving a small mass orbiting a supermassive black hole can be used to place a bound on the rest-mass of the graviton, thereby extending work by Will [14] and Cutler et al. [15] that assumed circular orbits.

The session was rounded out with talks by Jan Broekaert on a scalar gravity theory, Bertrand Chauvineau on the large ω limit of Brans-Dicke theory, Valdir Bezarra on gravitomagnetism in scalar-tensor theories, and Jonathan Oppenheim on gravity as thermodynamics. Finally, contributors who were assigned poster presentations were each given a one-minute, one-viewgraph opportunity to promote their posters.

References

1. É. É. Flanagan, Phys. Rev. Lett. **92**, 071101 (2004) (gr-qc/0308111)
2. É. É. Flanagan, Class. Quantum Gravit. **21**, 417 (2003) (gr-qc/0309015)
3. S. Capozziello, S. Carloni, A. Troisi, Recent Research Developments in Astronomy & Astrophysics"-RSP/AA/21-2003 (astro-ph/0303041)
4. M. E. Soussa and R. P. Woodard, Phys. Lett. **B578**, 253 (2004) (astro-ph/0307358)
5. J. D. Bekenstein, Phys. Rev. D, **70** 085039 (2004); erratum **71** 069901 (2005).
6. R. Maartens, Living Rev. Relativity **7**, (2004) 7, [Online article]: cited on 1 Oct. 2004, http://www.livingreviews.org/lrr-2004-7
7. M. Bouhmadi-López and D. Wands, Phys. Rev. D **71** 024010 (2005).
8. G. Allemandi, M. Francaviglia, M. Raiteri, Class. Quantum Gravit. **20**, 5103 (2003) (gr-qc/0308019)
9. T. Jacobson and D. Mattingly, Phys. Rev. D **64**, 024028 (2001) (gr-qc/0007031)
10. C. Eling and T. Jacobson, Phys. Rev. D **69**, 064005 (2004) (gr-qc/0310044)
11. J. F. Barbero G. and E. J. S. Villaseñor, Phys. Rev. D **68**, 087501 (2003) (gr-qc/0307066)
12. M. Soffel, et al. Astron. J. **126**, 2687 (2003)
13. S. Kopeikin and I. Vlasov, Phys. Reports **400** 209-318 (2004).
14. C. M. Will, Phys. Rev. D **57**, 2061 (1998) (gr-qc/9709011)
15. C. Cutler, W. A. Hiscock and S. L. Larson, Phys. Rev. D **67**, 024015 (2003) (gr-qc/0209101)

B1: RELATIVISTIC ASTROPHYSICS

LUIGI STELLA
Osservatorio Astronomico di Roma,
Via Frascati 33,
00040 Monteporzio Catone (Roma), Italy

Session B1 hosted a number of talks on recent studies and results that provide new exciting perspectives in the field of Relativistic Astrophysics and Cosmology. A brief summary of the session is presented here.

The discovery of the double pulsar relativistic binary system PSR J0737-3039 is certainly amongst the most important in recent years. This binary system consists of a millisecond pulsar with a spin period of 22.7 ms orbiting another radio pulsar with a spin period of $\sim 2.8\ s$. Its 2.4 hr orbital period, the shortest known in any double neutron star systems, will enable very accurate testing of theories of gravity and provide important new constraints on the physics of ultradense matter. Moreover, as a results of the discovery of such a binary, the estimated rate of merging of double neutron star systems has increased by a factor of ~ 10, therefore opening a new perspective in the detection of gravitational wave signals with experiments of the present generation. **Possenti** gave an account of the remarkable results on this unique system a year after its discovery. The very high inclination of PSR J0737-3039 ($i \geq 89\ deg$) appears to be especially promising also for investigating the interaction between the beam of one pulsar and the other pulsar's magnetosphere, thus shedding light on various aspects of the physics of radio pulsar beams, winds and magnetospheres.

Barret reviewed the very short term variability properties of accreting fast spinning neutron star and black holes in X-ray binary systems. He then surveyed the prospects of using fast X-ray timing to probe strong gravity fields around collapsed objects and constrain the equation of state of dense matter in neutron stars. A common and intriguing phenomenon is the presence of fast Quasi Periodic Oscillations, QPO, which reach frequencies as high as ~ 450 and $\sim 1300\ Hz$ in accreting black hole and neutron star binaries, respectively. Several QPOs at different frequencies are active at the same time in many sources. Their frequency changes show distinctive behaviours that are well reproduced across different source with vastly different properties. QPO models differ with respect to their detailed predictions, but all envisage that these signals originate from matter in the close vicinity and their frequencies are related to the fundamental frequencies of motion, close to the marginally stable orbit. The potential of these signals for "in situ" measurements of strong field gravity effects is currently humpered by the limited throughput of the instrumentation, which permits to study QPO only in the frequency domain and by Fourier transforming data interval containing hundreds of QPO "trains" (*i.e.* the time span over which QPOs remain nearly coherent). A factor of ~ 10 increase in the effective area of the instrumentation for X-ray timing (i.e. $\sim 10 - 20\ m^2$) would afford the study of the signal of individual QPO "trains" in the time domain, thus yielding crucial information on the strong gravity fields in the close vicinity of black holes and neutron stars. The X-ray Evolving Universe Spectroscopy mission (XEUS), currently discussed by the European Space Agency, combined with a small dedicated fast X-ray timing detector in the focal plane (XTRA: XEUS Timing for Relativistic Astrophysics) will afford the timing sensitivity required for these studies.

Some 15 accreting neutron stars in X-ray binary systems were discovered in recent years to possess spin periods in the $1.5 - 4\ ms$ range. As a result of the strong gravitational field of the star and its fast rotation, any radiation emitted from their surface must be redshifted by gravitational and transverse Doppler effects as well as relativistically Doppler-shifted. **Morsink** presented detailed calculations of the spectral line profiles observed at infinity for such neutron stars. These calculations include all relativistic effects, red/blue-shifting, gravitational lensing, frame-dragging as well as photon propagation in the space-time of the rotating neutron star. The line profiles are, in general, much broadened and shifted by the combination of all these effects. A comparison with the line profiles recently revealed in the X-ray binary EXO 0748-676, yields constraints on the neutron star spin.

The ejection of relativistic jets of plasma from accreting compact objects was long believed to be characteristic of black holes. The extremely powerful jets emanating from massive black holes ($\sim 10^6 - 10^9\ M_\odot$) in the nuclei of galaxies are a common phenomenon. The long timescales involved prevent any detailed study of the kinematics and evolution of the plasma of these jets. Indeed variability information has been obtained only in few cases, those in which the jets point along our line of sight. The so-called microquasars, accreting stellar mass black holes orbiting a mass donor companion of which the unique system SS433 is, at the same time, the prototype and the exception, have opened a new perspective in the study of relativistic jets. Indeed, as result of their much smaller scale, ejection episodes can be followed individually on timescales from hours to weeks, and their properties related to other properties of the accreting black holes. It was only recently that a number of accreting neutron star systems in X-ray binaries were indisputably shown to emit highly variable relativistic jets, which hold clear analogies with the jets of microquasars. **Fomalont** presented a detailed account of the very high resolution (VLBI) radio observations of the jets of Sco X-1, the brightest persistent X-ray source in the sky, that contains an old, presumably fast spinning neutron star. The flux evolution and expansion of the radio lobes of Sco X-1 was monitored for several years, yielding an average speed of $\sim 0.45\ c$ and a $\sim 45\ deg$ angle to our line of sight. The lifetime of lobe pairs is limited to a few days and the ejection velocity changes from one pair to another (range of $\sim 0.3 - 0.6\ c$). An extensive multiwavelength observational campaign from soft gamma ray to radio frequencies was carried out recently, in order to shed some light on the interplay of high energy (X and gamma ray) variability and the ejection of radio lobes.

The gravitational nano-lensing taking place in binary systems consisting of a stellar mass black hole and a radio pulsar was the subject of the presentation by **Hobill**. It was shown that for black holes with masses $> 20 M_{\odot}$ the expected orbital-phase dependent changes in the pulse arrival times, waveforms and intensity, as well as the distribution of angular image positions (on the order of milli-arcseconds, *i.e.* nano-radians), can provide unique signatures of this type of relativistic binary systems, and thus lead to their identification.

The initial mass function, IMF, describes the initial size distribution of stellar object and is a key ingredient for a variety of astrophysical themes, ranging from stellar evolution to the history of metal production in the universe. **Yunes** described some recent arguments aimed at constraining the shape of the IMF for very small values ($<< 1/10\ M_{\odot}$) of the mass. The result is that the IMF is expected to have a power law shape also at its low mass end, with and index of opposite sign to the high-mass index.

De Felice presented some new results in the study of the motion of magnetised particles in the close vicinity of a Kerr black hole. It is found that the spacetime curvature together with the electromagnetic field give rise to stable circular orbits inside the ergosphere. Particles on these orbits are characterised by a negative total energy and angular momentum. By extension, it is deduced that a belt of magnetised matter in such orbits inside the ergosphere of a stellar mass black hole can possess a binding energy as high as $\sim 10^{54}$ ergs, of potential interest for astrophysical applications, gamma ray bursts in particular.

The gravitational radiation rocket effect arises from the linear momentum carried away by gravitational waves, which imparts a kick to the radiating system. **Favata** presented the results obtained for two coalescing black holes in the limit of a very high mass ratio and by using perturbation theory. It is found that relativistic effects reduce the values of the kick velocities, compared to Newtonian estimates. In most cases values in the $\sim 10 - 100\ km/s$ are expected, whereas the highest velocities are $< 500\ km/s$. These values indicate that coalescing black holes systems are unlikely to be propelled away from galaxies as a result of the the radiation rocket effect.

Several presentations addressed a variety of aspects of the physics of the gravitational collapse of massive star cores, one of the most promising sources of high frequency gravitational radiation that instrumentation of the present generation should be able to detect. Issues related to the collapse of rotating stars received a great deal of attention. **Ott** reviewed the considerable progress made in recent years toward reliable estimates of the gravitational wave signal produced by core collapse events. He presented also new results from a set of fully general-relativistic three-dimensional simulations of rotating core collapse.

The gravitational waveforms produced when a rotating neutron star collapses to a Kerr black hole were the subject of the talks by **Ott** and **Baiotti**. The neutron star is modelled as an initially uniformly rotating polytrope. The dynamics of the matter, but also that of the trapped surfaces, of both the apparent and event horizons formed during the collapse are studied carefully, yielding accurate measurements of the black hole mass and spin, and the energy lost to gravitational radiation.

Saijo presented the results for the rotational core collapse of a rapidly rotating relativistic star by means of a 3+1 hydrodynamical simulations in conformally flat spacetime of general relativity. By evolving a differentially rotating equilibrium star of $T/W \sim 0.15$ towards core collapse, the threshold of dynamical bar instability is exceeded, The collapse of a spheroidal star (and perhaps also that of a toroidal star) leads to the formation of a bar. The strength of the dynamical bar instabiliy and, thus its associated gravitational wave signal, are found to depend on the equation of state of dense matter.

Another group of talks dealt with rotating neutron star models, especially from the point of view of non-axisymmetric oscillatory modes and production of intense gravitational radiation signals. R-modes in fast rotating neutron stars are unstable to the emission of gravitational radiation and might give rise to detectable signals. **Brink** addressed one of the crucial issues in R-mode research, namely the saturation amplitude of these modes. Detailed calculations are carried out for the two proposed mechanism for nonlinear mode saturation. The result is that the r-mode saturates at small amplitudes, $< 10^{-4}$.

Non-axisymmetric oscillation modes of non-Keplerian tori around a Schwarzschild black hole were the subject of **Yoshida**'s talk. As a result of their angular velocity distribution, the tori have finite extents and the trapped oscillations are possible. Using an incompressible torus model (the vertical structure of which is neglected), the spectrum of the discrete and continuum trapped oscillations is studied.

Watts discussed two phenomena that are specific to fast differentially rotating neutron stars, dynamical shear instabilities, and the existence of a co-rotation band, a frequency band in which the mode pattern speed matches the local angular velocity. Based on a simple model, dynamical shear instabilities are found that arise where modes cross into the co-rotation band, if the degree of differential rotation exceeds a certain threshold.

Magnetospheric Alfven waves are likely the origin of the bursts and flares of magnetars, the neutron stars with extremely high magnetic fields ($B \sim 10^{14} - 10^{15}\ G$) that are found in Soft Gamma Repeaters and Anomalous X-Ray Pulsars. **Kojima** studied in detail the propagation and transmission of these waves through the star's crust to the magnetosphere.

Haskell discussed the possibility that fast spinning accreting neutron stars in low mass X-ray binaries are conspicuous sources of gravitational radiation. Mechanisms such as accretion induced asymmetries, toroidal magnetic fields or unstable r-modes are considered which give rise to different predictions with respect to the spin-up/spin-down behaviour of these sources.

Hosoya discussed the effective evolution of an inhomogeneous cosmological model, as described in terms of spatially averaged variables. The time–evolution of 'effective information' was presented together with the conjecture that the information content of the Universe, measured by relative information entropy of a cosmological model containing dust matter, is increasing.

Clarkson reported on an investigation of the generation of electromagnetic radiation by gravitational waves interacting with a strong magnetic field in the vicinity of a vibrating Schwarzschild black hole, an effect that may play a role in gamma-ray bursts and supernovae. It is found that the black hole ringdown process can produce substantial amounts of electromagnetic radiation from a dipolar magnetic field in the vicinity of the photon sphere.

Singh examined the dynamical behaviour of an uncharged classical spinning particle in orbit around a star undergoing gravitational collapse, as governed by the Dixon equations of motion in Vaidya space-time. By assuming various central mass functions, the characteristics of the trajectories and gravitational wave output generated by the Dixon particle are evaluated.

Dunsby presented the exact 1+3 covariant dynamical fluid equations for a multi-component plasma, together with Maxwell's equations, in such a way as to make them suitable for a gauge-invariant analysis of linear density and velocity perturbations of the Friedmann-Robertson-Walker model. For a two component plasma where thermal effects are neglected, a new mode representing high-frequency plasma oscillations is found. Moreover, velocity perturbations can lead to the generation of cosmic magnetic fields of sufficient strength to be plausible seed field candidates.

Scott described GRworkbench, a visual, numerical software tool developed at the Australian National University for exploring analytical spacetimes in General Relativity. The claimed possibility of measuring the mass of the Milky Way by using a small interferometer on the surface of the Earth was investigated and found to arise from an artefact of the approximations used in the analysis. A precise estimate of the effect was obtained by using GRworkbench, without making any simplifying assumptions.

Berti presented a comparison of three different models for the spacetime of a rotating neutron star: the Hartle-Thorne slow-rotation approximation, the exact analytic vacuum solution of Manko *et al.*, and an accurate numerical solution of the full Einstein equations. The Hartle-Thorne equations are integrated for five different equations of state, keeping terms up to second order in the stellar angular velocity. These models are then matched to numerical solutions of the Einstein equations with the identical values of the gravitational mass and angular momentum. Concerning the exterior vacuum spacetimes, the Hartle-Thorne approximation and the Manko analytic solution are compared to the numerical models using Newman-Penrose coordinate-independent quantities. The main conclusion is that the Hartle-Thorne approximation is very reliable for most astrophysical applications.

Chaisi presented new solutions to the anisotropic Einstein field equations that are important in the study of very dense stellar structures. Two new algorithms are generated, which produce a new anisotropic solution to the Einstein field equations from a given isotropic solution. For any new anisotropic solution generated with the help of these algorithms, the original isotropic seed solution is regained as a special case. Two examples of known isotropic solutions are used as demonstrations of the way the algorithms work.

B.2(I): COSMOLOGY (I). EARLY UNIVERSE

MISAO SASAKI
Yukawa Institute for Theoretical Physics
Kyoto University
Kyoto 606-8502, Japan

1. Introduction

There were about 40 contributions to Session B.2(i), Cosmology (i) Early Universe. Topics discussed by a dominant fraction of the contributions were either braneworld cosmology or inflation. For completeness, I first list authors and titles in accordance with the program of the session. Then, I summarize the contributions by classifying them to those on the braneworld, inflation and others, respectively.

Oral presentations:

Ruth Lazkoz	Anisotropy and inflation in Bianchi I brane worlds
Peter Dunsby	Is the braneworld born isotropic?
David Matravers	Comments on a Braneworld Exact Solution
Maria Rodriguez-Martinez	Defining perturbations on submanifolds
Tsutomu Kobayashi	Bulk Inflaton Shadows of Vacuum Gravity
Kiyotomo Ichiki	Causal Structure and Gravitational Waves in Brane World Cosmology
Tetsuya Shiromizu	Spacetime dynamics and baryogenesis on the brane
Robert Mann	Search for a Fingerprint of Planck-Scale Physics in the CMB
Oriol Pujolas	Stress tensor in the inflationary brane world
Takashi Torii	Brane cosmology with curvature corrections
Masato Minamitsuji	Effective Gravity on the de Sitter-Einstein Gauss-Bonnet brane
Anthony Aguirre	The past and future of eternal inflation
David Wands	Adiabatic and non-adiabatic perturbations from inflation
Hsien-chung Kao	Large N Cosmology
Richard Woodard	Exact Mode Functions for Single Scalar Cosmology
Bojan Losic	Backreactions in de Sitter spaces
Paulo Vargas Moniz	FRW quantum cosmology with a generalized Chaplygin gas
Tomislav Prokopec	Photon mass in inflation and nearly minimal magnetogenesis
Dmitri Grigoriev	Late Reheating, Quark-Gluon Plasma and Baryogenesis
Ken-ji Hamada	Do we see quantized gravity in CMB anisotropies?
Yasusada Nambu	Evolution of Non-linear Fluctuations in Preheating after Inflation
Tomohiro Harada	The evolution of primordial black holes with with a massless scalar field and the implications for gravitational memory
Ilia Musco	Hydrodynamical simulations of primordial black hole formation

Poster presentations:

Tiago Charters	Preheating for field interactions of a general form
Sri Sriranjan Banerji	The Model of a Spherical Void Collapsing in the Universe
Takashi Hiramatsu	Evolution of Gravitational Waves from Inflationary Brane-world
David Hobill	Cosmic Microwave Background Radiation anisotropiesin Bianchi-IX cosmologies
Giovanni Imponente	Inhomogeneous de Sitter Solution with Scalar Field and Perturbations Spectrum
Hideki Ishihara	Gravitational Waves in Brane Cosmologies
Piret Kuusk	A cosmological model for a two-brane world
Hideki Maeda	Kink instability and stabilization of the Friedmann universe
Abd El Fady Morcos	Formation of condensations in some cosmological models
David Mulryne	Inflationary Cosmology and Quantization Ambiguities in Semi-Classical Loop Quantum Gravity
Teruaki Suyama	Primordial Black Holes production during preheating
Izumi Tanaka	Quantum radiation by a moving brane
ChulMoon Yoo	The condition for higher-dimensional black hole formation
Tetsuyuki Yukawa	Anisotropies of the cosmic background curvature from the simplectic quantum gravity

2. Braneworld Cosmology

Braneworld cosmology has become one of the most topical themes in the GR community for the past few years. As a simple realization of braneworld, the model proposed by Randall and Sundrum (RS) [1] has attracted much attention because of its interesting feature that gravity is localized on the brane not because of compactification but by warping of the extra dimension. This model is a solution of the 5-dimensional Einstein equations with a negative cosmological constant, where a Minkowski brane is embedded in the 5-dimensional anti-de Sitter (AdS) bulk. Subsequently, cosmology based on this RS type braneworld scenario has been extensively studied.

Lazkoz and Matravers discussed spatially anisotropic models of the braneworld. Dunsby pointed out that the braneworld tends to be born more isotropic than in the conventional 4-dimensional Einstein gravity, hence it may explain the observed large scale isotropy of the universe.

Cosmological perturbations in the braneworld have been one of the biggest issues to be clarified. Rodriguez-Martinez presented a reformulation of the cosmological perturbations in the braneworld. Using a compactification of extra dimensions, Kobayashi gave an interesting method to obtain exact solutions of the cosmological perturbation. Ishihara presented an analysis of gravitational wave modes in the bulk. Ichiki and Hiramatsu presented numerical computations of the evolution of gravitational waves in the braneworld.

Corrections to or modifications of Einstein gravity in the 5 dimensional bulk have been also discussed. Kuusk discussed a particular two-brane cosmological model. Torii and Minamitsuji reported the case of Einstein Gauss-Bonnet gravity in 5 dimensions, which exhibits an interesting short distance behavior.

Pujolas pointed out that the backreaction of the quantum effect in the bulk may be crucially important, especially near the branes. Tanaka discussed the quantum radiation by a moving brane.

Related to observational signatures, Shiromizu pointed out an interesting possibility of baryogenesis in the braneworld, which may be called the 'geometrical baryogenesis'.

3. Inflation and CMB

Inflation has been almost a unique theoretical model that can explain the present large scale structure of the universe. This was strengthened by the WMAP first year data that appear to be perfectly consistent with predictions of the inflationary cosmology [2]. Yet, the WMAP data in details also suggest possible deviations of the primordial spectrum from a simple scale-invariant one that may not be explained easily within the standard slow-roll inflation.

Wands discussed general predictions of inflation on the primodial perturbation spectrum including isocurvature perturbations. Hamada and Mann discussed possible signatures of Planck-scale physics in the CMB. Kao discussed an $SO(N)$ symmetric model. Woodard presented some elaborations of the standard inflationary theory. Imponente presented an inhomogeneous solution on large scales. Losic discussed possible importance of quantum backreaction to the dynamics of inflation. Prokopec presented a scenario of magnetogenesis during inflation. Auguirre discussed causal structure of eternal inflation.

Preheating of inflation has been also a matter of interest because of possible non-linear effects associated with the parametric particle production. Nambu discussed possible appearance of non-linear fluctuations at preheating. Related to this, Suyama discussed black hole formation during preheating. Charters discussed general properties of preheating. Grigoriev presented a scenario of low temperature reheating to explain baryogenesis.

4. Other Issues

There are of course many other issues in the early universe cosmology. In our session, in addition to those summarized in the previous sections, there were presentations on quantum cosmology, primordial black holes, topological defects and nonlinear cosmological structures.

Mulryne discussed creation of an inflationary universe in the loop quantum gravity. Yukawa discussed CMB anisotropies in the simplectic quantum gravity. Harada discussed gravitational memory of a Brans-Dicke-like charge of primordial black holes. Musco presented a hydrodynamical simulation of primordial black hole formation. Yoo discussed a hyper-hoop conjecture on black hole formation in higher dimensions.

Maeda investigated stability of a kink mode in scalar field dominated universes. Moniz discussed cosmologies with Chapligin gas. Hobill considered CMB anisotropies in an anisotropic universe. Morcos discussed formation of condansations. Banerji presented a model for a collapsing spherical void.

5. Conclusion

Thanks to active participation of the audience at the session B.2(i), the workshop was very successful. All the speakers were very cooperative and spared us enough time for discussions.

We learned a lot: There are issues that have been resolved, but there are many more remaining issues that need to be resolved.

In particular, if we take the reality of braneworld or higher dimensions seriously, we definitely are in need of specific, testable observational or experimental signatures of extra-dimensions or Planckian physics, perhaps in CMB anisotropies and polarizations or in high energy collider experiments.

At the same time, of course we need more quantitative pieces of evidence for inflation. However, there still seems to be some purely theoretical issues in the inflationary cosmology. For example, controversies on the effect of quantum backreaction need to be settled in order to give quantitatively unambiguous predictions.

It is hoped that we will hear encouraging reports on these issues at the next GR meeting.

Acknowledgments

I would like to thank the organizers of GR17 for warm hospitality given to us during the conference. This work was supported in part by Monbukagakusho Grant-in-Aid for Scientific Research No. 14047214.

References

1. L. Randall and R. Sundrum, Phys. Rev. Lett. **83**, 4690 (1999) [arXiv:hep-th/9906064].
2. H. V. Peiris *et al.*, Astrophys. J. Suppl. **148**, 213 (2003) [arXiv:astro-ph/0302225].

B2(II): COSMOLOGY (II). LATE UNIVERSE AND GRAVITATIONAL LENSES

DANIEL E. HOLZ

Theoretical Division, Los Alamos National Laboratory
Los Alamos, NM 87544 USA;
Department of Astronomy and Astrophysics
University of Chicago
Chicago, IL 60637 USA

The "Late Universe and Gravitational Lenses" session featured a broad spectrum of talks, with topics ranging from the cosmic microwave background to the density profiles of dark matter halos, and from perturbation theory to cosmic topology. The common thread through all of the talks was the way general relativity impacts our current understanding of cosmology. As GR forms the underpinning of our cosmological standard model, seemingly abstract and abstruse relativity questions can often yield important cosmological insights.

1. Jiri Bicak: Cosmological Perturbation Theory

The session began with a talk by Jiri Bicak, on cosmological perturbation theory and the effect of toroidal motions beyond the cosmological horizon on the local inertial frame. This was based on work by the speaker, in collaboration with Donald Lynden-Bell and Joseph Katz. An abstract of their work:

We study linear perturbations of Friedmann–Robertson–Walker (FRW) Universes with arbitrary curvature and cosmological constant in a general gauge without decomposition into harmonics. Desirable gauges are selected as those which best embody Mach's principle: in these gauges local inertial frames can be determined instantaneously via the perturbed Einstein field equations from the distributions of energy and momentum in the Universe. In closed spherical Universes, integral gauge conditions can be imposed to eliminate motions generated by the conformal Killing vectors. The meaning of Traschen's integral constraint vectors is elucidated. For all three types of FRW Universes the Machian gauges admit much less residual freedom than the synchronous or harmonic gauge.

The general formalism has been used recently to study toroidal motions of the cosmic fluid. In Bicak, Lynden-Bell, & Katz (2004) [2] we give explicit expressions for the axisymmetric metric perturbations of the FRW Universes caused by such motions. The perturbations are decomposed in vector spherical harmonics on 2-spheres, but the radial dependence is left general. Solutions for general odd-parity l-pole perturbations are found for either prescribed angular velocities or angular momenta. In particular, in the case of closed Universes the solutions require a special treatment of the Legendre equation.

In Bicak, Lynden-Bell, & Katz (2004) [1] we show that if perturbations beyond the horizon have the velocities prescribed everywhere then the dragging of inertial frames near the origin is suppressed by an exponential factor. However if perturbations are prescribed in terms of their angular momenta there is no such suppression. We resolve this paradox and in doing so give new explicit results for the dragging of inertial frames in closed, flat and open FRW Universes with and without a cosmological constant.

2. Sergei Kopeikin: Quadratic Cosmological Perturbations

The next talk was also on the subject of cosmological perturbations. It was by Sergei Kopeikin, who stood in for his collaborator Alexander Petrov (who sadly was unable to make it to the conference). He discussed a decoupled system of equations for quadratic cosmological perturbations on an FRW background. The abstract of their work:

The standard treatment of cosmological perturbations, beginning from the work by Lifshitz, [3] is based on the gauge-invariant decomposition of the metric perturbations $h_{\mu\nu}$ in scalar, vector, and tensor parts which can be analyzed independently in the Fourier domain. In some recent papers [4, 5] a new gauge-invariant approach for description of linear perturbations on FRW background was presented. It allows one to decouple the system of differential equations for cosmological perturbations for each of the metric tensor components of $h_{\mu\nu}$. A new gauge was introduced that allowed to solve the equations for linear cosmological perturbations in terms of the standard retarded potentials without performing a Fourier transform.

The present work generalizes our previous research and includes a covariant analysis of quadratic cosmological perturbations around a spatially flat FRW Universe. We derive a system of differential equations for quadratic perturbations and prove, by making use of the general relativistic technique of exact gauge transformations, [6] that the equations are gauge-invariant. Extending results of earlier papers [4,5] we find a gauge in which the quadratic system of cosmological perturbations is also decoupled, and analyze its residual gauge freedom. Results presented in this paper may find useful applications in the analysis of cosmological gravitational waves—their generation by isolated astronomical systems like supermassive black holes, and propagation of the waves across the Universe which is most likely spatially-flat.

3. Daniel Holz: Gravitational lensing of standard candles

The following talk was given by Daniel Holz (the session chair, and the present author). This was the third talk on cosmological perturbations, though with a different emphasis: I explored how perturbations to a perfect FRW Universe cause gravitational lensing effects on distant standard candles. The talk was based on a number of papers, with collaborators Bob Wald, Xuelei Chen, Neal Dalal, Josh Frieman, and most recently, Eric Linder. The abstract:

As is well known, the Universe is not described by the Friedmann–Robertson–Walker (FRW) metric of general relativity. Although this metric works extraordinarily well on the largest scales, it makes the critical assumption that the Universe is homogeneous and isotropic. The Universe we live in is exceedingly lumpy: galaxies, stars, and planets all represent tremendous overdensities of matter, compared to the background average density of the Universe. The gravity from these overdensities causes gravitational lensing: light from distant sources deviates from the pure FRW null geodesics. Robert Wald and I have developed a formalism to calculate lensing statistics in inhomogeneous Universes, based upon a careful analysis of the assumptions needed for a physically reasonable model of a globally FRW, but locally grossly inhomogeneous, Universe. [7] The important qualitative features of the magnification distributions can be summarized: Any source in an inhomogeneous Universe is seen significantly brightened by a few "observers", while the remaining vast majority of observers see the source slightly dimmed (to conserve photon flux).

Because the distance-redshift relation is our most direct and promising probe of the dark energy, there has been an explosion of observational interest in high-redshift supernovae as standard candles (e.g. ACS/GOODS, SNAP/JDEM, LSST). The "noise" due to lensing lies just beneath the current estimates of supernova intrinsic noise. As our understanding of the supernovae improves (from ground-based, nearby surveys, as well as from future high-redshift data), gravitational lensing could well dominate the error budget. In addition, because the lensing is emphatically non-Gaussian, without a careful and thorough understanding of the effects, the primary analysis and results from these missions could be compromised. [8]

In a very recent paper with Eric Linder, [9] we investigate the degree to which gravitational lensing degrades the cosmological measures that underpin a SNAP-like mission, and thereby estimate how many "extra" supernovae are needed to compensate for the lensing. For sufficient numbers of sources observed at high redshift, the average brightness will tend towards the true, unlensed, brightness. For example, the "degradation factor" due to lensing is a factor of three reduction in the effective number of usable supernovae at $z = 1.5$ (for sources with intrinsic flux dispersion of 10%). Gravitational lensing is a major source of noise for high-redshift standard candle measurements, and will need to be thoroughly understood for future missions to be successful.

4. Noriyuki Kogo: Reconstructing the Primordial Spectrum

We shifted gears in the following talk, jumping to the early Universe. Noriyuki Kogo discussed a reconstruction of the primordial power spectrum at the time of decoupling, based on his work with Makoto Matsumiya, Misao Sasaki, and Jun'ichi Yokoyama. This was done by using a "Cosmic Inversion Method" on the recent data from the Wilkinson Microwave Anisotropy Probe. His abstract:

We reconstruct the primordial power spectrum of curvature perturbation, $P(k)$, from the angular power spectrum of the CMB temperature fluctuations, C_ℓ, obtained by the Wilkinson Microwave Anisotropy Probe (WMAP). [10] We use the cosmic inversion method proposed by Matsumiya et al. [11, 12] In contrast to conventional parameter-fitting methods, our method can reproduce and investigate small features in $P(k)$, with the resolution up to $\Delta\ell \simeq 5$. Although we obtain a complicated oscillatory $P(k)$, similar oscillatory features are found in simulations using artificial CMB data with the same observational errors as WMAP assuming a scale-invariant $P(k)$. Therefore, we investigate the statistical significance of nontrivial features by quantifying the deviations from the scale-invariance of the reconstructed $P(k)$. As a result, we find that there are possible deviations from scale-invariance around $k \simeq 1.5 \times 10^{-2}\text{Mpc}^{-1}$ and $k \simeq 2.6 \times 10^{-2}\text{Mpc}^{-1}$. We also discuss the extension of this method to the

CMB polarization, which is currently under study.

5. David Hobill: CMB anisotropies in Bianchi-IX Cosmologies

The following talk was also concerned with the Cosmic Microwave Background. David Hobill discussed his work with Wanwen Guo on CMB anisotropies in Bianchi IX cosmologies. His abstract:

Using the equations for the null geodesics in Bianchi-IX cosmologies, the temperature distributions of CMB photons is computed in chaotic Bianchi-IX cosmologies. It is found that the measured fluctuations of the $\ell = 2, 3, 4, 5$ modes place bounds on the initial values for these cosmologies assuming that the low ℓ modes are due to the evolution of large scale gravitational anisotropies. It is shown that Bianchi-IX cosmologies that satisfy the bounds on the magnitude of fluctuations in the quadrupole and octupole modes also satisfy the bounds placed upon the shear and Weyl parameters associated with all Bianchi models. This is not always true as can be seen in the case of Bianchi VII_0 models.

6. Bruno Mota: Cosmic Topology for Very Nearly Flat Universes

In a final talk relating to the Cosmic Microwave Background, Bruno Mota discussed the local nature of cosmic topology for very nearly flat Universes. This was work in collaboration with G. I Romero, M. J. Rebouças, and R. Tavakol. His abstract:

Recent studies of the detectability of cosmic topology of nearly flat Universes have often concentrated on the range of values of Ω_0 given by current observations. We have studied the consequences of taking a range of bounds satisfying $|\Omega_0 - 1| \ll 1$, which include those expected from future observations such as the Planck mission, as well as those predicted by inflationary models. [13] We show that in this limit, a generic detectable non-flat manifold is locally indistinguishable from either a cylindrical ($\mathbb{R}^2 \times \mathbb{S}^1$) or toroidal ($\mathbb{R} \times \mathbb{T}^2$) manifold, irrespective of its global shape, with the former being more likely. Importantly this is compatible with some recent indications of the alignment of the quadrupole and octupole moments, based on the analysis of the first year WMAP data. It also implies that in this limit an observer would not be able to distinguish topologically whether the Universe is spherical, hyperbolic or flat. By severely restricting the expected topological signatures of detectable isometries, our results provide an effective theoretical framework for interpreting cosmological observations, and can be used to confine the parameter spaces which realistic search strategies, such as the 'circles in the sky' method, need to concentrate on.

7. Charles Hellaby: Cosmic Geometry from Observations

We then moved from topology to geometry, and a discussion of observational limitation to the perturbations discussed in the first few talks of the session. Charles Hellaby discussed how to determine the cosmic geometry from observations. In particular, he was concerned with quantifying the inhomogeneity present in our Universe. His abstract:

Current and planned galaxy surveys will vastly increase the amount of cosmological data available for analysis. Already in recent years there has been a dramatic improvement in the number of measured cosmological parameters and the accuracy of their values. Properties of the matter distribution have been well studied, but always with the assumption of a homogeneous background metric.

The general assumption that the Universe has a Robertson-Walker metric on the very large scale has served cosmology very well, and is implicit in many calculations. But this makes verifying homogeneity rather tricky as there is a distinct danger of a circular argument. To do this consistently requires the use of a non-homogeneous metric. With all the new data, we will soon be in a position to quantify homogeneity—the degree of lumpiness on each scale—rather than assuming it. A major problem is that the effects of cosmic evolution, source evolution and inhomogeneity are all mixed together on the null cone.

There has been a fair bit of theoretical work on how to determine the metric from standard observations, but implementation has not been attempted. This is clearly a very big task, and will take years to develop into a rigorous algorithm generating believable results. We describe the beginnings of such a procedure, necessarily simple at first, and look at the relationship between a continuous metric and discrete data.

8. Akio Hosoya: Information Entropy in Cosmology

In a similar spirit to the previous talk, Akio Hosoya discussed information entropy in cosmology. This work, in collaboration with Thomas Buchert and Masaaki Morita, attempts to determine an entropy measure for the inhomogeneity in the Universe. The abstract:

The effective evolution of an inhomogeneous cosmological model may be described in terms of spatially averaged variables. [14] We point out that in this context, quite naturally, a measure arises which is identical to a fluid model of the 'Kullback–Leibler Relative Information Entropy', expressing the distinguishability of the local inhomogeneous mass density field from its spatial average on arbitrary compact domains. We discuss the time-evolution of 'effective information' and explore some implications. We conjecture that the information content of the Universe—measured by Relative Information Entropy of a cosmological model containing dust matter—is increasing.

9. Tonatiuh Matos: Flat Central Density Profiles from Scalar Field Dark Matter Halo

In the final talk of the session, Tonatiuh Matos discussed a very particular form of cosmological inhomogeneity: the cosmological dark matter halo. He explored the density profiles formed by scalar field dark matter. His abstract:

The scalar field endowed with a cosh scalar field potential behaves exactly in the same way as cold dark matter (CDM) in the region where the scalar field oscillates around its minimum. Also, in the linear regime, the scalar field dark matter (SFDM) hypothesis predicts the same structure formation as the cold dark matter one. This means that CDM and SFDM are equivalent from the cosmological point of view. The free parameters of the SFDM model can be fixed using cosmological observations. In previous work, we showed by solving the Einstein Klein Gordon equations that if we use such parameters, the scalar field collapses forming stable objects with a mass around $10^{12} M_{\odot}$. [15] In the present work we use analytical solutions of the flat and weak field limit of the Einstein-Klein-Gordon equations. With these solutions we show that the scalar field density profile corresponds to a halo with an almost flat central density and that this halo coincides with the CDM model in a large outer region. Such a result could solve the problem of cuspy DM halos in galaxies without extra hypotheses, adding to the viability of the SFDM model.

10. Summary

The session was quite lively and engaging, with a wide range of discussion, and many incisive questions. It is apparent that there are many fascinating issues in general relativity having to do with our understanding of the "late" Universe. Of course, this is precisely the Universe we now occupy, and thus these questions take on added relevance. In addition to the animated talks and discussion, there was also a very diverse array of posters, covering an immensely broad range of subjects.

Acknowledgments

The author would like to thank Josh Frieman, the co-chair of the session, for his help in organizing a lively and engaging session. In addition, many thanks are due to the organizers of GR17, and to Brien Nolan in particular, for an immensely successful meeting.

References

1. J. Bicak, D. Lynden-Bell, J. Katz, *Phys. Rev. D* **69**, 064012 (2004)
2. J. Bicak, D. Lynden-Bell, J. Katz, *Phys. Rev. D* **69**, 064011 (2004)
3. E.M. Lifshitz, *J. of Phys. (Moscow)* **10**, 116 (1946)
4. S. Kopeikin, J. Ramirez, B. Mashhoon & M. Sazhin, *Phys. Lett. A* **292**, 173 (2001)
5. J. Ramirez & S. Kopeikin, *Phys. Lett. B* **532**, 1 (2002)
6. L. P. Grishchuck, A. N. Petrov, & A. D. Popova, *Commun. Math. Phys.* **94**, 379 (1984)
7. D. E. Holz & R. M. Wald, *Phys. Rev. D* **58**, 063501 (1998)
8. D. E. Holz, *Astrophys. J. Lett.* **506**, L1 (1998)
9. D. E. Holz & E. V. Linder, submitted to *Astrophys. J.* (2004) (astro-ph/0412173)
10. N. Kogo, M. Matsumiya, M. Sasaki, & J. Yokoyama, *Astrophys. J.* **607**, 32 (2004)
11. M. Matsumiya, M. Sasaki, & J. Yokoyama, *Phys. Rev. D* **65**, 3007 (2002)
12. M. Matsumiya, M. Sasaki, & J. Yokoyama, *J. Cosmol. Astropart. Phys.* **02**, 003 (2003)
13. B. Mota, G. I. Gomero, M. J. Rebouças, & R. Tavakol, *Class. Quant. Grav.* **21**, 3361 (2004)
14. A. Hosoya, T. Buchert, & M. Morita, *Phys. Rev. Lett.* **92**, 141302 (2004)
15. M. Alcubierre et al., *Class. Quant. Grav.* **19**, 5017 (2002)

B.2(III): COSMOLOGY (III). DARK ENERGY AND THE COSMOLOGICAL CONSTANT

VARUN SAHNI
Inter-University Centre for Astronomy and Astrophysics
Post Bag 4
Pune 411007
India
e-mail: varun@iucaa.ernet.in

The workshop commenced with presentations on the cosmic equation of state and its relevance for dark energy model building.

The equation of state of dark energy is clearly one of the most important issues in modern cosmology. While the cosmological constant has $w = p/\rho = -1$, the equation of state in dynamical dark energy models can vary significantly with time, and its clearly a matter of the utmost importance to determine whether or not such a time-variation of the equation of state is actually suggested/supported by the data.

Pier-Stefano Corasaniti emphasised that recent developments in observational cosmology brings the dark energy scenario into a new phase of constrained phenomenology. Of particular importance is to determine whether dark energy manifests a dynamical behavior or not. And the answer to this question would allow to falsify or confirm the standard cosmological constant hypothesis. Addressing such an issue however demands a fully consistent analysis of CMB, large scale structure and SN Ia data with a multi parameters freedom valid at all redshifts. Corasaniti analyzed the WMAP, 2dF and latest SN Ia data to constrain in addition to the standard cosmological parameter, four dark energy parameters which parametrized the dark energy equation of state in terms of its present value w_0, its value at high redshift w_m, the redshift z_t corresponding to the transition from w_m to w_0 and the duration Δ of the transition. He reported no new degeneracies with the standard cosmological parameters despite the increased freedom in dark energy. The best fit model showed a strong variation at $z < 1.5$. It was interesting that the CMB strongly constrains the value of the equation of state at high redshifts for models with $z_t < 10$. In such a case $w_m < -0.1$. The present equation of state of dark energy was found to be $w_0 < -0.8$ at two sigma. Nonetheless the data are still fully consistent with, and arguably prefer, a cosmological constant. Phantom models were also discussed and in this case the best fit crossed $w = -1$ a result which, if confirmed, would be a clear signal for radically new physics.

Ujjaini Alam reported results of a reconstruction exercise in which properties of dark energy were reconstructed from supernova observations using a model independent ansatz. The results obtained indicated that an evolving form of dark energy was favoured by supernova observations provided no additional priors were imposed. In this case the best-fit equation of state evolved from $w \lesssim -1$ at $z \simeq 0$ to $w \simeq 0$ at $z \sim 1$. The results were robust to changes in the value of the matter density and remained in place for $0.2 \lesssim \Omega_m \lesssim 0.4$. However if one imposed the additional prior that dark energy satisfied the weak energy condition $\rho + p \geq 0$ then the evolution of dark energy became much weaker and a cosmological constant agreed very well with the data. The inclusion of CMB results obtained by the WMAP experiment did not lead to significant changes in the above picture. WMAP+SNe data preferred evolving dark energy to a cosmological constant for higher values of the matter density ($\Omega_m \simeq 0.38$) and lower values of the Hubble parameter ($h \simeq 0.6$). If strong priors were imposed on these parameters ($\Omega_m \simeq 0.3, h \simeq 0.6$) then a cosmological constant was in excellent agreement with observations. An accurate knowledge of the cosmological parameters Ω_m, h was clearly an important precursor to determining the cosmic equation of state of dark energy.

Different data sets, carry important (sometimes orthogonal) information about the properties of dark energy. *Jochen Weller* presented new constraints on the equation of state of dark energy from a combined analysis of supernovae, CMB and X-ray cluster observations. Weller emphasised the existence of degeneracies in the CMB on degree and subdegree scales for different values of the equation of state of dark energy, and also the fact that this degeneracy was broken by the integrated Sachs-Wolfe (ISW) effect. However the ISW effect, which was important on large angular scales was subject to a large cosmic variance, and therefore was, on its own, a difficult probe of dark energy. Weller also emphasised the importance of perturbations in the dark energy fluid and that, in the case of cluster observations, the baryon gas fraction was a powerful complementary constraint on the matter content.

After the preceeding discussion on dark energy, which was based on model independent arguments, the discussion began to focus around concrete models of dark energy, commencing with the Cardassian approximation. In this approach, the expansion rate of the Universe at late times is modified according to the form:

$$H^2 = \frac{8\pi G}{3}(\rho + \rho^n) \tag{1}$$

W. Godlowski discussed observational constraints from type Ia supernovae on the spatially flat Cardassian model, and its extension with the curvature term included. The Cardassian model, which is an alternative to LCDM, was tested against the Perlmutter SN Ia data as well as the Knop and Tonry samples. In the non-flat Cardassian model Godlowski obtained the matter density value ($\Omega_{m,0} \geq 0.3$), while for the flat Cardassian model a high density universe with $\Omega_{m,0} \simeq 0.5$ was preferred. For the high density universe Godlowski also found a negative value of n which can be interpreted as the effect of a phantom fluid. It is interesting that a nearly flat universe is preferred from Sn observations, which is not the case for LCDM for which Sn observations appear to prefer a closed universe.

An inverse problem approach to the dark energy equation of state was presented by *Ruiz–Lapuente*. This approach quantifies the degree of information on dark energy contained in samples of cosmological distance indicators without specifying any a priori shape of the function w(z).

The present sample of SNe Ia at high z is found to be compatible with the cosmological constant although neither an evolution of the equation of state nor a running of the cosmological constant can be discarded. The z > 0.8 region of w(z) is very poorly sampled at present. It is found that a reliable reconstruction of the continuous function w(z) (without constraining the functional form) will not be feasible from data expected in future experiments like SNAP (even using priors on Ω_m). The present value of the index of the equation of state w_0 and its first derivative w_a can be well retrieved from future samples. It is found that combining SNe Ia at high–z plus another different cosmological distance indicator that samples well z > 1.5 (for instance FR IIb radio galaxies), the knowledge of w_0 and w_a improves up to 50%. The mean index for the first derivative of the equation of state I_{wa} has a value of ~ 0.7 in the Perlmutter et al (1999) sample and of ~ 0.95 in the Riess et al. (2004) sample, both values indicating that there is not enough knowledge yet on w_a in those samples, while simulated data sets for SNAP show that the future data from this mission would allow a retrieval of w_a with I_{wa} close to 1. Reliablity of the result would only depend on the adequacy of the formulation of the dark energy up to a first order expansion in the derivatives of w(z).

Kei-ichi Maeda and *Marcos Maia* discussed braneworld models and their possible relation with dark energy.

Kei-ichi Maeda studied the so-called cosmological constant problem in the context of a braneworld scenario. He emphasised that since the four dimensional gravity in a brane world was modified, the constants which appeared in the effective Friedmann equation would not be the same as those in the local equation on a small scale. This difference may explain the present value of dark energy. As a concrete example, Maeda discussed the induced-gravity brane model first proposed by Dvali, Gabadadze and Porrati. Generalizing this model, Maeda showed that an effective cosmological constant on the brane can be extremely suppressed in contrast to the case of the Randall-Sundrum model even if a bulk cosmological constant and a brane tension are not fine-tuned.

Marcos Maia drew attention to the fact that despite the good observational indications and theoretical reasons for the existence of dark energy, its fundamental nature still remained obscure. In his talk he gave a possible geometrical explanation for such a dark energy component within the context of brane-world theory. In the particular case of the Friedmann-Robertson-Walker universe regarded as brane-world embedded into a five-dimensional constant curvature bulk, without the Z_2 symmetry, Maia derived the modified Friedman's equation, showing that the effect of the extrinsic curvature on the accelerating universe is compatible with phenomenology of a dark energy model.

Salvatore Capozziello showed that a quintessence model could be constructed by taking into account higher order curvature invariants into the effective action of the gravitational field. Such an approach is naturally related to fundamental theories of quantum gravity which predict higher order terms in loop expansion of quantum fields in curved space-times. Capozziello showed that using this approach it was possible to reproduce results which were able to fit high redshift supernovae and WMAP observations.

One of the central questions of modern cosmology is whether the early acceleration of the Universe (inflation) and the present, late-time acceleration, could be related. In this context *Gianluca Allemandi* discussed the importance of investigating modified and alternative theories of gravity, which are able to explain both present acceleration and early time inflation of the universe, without the need of introducing of dark energy. Alternative gravitational theories (in minimal interaction with matter), defined by means of Lagrangians depending on analytical functions of the Ricci scalar invariant or higher curvature invariants, were studied in a Palatini formalism in view of their possible cosmological applications. The Palatini formalism prevents field equations from instability problems and it has been proven to fit the solar system experiments at an acceptable level. Structural equations for the corresponding spacetimes were solved a modified Hubble parameter and deceleration parameter were obtained. Allemandi showed that particular examples of these theories can provide, under suitable hypotheses, a theoretical explanation to the present acceleration of the universe and as well as early time inflation.

Marek Demianski presented results of investigation of two classes of minimally coupled scalar field models of quintessence with exponential potentials for which exact solutions of the Einstein equation are known based on a joint work with E. Piedipalumbo and C. Rubano. These models are very sturdy depending essentially only on one parameter - the Hubble constant. Theoretical predictions of these models have been confronted with recently released observational data: the power spectrum of the CMBR anisotropy, as measured by WMAP, the publicly available data on type Ia supernovae, and the parameters of large scale structure determined by the 2-degree Field Galaxy Redshift Survey (2dFGRS). Both models nicely reproduce the observational data but they have different behaviour at $t \rightarrow \infty$. In one model $\Omega_m \rightarrow 0$ with $t \rightarrow \infty$, so it asymptotically becomes the de Sitter model, in the other model $\Omega_m \rightarrow$ constant with $t \rightarrow \infty$, so even asymptotically the average density of matter does not tend to zero and the scale factor $a(t) \sim t^{2/3}$, though the average density of dark energy is different form zero. It turns out that on the basis of present day observational data it is not possible to uniquely determine the future evolution of the universe.

Current observations of type Ia supernovae allow an equation of state of dark energy which is supernegative, $w < -1$. Such equations of state arise in so-called Phantom dark energy models as well as in Braneworld models of dark energy. *José Jiménez Madrid* and *Valerio Faraoni* discussed the future evolution of the Universe in the presence of matter which violated the weak energy condition and led to $w < -1$.

In such a case the future of the Universe is unusual in the sense that the universe encounters a Big Rip singularity at which all curvature invariants together with the dark energy density diverge. The occurrence of a Big Rip singularity which occurs at a finite time in the future was investigated in the context of superquintessence by Faraoni. (Superquintessence is dark energy with an effective equation of state parameter $w < -1$ and $\dot{H} > 0$.)

Perhaps the simplest classical model of superquintessence is based on a single nonminimally coupled scalar field, which is allowed to be massive and to self-interact through a quartic potential. This model exhibits Big Rip solutions and its phase space is a 2-dimensional manifold composed of two sheets attached to one another and embedded in a higher-dimensional space. Faraoni demonstrated that a stability analysis yields the result that the Big Rip solutions are attractors in phase space.

José Jiménez Madrid considered two situations: in the first, phantom energy was described by a perfect fluid satisfying the equation of state $P = (\beta - 1)\rho$ (with $\beta < 0$ and constant). In this case the universe reaches a "Big Rip" independently of the spatial geometry of the FRWL universe. In the second situation, the dark energy is described by a phantom (generalized) Chaplygin gas which violates the dominant energy condition. Contrary to the previous case, for this material content a FRWL universe would never reach a "Big Rip" singularity (indeed, the geometry is asymptotically de Sitter). Madrid showed how this dark energy model could be described in terms of scalar fields, corresponding to a minimally coupled scalar field, a Born-Infield scalar field and a generalized Born-Infield scalar field.

There is already a vast amount of literature pertaining to dark energy models. Distinguishing between them is clearly a paramount necessity and for this purpose Sahni, Saini, Starobinsky and *Alam* (2003) suggested the Statefinder parameters

$$r \equiv \frac{\dddot{a}}{aH^3} = 1 + \frac{9w}{2}\Omega_X(1+w) - \frac{3}{2}\Omega_X \frac{\dot{w}}{H} , \tag{2}$$

$$s \equiv \frac{r-1}{3(q-1/2)} = 1 + w - \frac{1}{3}\frac{\dot{w}}{wH} . \tag{3}$$

The Statefinder parameters are based on higher derivatives of the expansion factor and therefore usefully supplement the Hubble parameter and the deceleration parameter. They can be applied to quintessence models on the one hand, and dark energy arising from modified gravity theories such as braneworld models and scalar-tensor theories, on the other.

Diego Pavon discussed the behaviour of the Statefinder parameters in a class of dark energy models which involved interactions between dark matter and dark energy. Such models are interesting because, by allowing energy transfer between the different components, they alleviate the 'coincidence problem' which plagues most models of dark energy. (The coincidence problem in standard LCDM singles out the present epoch as being somewhat special since the dark matter and dark energy densities are almost equal at precisely the present epoch.) Pavon showed that interactions between dark matter and dark energy explicitely affect the third time derivative of the expansion factor $\dddot{a}$; neither $\dot{a}$ nor $\ddot{a}$ show explicit dependence on the interaction. As a result the statefinder r contains important information for this class of dark energy models and the statefinder pair r, s can be used as a discriminator between interacting and non-interacting models of dark energy. Interacting dark energy allow scaling solutions for which $\rho_m/\rho_{\rm DE} \propto a^{-\xi}$, where ξ is a constant whose value lies in the range [0,3]. These solutions are of interest because the alleviate the coincidence problem. Pavon concluded that the statefinder parameters would prove a useful tool for discriminating between different models of late-time acceleration when appropriate observational techniques become available.

The importance of higher derivatives of the expansion factor was also emphasised by *Matt Visser*. Visser emphasised that the value and derivatives of the scale factor determined at the current epoch place constraints on the value and derivatives of the cosmological equation of state at the current epoch. Determining the first three Taylor coefficients of the equation of state at the current epoch requires a measurement of the the second, third, and fourth derivatives of the scale factor with respect to time. Visser showed that it was possible to expand the physical distance $D = c \int dt$ as follows:

$$D = \frac{c\,z}{H_0}\left\{1 - \left[1 + \frac{q_0}{2}\right] z + \left[1 + q_0 + \frac{q_0^2}{2} - \frac{j_0}{6}\right] z^2 \right. \tag{4}$$

$$\left. - \left[1 + \frac{3}{2} q_0 (1 + q_0) + \frac{5}{8} q_0^3 - \frac{1}{2} j_0 - \frac{5}{12} q_0 j_0 - \frac{s_0}{24}\right] z^3 + O(z^4)\right\},$$

where the 'jerk' is

$$j(t) = +\frac{1}{a}\,\frac{d^3 a}{dt^3}\,\left[\frac{1}{a}\,\frac{da}{dt}\right]^{-3} ; \tag{5}$$

and the 'snap' is

$$s(t) = +\frac{1}{a}\,\frac{d^4 a}{dt^4}\,\left[\frac{1}{a}\,\frac{da}{dt}\right]^{-4} . \tag{6}$$

Note that j is directly related to the statefinder 'r'. This simple calculation demonstrates that the jerk shows up at third order in the Hubble law, and the snap at fourth order.

A radically different approach to dark energy was presented by *Marie-Noelle Celerier*. Celerier emphasised that the assumption of large scale spatial homogeneity of the universe lies at the basis of the interpretation of the Sn observations in terms of cosmological acceleration. This assumption proceeds from a hypothesis brought to the status of cosmological principle of which the justification is based on two arguments: 1) The isotropy, or quasi-isotropy, of the temperature of the Cosmic Microwave Background Radiation (CMBR) around us. 2) The Copernican assumption that, as our location must not be special, the isotropy we observe on a surface of our past light cone must be observed identically from any point of the universe. Since a matter distribution which is seen isotropic from everywhere implies homogeneity, the cosmological principle follows. However, of the two above arguments, only the first is observation grounded. The second, purely philosophical, has never been verified. Celerier proposed a method, which, applied to a statistically pertinent sample of high-redshift SNIae, could be used as a test of the cosmological principle and as a consequence as a test of the cosmological constant assumption (which follows from the use of the FLRW magnitude-redshift relation to analyse the data). This test could be applied provided sufficiently accurate and large samples of SNIe were available (for instance from the SNAP project).

In addition to the talks summarised above, poster presentations on the following topics were also made.

Christian Boehmer considered Einstein-Cartan theory with cosmological constant and constructed the Einstein static universe with torsion, thereby generalising the Einstein cosmos to Einstein-Cartan theory.

Jack Sarfatti derived a generalized ODLRO local inflation field from an instability in the globally flat vacuum of quantum electrodynamics. He claimed to show that Einstein's general relativity equations emerge from the modulation of the macro-quantum coherent phase of this inflation field and that Dark matter, although w = -1 looks like $w \simeq 0$ CDM for distant observers.

Daksh Lohiya showed that a strictly linear evolution of the cosmological scale factor provides an an excellent fit to a host of cosmological observations. He discussed the concordance of such an evolution in relation to minimal requirements for large scale structure formation and cosmic microwave background anisotropy along with the overall viability of such models and conjectured that a linear evolution would emerge naturally either from General Relativity itself or from a General Relativistic theory of a non-minimally coupled scalar field theory.

Zoltán Perjés discussed spatially homogeneous and isotropic cosmological models, with a perfect fluid matter source and non-vanishing cosmological constant and showed that the problem is solved for all C^∞ perturbations and pressure-free matter. *Perjés* and *Czkinner* also obtained the complete solution of the equations governing linear perturbations of the space-time and the variation of energy density and computed the Sachs-Wolfe fluctuations of the temperature of the cosmic background radiation for the relatively growing modes.

Takao Fukui discussed the Magnitude-Redshift Relation in Cosmological Models with a Variable Cosmological Term such as generalized scalar-tensor (GST) cosmology. He showed that GST fits the observational supernovae magnitude-redshift relation quite well.

Tomoko Uesugi investigated braneworld cosmology based on the D-brane where the brane is described by a Born-Infeld action and the gauge field. He showed that the equation of state of the gauge field on the brane is radiation-like in low energy scales and almost dust-like fluid in high energy scales.

Antonio Segui discussed a toy model for the cosmological coincidence problem by means of a mechanical model made with a percolating network of black holes.

R. Aldrovandi, J. P. Almeida and J. G. Pereira claimed that new physics is required by the Cosmological Constant. They claimed that the usual spacetime translations of the Poincaré group should be replaced by the mixed translation+conformal transformations of the de Sitter group, leading to obvious alterations in elementary concepts such as time, energy and momentum. Although negligible at small scales, such modifications may have important consequences both in the large and for the inflationary picture of the primaeval Universe.

R. Aldrovandi, J. P. Almeida and J. G. Pereira also discussed Conformal Symmetry and the Cosmological Term.

Paulo Vargas Moniz discussed implications of the de Broglie - Bohm (dBB) quantum mechanics towards the description of the late universe in superstring cosmology. He claimed that the presence of a dilatonic potential induces effects similar to dark energy and extra dimensions.

Z. Stuchlik and P. Slany discussed Relativistic discs orbiting the Kerr–de Sitter black holes. They showed that the presence of $\Lambda > 0$ leads in the existence of an outer marginally stable orbit located under the so-called static radius where the gravitational attraction is just balanced by the cosmic repulsion. For central supermassive black holes ($10^6 M_\odot$–$10^9 M_\odot$) the dimensions of the disc-like structures are smaller but comparable to the dimensions of elliptical galaxies, indicating some role of $\Lambda > 0$ in the creation processes of the galaxies.

Zdeněk Stuchlík and Jan Schee discussed the role of the observed relict vacuum energy on the fluctuations of Cosmic Microwave Background Radiation in the framework of the Einstein–Strauss–de Sitter vacuole model.

Remo Garattini discussed a spacetime foam model for the cosmological constant. In this model spacetime foam is made by N wormholes in a semiclassical approximation. The renormalized Casimir-like energy of the quantum fluctuation of such a model via an induced cosmological constant was examined and implications on the Bekenstein-Hawking entropy and the cosmological constant were discussed.

B3: GRAVITATIONAL-WAVE SOURCES: SOURCE SCIENCE AND STATISTICAL METHODS

ALESSANDRA BUONANNO

Groupe de Gravitation et Cosmologie (GReCO)
Institut d'Astrophysique de Paris
98$^{\rm bis}$ Boulvard Arago, 75013 Paris, France
e-mail: buonanno@iap.fr

We summarize recent progresses in understanding physical mechanisms responsible for gravitational-wave emission in a variety of astrophysical sources, and statistical strategies which have been developed to detect gravitational waves both with ground- and space-based detectors.

1. Introduction

The last few years have been marked by the first scientific runs of long-baseline ground-based laser-interferometer gravitational-wave detectors (10–10^3 kHz). TAMA [1] operated at design sensitivity, LIGO and GEO [2] have completed three science runs (although not at design sensitivity) yielding new upper limits on event rates for various classes of astrophysical sources. They expect to reach full sensitivity in 2005. VIRGO [3] is in its commissioning phase and plans to begin science operations quite soon. The space-based detector LISA, an ESA and NASA joint mission, should fly in 2013 and, being sensitive to lower GW frequencies (10^{-4}–10^{-1} Hz), will provide us with information complementary to the ones of ground-based detectors. Approaching sensitivities for which gravitational waves could be finally detected makes it more and more crucial to predict accurate gravity-wave signals and develop robust statistical methods to extract them. This session gathers talks discussing recent advances in this research area.

2. Which are the most likely compact binaries to be first observed?

Inspiraling compact binaries made of neutron stars and/or black holes are among the most promising (or, at least, are among the best-understood) candidates for the production of gravitational waves detectable by both ground- and space-based observatories. The data analysis for these sources relies on cross correlating the output data stream with theoretical templates predicted by post-Newtonian calculations. Those templates depend on various binary parameters, such as masses, spins, eccentricity, etc., and it is very desirable to understand in advance what are the most likely compact binaries that we should first observe and what are the most likely range of parameters for which we should expect the first detection.

For many years various groups have been investigating Galactic inspiral event rates of compact binaries, either empirically [4], using information from observations of the six double neutron star binaries [5], or theoretically, through population synthesis codes [6]. Such estimates of event rates will be widely used in appraising the gravitational inspiral detectability for a given instrument sensitivity.

Tomek Bulik [7] presented recent results on various features of binary populations obtained using the StarTrack population synthesis code. Bulik and collaborators calculated the distribution of masses and mass ratios by weighting the probability distribution with the volume of observability, which for an inspiraling Newtonian signal scales as $\mathcal{M}^{5/2}$ [$\mathcal{M} = M\eta^{3/5}$ being the chirp mass, with $M = m_1 + m_2$ the total mass and $\eta = m_1 m_2/M^2$]. By assuming neutron stars with mass of $1.2 - 3M_\odot$ and black holes with mass of $3 - 16M_\odot$, they found that in the observability weighted distribution of double neutron star binaries there are two peaks: one for nearly equal-mass systems, and one for systems consisting of a low and a high mass neutron star, i.e. with mass ratio $0.6 - 0.7$. The observability weighted distribution of black-hole/neutron-star binaries is concentrated on systems with mass ratio $0.3 - 0.5$, while for double black-hole binaries the observability weighted distribution is dominated by massive, nearly equal-mass binaries with mass ratio 0.8. Therefore, high equal-mass black holes are likely to be the first compact objects to be observed.

Unlike the binary neutron star merger rate, which has been both theoretically and empirically constrained, the binary black hole merger rate has only been estimated by purely theoretical population synthesis calculations. Population synthesis codes (Monte Carlo simulations) consider the complete formation history of double compact objects through long sequences of binary evolution phases. Since our current understanding of single and binary star evolution is incomplete, the population synthesis codes are affected by uncertainties, generally parametrized by empirical constraints. Due to the high computational cost of Monte Carlo simulations it is very hard to cover a large part of the multi-dimensional parameter space and, as a consequence, rate predictions can be uncertain by several orders of magnitude. Richard O'Shaughnessy [8] discussed a recent approach, based on *genetic* algorithms, which are ideal for the exploration of highly multidimensional spaces. They systematically mapped theoretical predictions for neutron star/neutron star and black-hole/black-hole merger rates over many of the most significant parameters which enter population synthesis models and employed these maps, along with the empirically-determined neutron-star/neutron-star merger rate distribution as a model constraint to derive a semi-empirical black-hole/black-hole merger rate distribution. They found that it is possible to fit inspiral rate to an acceptable accuracy (currently 50%, but smaller values can be obtained if the original simulations were more accurate). They also described how these computations can be generalized to include additional observational constraints other than the neutron-star/neutron-star merger rate.

Intermediate-mass black-hole binaries have gained a lot of attention during the last years, both as sources of low-frequency ground-based detectors and of space-based detectors. These are binaries typically containing one black hole in the mass range 100 – 1000 $M_\odot$.

Matthew Benacquista gave a talk on preliminary results [9] of event rates of intermediate-mass black holes in the scenario in which they formed by runaway growth of a star through stellar collisions in star clusters [10]. The massive stars then collapse to form either an intermediate mass black hole or a seed black hole for further growth through black hole mergers. He found that the expected birthrate of these intermediate mass black holes is between 10^{-7} to 10^{-6} per year per galaxy [11]. Gravitational radiation from the ringdown of these newly formed black holes will be in the frequency range of ground-based detectors (10 to 100 Hz), especially VIRGO and Advanced LIGO.

Tomek Bulik [12] presented another possible scenario of formation of intermediate mass black holes in which they originated from the evolution of first population of massive metal-free binary stars. Due to the low metallicity, those stars are allowed to form with large initial masses and they evolve without significant mass loss. Evolution at zero metallicity, therefore, may lead to the formation of massive remnants. In particular, black holes of intermediate-mass (100-500 $M_\odot$) are expected to have formed in the early Universe, in contrast to the much lower mass stellar black holes (10 $M_\odot$) being formed at present time. Assuming that some of these Population III stars have formed in binaries, Bulik and collaborators found that a significant fraction of such binary black holes coalesces within Hubble time. The burst of gravitational waves emitted at coalescence and during the ringdown phase could be observed by ground-based detectors of first generation. For second-generation detectors, such as Advanced LIGO, the detection rate of such mergers is at least several events per year with signal-to-noise ratio larger than ~ 10.

Cliff Will [13] discussed event rates for intermediate-mass black holes for the space-based interferometer LISA. He compared and contrasted his estimates with earlier estimates by Miller [14]. For one-year integration leading up to the innermost stable circular orbits, and a signal to noise ratio of 10, he estimated a detection rate of only 1 per million years for $10M_{\odot}/100M_{\odot}$ binaries. The estimate uses the method of parameter estimation via matched filtering, incorporates a noise curve for LISA established by the LISA Science Team that is available online, and employs an intermediate-mass black holes formation rate model used by Miller [14]. Will found that the detectable distance is relatively insensitive to LISA arm lengths or acceleration noise, but is roughly inversely proportional to LISA position errors, and varies roughly as $T^{1/2}$, where T is the integration time in years. He also showed that, while all intermediate-mass black-hole systems in this mass range may be detected in the Virgo cluster up to 40 years before merger, none can be detected there earlier than 400 years before merger.

3. Sources for kilometer-scale ground-based detectors

3.1. *Compact binaries*

One of the most challenging problem in black-hole binaries is the description of the dynamics when the two bodies enter the regime of strong gravitational, non-linear interactions. This problem can be tackled by solving the Einstein equations numerically. Despite big efforts and many advances during the last years, the numerical relativity community has not been able to simulate a single binary black-hole orbit, yet. In fact after rather short evolution times numerical simulations of black-hole binaries become unstable.

Bernd Brügmann [15] presented recent results of numerical simulations of binary black hole systems which use fixed mesh refinement and comoving coordinates. For the first time the simulations last for about one orbital period and concern black holes close but well outside the innermost stable circular orbit. The apparent horizons do not merge after one orbital period has passed, as indicated by the absence of a common event horizon. While such orbit simulations have not been possible before, there are still several problems that remain to be solved before being able to evolve for a longer time and extract accurately the gravitational-wave signal.

Many results have been recently obtained by describing the two-body dynamics during the inspiral phase with quasi-equilibrium circular configurations [16]. Uryu [17] reviewed numerical solutions for binary compact objects in close circular orbit which can be used as a model for the final stage of inspiral. Such solutions could describe an intermediate phase between a post-Newtonian inspiral and a binary merger and could be used to extract waveforms that connect the gravity-wave signal emitted from inspiral and merger phases. These numerical solutions could be also used as data sets for merger simulations, since they yield orbits with small eccentricity and may have less spurious radiation. More specifically, Uryu discussed results for: (1) binary neutron star solutions, using various nuclear equations of state, obtaining the gravitational-wave frequency just before merger, (2) new binary black hole initial data, (3) new formulations for modeling binary compact objects that incorporate generic spacetime metrics with helically symmetric or waveless approximations, (4) toy models for binaries with helical symmetry in linearized gravity.

The best theoretical predictions for the gravitational-wave signal emitted by inspiraling compact binaries have been derived using post-Newtonian calculations. If the compact body carries a large spin, and if the spin is not aligned with the (orbital) angular momentum, spin-orbit, spin-spin, quadrupole-monopole interactions may affect the waveform and, if this is the case, their effects should be consistently included in building accurate templates. If the compact object has a strong magnetic field — neutron stars with strong magnetic fields are known to exist (magnetars) — the magnetic couplings could also affect the dynamical evolution of the binary.

Mátyás Vasuth [18] discussed the effects of magnetic dipole fields of neutron stars in a binary system moving along eccentric orbits, extending work done by Ioka and Taniguchi [19] for binaries moving along circular orbits. The interaction of magnetic dipole moments contributes to the equations of motion and gravitational radiation at the second post-Newtonian order, if magnetic fields are rather strong ($\gtrsim 10^{16}$ Gauss). [Current observations are compatible with magnetic fields in magnetars of 10^{15} Gauss.] The secular contributions to the energy and orbital angular momentum losses for eccentric orbits is obtanied by parametrizing the radial motion and averaging over a radial period. They worked out the secular evolution of the relative orientation of the orbital angular momentum and spins and found that those equations, augmented by the evolution equations for the angles characterizing the orientation of the dipole moments, form a closed first order differential system. It rests to be investigated the actual impact of large magnetic dipoles on the phasing of the gravitational wave for realistic strong magnetic fields.

3.2. *Pulsating neutron stars*

Rotating stars possess a class of circulation modes, r-modes, that are driven toward instability by gravitational radiation reaction [20]. In hot, rapidly rotating young neutron stars these instabilities might be so strong to dominate over viscous dissipation. When r-modes achieve sufficient amplitude, the neutron star is spun down as angular momentum is lost due to gravitational-wave emission. During the last years, various calculations showed that the r-mode instability could not survive when all the complicated physics that occurs in neutron stars is taken into account — relativistic effects, equation-of-state effects, solid crust effects, magnetic fields, etc. Moreover, the amount of GWs emitted, which is determined by the maximum r-mode amplitude, can be limited by magnetic-field effects or by transfer of energy to other modes through nonlinear hydrodynamical couplings [21].

Recently, Wagoner [22], based on the results of Lindblom and Owen [23], proposed that accreting neutron stars in low-mass X-ray binaries can emit gravitational waves for long periods of time through the r-mode instability if the equation of state favors hyperons at high densities. If the bulk viscosity of the hyperon core has the right temperature dependence, an accreting star can steadily radiate its accreted angular momentum as gravitational waves without suffering thermal runaway. Ben Owen [24] presented preliminary results in which they evaluated the bulk viscosity for a variety of microphysics and determined under which conditions the star is a persistent source of gravitational waves. If r-modes are observed in an accreting neutron star, their results could be used to place constraints on the microphysics.

Leonardo Gualtieri [25] reviewed work on non-adiabatic, non-radial oscillations of relativistic stars in the frequency domain by including consistently the effects of thermal diffusion and computing how those effects could alter the gravitational wave emission. When a general equation of state depending on temperature is used, the perturbations of the fluid result in heat flux, that is coupled with the geometry through the Einstein field equations. They found that in some cases, the frequencies and damping times of the quasi-normal modes (and of the corresponding gravitational emission) are substantially altered with respect to the standard, adiabatic case. This happens especially for $p-$modes. In the semi-transparent regime, the damping time of the gravitational waves is found to be similar to the thermal damping time predicted by simple estimates. Thermal diffusion is expected to be relevant in hot, newly born proto-neutron stars, and in other astrophysical scenarios.

Another promising class of neutron-star modes, the f-modes, is discussed by Karino and collaborators [28]. It is known that f-mode oscillation becomes dynamically unstable when $T/|W|$ is sufficiently large, T and W being the rotational kinetic energy and the gravitational potential energy of the star, respectively. It is generally believed that the classical criterion of the instability of $m = 2$ f-mode (bar-mode) is $T/|W| > 0.27$, at least if the simple stellar model of Maclaurin spheroids is used. Karino [28] studied the stability of those modes in stars with strong differential rotations, by using linear-stability–analysis methods for a broad region of parameters. As a result, they found that under strong differential rotations, at least one unstable mode exists even when the stellar rotation is very slow (say, $T/|W| < 0.1$). Such an unstable mode has the f-mode-like features, but classical discussions of stability can not be applied and the mechanism responsable for the instability is still unknown. The existence of unstable modes in slowly rotating compact stars is significantly important from the point of view of gravitational-wave astrophysics, since unstable rotating stars will emit quasi-periodic gravitational waves. The event rate of unstable-stars emitting GWs will dramatically increase.

3.3. *Rotating neutron stars*

Another important class of astrophysical sources are rotating neutron stars or pulsars. It is well known [26] that all-sky searches for unknown gravitational-wave pulsars are computationally bound, in the sense that their efficiency is limited by the available computer power. It is therefore important to find the most efficient possible search algorithm. Krishnan [27] discussed preliminary results for strategies to look for pulsars with unknown sky-positions, frequencies, and spindowns for a large range of these parameter values. The fully coherent searches over these large parameter space volumes are not feasible, and therefore sub-optimal, incoherent, and inexpensive methods are needed. They considered a general multi-stage hierarchical search which consists of an arbitrary number of incoherent stages with a final coherent follow-up stage. The search strategy is based on reducing the number of signal candidates using sub-optimal methods so that the final follow-up stage can analyze a small number of candidates. Using the stack-slide method as an example, they showed how such searches can be optimized to make the best possible use of given computational resources.

3.4. *Gravitational waves from accretion of matter onto non-rotating compact objects*

Nagar [29] summarized results obtained with numerical simulations of gravitational waves emitted from accretion of matter onto non-rotating compact objects (both neutron stars and Schwarzschild black holes). Assuming axisymmetry, they studied non-spherical polar metric perturbations of the compact body due to infalling matter. They adopted a *hybrid* procedure in which they evolved numerically the inhomogeneous Zerilli-Moncrief equation, when the compact object is a black hole, and the linearized Einstein equations describing matter and metric perturbations when the compact object is a star, but they coupled those equations to a fully nonlinear hydrodynamics code that calculates the motion of accreting matter. Both self-gravity of the fluid's accreting layers and radiation reaction effects are neglected. The initial matter distribution, which is initially at rest and is located at a finite distance from the object, is shaped in the form of radially extended quadrupolar shells either of dust or obeying a perfect fluid equation of state. In the case of black holes, they found that the energy spectrum of gravitational waves is far from having only one clear, monochromatic peak at the frequency of the fundamental quasi-normal mode; rather, it exhibits a complex pattern, with distinctive interference fringes related to the extended spatial distribution of the accreting matter. Depending on the parameters characterizing the shells (i.e., its radial extension and its initial position), the energy is mainly emitted at frequencies lower than that of the fundamental quasi-normal mode. They also compared with the well-known case of point-like particles plunging radially into the black hole, and find that the energy emitted in gravitational waves is roughly two order of magnitude smaller than the point-particle values. This analysis illustrates that gravitational-wave signals driven by accretion onto a Schwarzschild black hole is influenced more by the details and dynamics of the process, and by the external distribution of matter, than by the quasi-normal mode structure of the object. Similar results are found when the black hole is replaced by a relativistic star, with shells involving perfect fluid instead of dust; in this case the contribution of the stellar fundamental mode stands clearly in the energy spectrum. Using the same approach, Nagar discussed also preliminary results where the shell of dust around the black hole is replaced by a centrifugally supported, axially symmetric, thick accretion disk (torus) of perfect fluid, which eventually accretes onto the black hole due to the "runaway" instability driven by the mass loss through the cusp of the outermost closed equipotential surface (the toroidal equivalent of the Roche lobe).

4. Sources for space-based detectors

Binary systems composed of compact objects and supermassive black holes at galactic nuclei are expected to be important sources for space-based detectors. The gravitational waves will carry detailed information on the geometry of supermassive black holes and test gravitational theory in strong field regime. Because of the small mass ratio, binary dynamics can be described using perturbation theory: the supermassive black hole is treated as a background and the solar-mass compact object as a source of metric perturbations. To calculate the waves emitted by such systems require an understanding of the motion beyond the test-mass approximation. In the massless limit the particle moves along a geodesic, but when the particle has a mass, it becomes a source of gravitational radiation that carries away orbital energy and angular momentum. Thus, the orbit deviates from a geodesic and this deviation is determined by solving the equation of motion with radiation-reaction effects (self-force) included [30]. This problem turned out to be rather challenging, but significant, encouraging progresses were made during the last years [30, 31].

A true geodesic around a rotating black hole is characterized by the energy, the angular momentum and the Carter constant. Whereas various investigations, which assume adiabatic evolution of the orbits, have lead to the determination of radiation reactions effects on the energy and the angular momentum, the evaluation of radiation-reaction effects on the Carter constant remains so far an unsolved problem.

Drasco [32] described a method to compute gravitational waves produced by point particles on generic geodesic orbits around a spinning (Kerr) black hole. He reviewed preliminary results for waveforms computed numerically using a code based on the perturbation formalism developed by

Teukolsky, Sasaki and Nakamura. They obtained the energy and angular momentum fluxes to infinity and down to the black hole horizon. Their results extend earlier work on circular and equatorial orbits to accomodate completely generic orbits and this extension was made possible by the recently obtained frequency-domain description of generic Kerr orbits described in Ref. [33]. Drasco also discussed the possibility of applying their method to compute the radiative changes to the Carter constant, determining the waveforms for realistic, inspiraling generic orbits in the adiabatic limit.

The possibility that the two-body dynamics of spinning compact objects could be chaotic has attracted the attention of the GW community [34–36] during the last years. If chaos affects orbits that are of astrophysical interest for GW detectors, the gravitational waveforms would come to depend sensitively on the initial conditions and the strategy of using match filtering techniques could be spoiled.

Kiuchi [37] discussed the issue of chaoticity for small mass ratio binaries, using Papapetrou equations. It is known [35, 36] that the dynamics of a spinning test particle orbiting around a rotating (Kerr) black hole is not integrable and can be chaotic. By using multipole expansion of the gravitational field, Kiuchi estimated the gravitational waves and its energy spectrum for some chaotic orbits. They observed that whereas the energy spectrum of non-chaotic orbits contains discrete characteristic frequencies, the one of chaotic orbits is continuous and contains several peaks. Although it is theoretically interesting to investigate the presence of chaos in the two-body dynamics, however, chaos will not affect GW detection. Infact Hartl showed [36] that despite the existence of chaotic solutions, physically realistic solutions to the Papapetrou equations are not chaotic. The chaotic solutions either do not correspond to realistic astrophysical systems, or involve a breakdown of the test-particle approximation leading to the Papapetrou equations (or both). As a result, the gravitational radiation from bodies spiraling into much more massive black holes (as detectable, for example, by LISA) should not exhibit any signs of chaos.

It is well known that the collapse of supermassive star produces supermassive black hole with a small amount of disk. To investigate gravitational waves in the accretion phase of the collapsing supermassive star, black hole perturbation approach is one of the best tools. Sotani [38] calculated the gravity-wave signal emitted from a disk composed of test particles which almost retains a circular orbit around a Schwarzschild black hole within the time-domain approach [39–41]. They discussed the possibility of extracting the location of the innermost stable circular orbit and the disk radius by combining information from the GW waveform and the power spectrum.

5. Sources for meter-scale ground-based detectors

Bar detectors have been operating for more than 30s years. They are cylindrical-shape bars whose mechanical oscillations can be driven by GWs ($\sim$ kHz). Five cryogenic resonant-bar detectors are in operation since 1990: ALLEGRO, AURIGA, EXPLORER, NAUTILUS and NIOBE [42]. Their sensitivities and bandwidths are expected to improve during the next years. Various sphere GW detectors have been recently proposed and some are under construction in Belgium, Brazil and Spain. de Araujo [43] reviewed the status of the Brazilian spherical antenna Schenberg, which is planned to detect high frequency gravitational waves ranging from 3.0 kHz to 3.4 kHz, and summarized the most probable sources: core collapse in supernova events; (proto)neutron stars undergoing hydrodynamical instability; f-mode unstable neutron stars, caused by quakes and oscillations; excitation of the first quadrupole normal mode of 4-10 $M_{\odot}$ black holes; coalescence of neutron stars and/or black holes; exotic sources such as bosonic or strange matter stars rotating at 1.6 kHz; and inspiralling of mini binary black holes.

I wish to thank the organizers of GR17 for inviting me to chair Workshop B3 and to participate in a very interesting meeting.

References

1. M. Ando *et al.* (TAMA Collaboration), *Phys. Rev. Lett.* **86**, 3950 (2001).
2. B. Abbott *et al.* (LIGO Scientific Collaboration), *Phys. Rev.* **D69**, 122001 (2004); **D69**, 082004 (2004); **D69**, 102001 (2004); **D69**, 122004 (2004).
3. F. Acernese *et al.* (VIRGO Collaboration), [arXiv:gr-qc/0406123].
4. R. Narayan, T. Piran and A. Shemi, *Astrophys. J.* **379**, L17 (1991); S.J. Curran and D.R. Lorimer, *Mon. Not. R. Astron. Soc.* **276** 347 (1995); S. Phinney, *Astrophys. J.* **380**, L17 (1991); V. Kalogera, R. Narayan, D. N. Spergel and J. H. Taylor, *Astrophys. J.* **556**, 340 (2001); V. Kalogera, in *Gravitational Waves, Third Edoardo Amaldi Conference*, S. Meshkov, ed. (AIP, New York, 2000); V. Kalogera, in *Astrophysical Sources for Ground-Based Gravitational Wave Detectors*, J. M. Centrella, ed. (AIP, New York, 2001).
5. M. Burgay *et al.*, *Nature* **426**, 531 (2003).
6. C.L. Freyer, A. Burrows and W. Benz, *Astrophys. J.* **496**, 333 (1998); S. F. Portegies-Zwart and L.R. Yungelson, *Astronomy and Astrophysics* **332** 173 (1998); H.A. Bethe and G.B. Brown, *Astrophys. J.* **506**, 780 (1998); K. Belczynski, V. Kalogera and T. Bulik, *Astrophys. J.* **572** 407 (2002)

7. T. Bulik, D. Gondek-Rosinska and K. Belczynski, *Mon. Not. R. Astron. Soc.* **352**, 1372 (2004).
8. R. O'Shaughnessy, V. Kalogera and K. Belczynski, *Astrophy. J.* **620** 385-389 (2005).
9. M. Benacquista, (in preparation).
10. S.F. Portegies-Zwart and S.L.W. McMillan, *Astrophys. J.* **576**, 899 (2002); S.F. Portegies-Zwart, S.L.W. McMillan and O. Gerhard, *Astrophys. J.* **593**, 352 (2002); M.C. Miller and E.J.M. Colbert, *Int. J. Mod. Phys.* **D13**, 1 (2004).
11. B.M.S. Hansen and M. Milosavljevic, *Astrophys. J.* **593**, 77 (2003).
12. K. Belczynski, T. Bulik and B. Rudak, *Astrophys. J.* **608** L45 (2004).
13. C.M. Will, *Phys. Rev.* **D69**, 1729 (2002).
14. M.C. Miller, *Astrophys. J.* **581** 483 (2002).
15. B. Brügmann, W. Ticky and N. Jansen, *Phys. Rev. Lett.*, **92**, 211101 (2004).
16. J.K. Blackburn and S. Detweiler, *Phys. Rev.* **D46** 2318 (1992); T.W. Baumgarte, G.B. Cook, M.A. Scheel, S.L. Shapiro and S.A. Teukolsky, *Phys. Rev. Lett.* **79** 1182 (1997); S. Bonazzola, E. Gourgoulhon and J.A. Marck, *Phys. Rev.* **D56** 7740 (1997); P. Marronetti, G.A. Mathews and J.R. Wilson, *Phys. Rev.* **D60** 087301 (1999).
17. M. Shibata, K. Uryu and J.L. Friedman, *Phys. Rev.*, **D70**, 044044 (2004).
18. M. Vasuth, Z. Keresztes, A. Mihaly and L.A. Gergely, *Phys. Rev.* **D68**, 124006 (2003).
19. K. Ioka and T. Taniguchi, *Astrophys. J.* **537**, 327 (2000).
20. N. Andersson, *Astrophys. J.*, **502** 708 (1998); J.L. Friedman and S.M. Morsink, *Astrophys. J.*, **502** 714 (1998); L. Lindblom, B.J. Owen and S.M. Morsink, *Phys. Rev. Lett.* **80**, 4843 (1998); N. Andersson, K.D. Kokkotas and B.F. Schutz, *Astrophys. J.*, **510** 846 (1999).
21. P. Arras, É.É. Flanagan, S.M. Morsink, A.K. Schenk, S.A. Teukolsky and I. Wasserman, *Astrophys. J.*, **591** 1129 (2003); A.K. Schenk, P. Arras, É.É. Flanagan S.A. Teukolsky and I. Wasserman, *Phys. Rev.* **D65**, 024001 (2002)
22. R.V. Wagoner, *Astrophys. J.* **578**, L63 (2002).
23. L. Lindblom and B. Owen, *Phys. Rev.* **D65** 063006 (2002).
24. M. Nayyar and B. Owen, (in preparation).
25. L. Gualtieri, J.A. Pons and G. Miniutti, *Phys. Rev.* **D70** 084009 (2004).
26. P. Brady, T. Creigthon, C. Cutler and B. Schutz, *Phys. Rev.* **D57**, 2101 (1998).
27. C. Cutler, I. Gholami and B. Krishnan, (in preparation).
28. M. Shibata and S. Karino, *Phys. Rev.* **D70** 084022 (2004).
29. A. Nagar, G. Diaz, J.A. Pons and J.A. Font, *Phys. Rev.* **D69**, 124028 (2004).
30. B.S. DeWitt and Brehme, *Ann. Phys.* **9**, 220 (1960); Y. Mino, M. Sasaki and T. Tanaka, *Phys. Rev* **D55**, 3457 (1997); Y. Mino, M. Sasaki and T. Tanaka, *Prog. Theor. Phys. Suppl.* **128**, 373 (1997); T.C. Quinn and R.M. Wald, *Phys. Rev.* **D56**, 3381 (1997); S. Detweiler and B.F. Whiting, *Phys. Rev.* **D67**, 024025 (2003); L. Barack and A. Ori, *Phys. Rev.* **D67**, 024029 (2003).
31. See, e.g, E. Poisson, *Living Review Relativity* **7** (2004).
32. S.A. Hughes, S. Drasco, É. É. Flanagan and J. Franklin, [arXiv:gr-

qc/0504015], to appear in *Phys. Rev. Lett.*; S. Drasco and S.A. Hughes, (in preparation).
33. S. Drasco and S. A. Hughes, *Phys. Rev.* **D69**, 044015 (2004).
34. J. Levin, *Phys. Rev. Lett.* **84**, 3515 (2000); J. D. Schnittman and F. Rasio, *Phys. Rev. Lett.* **87**, 121101 (2001); N. J. Cornish and J. Levin, [arXiv:gr-qc/0207016]; N. J. Cornish and J. Levin, *Phys. Rev. Lett.* **89**, 179001 (2002); M. D. Hartl and A. Buonanno, *Phys. Rev.* **D71** 024027 (2005).
35. T. Tanaka, Y. Mino, M. Sasaki and M. Shibata, *Phys. Rev.* **D54**, 3762 (1996); S. Suzuki and K. Maeda, *Phys. Rev.* **D55**, 4848 (1997); S. Suzuki and K. Maeda, *Phys. Rev.* **D55**, 023005 (2000);
36. M . D. Hartl, *Phys. Rev.* **D67**, 024005 (2003); **D67**, 104023 (2003).
37. K. Kiuchi and K. Maeda, *Phys. Rev.* **D70**, 064036 (2004).
38. H. Sotani, (in preparation).
39. M. Davis, R. Ruffini, W. H. Press and R. H. Price, *Phys. Rev. Lett.* **27**, 1466 (1971).
40. T. Tanaka, H. Tagoshi, and M. Sasaki, *Prog. Theor. Phys.* **96**, 1087 (1996).
41. K. Martel and E. Poisson, *Phys. Rev.* **D66**, 084001 (2002); K. Martel, *Phys. Rev.* **D69**, 044025 (2004).
42. P. Astone et al., *Europhys. Lett.* **12**, 5 (1990); G. Pallottino, in "Gravitational Waves, Sources and Detectors," p 159 (Singapore, World Scientific, 1997); E. Mauceli et al., *Phys. Rev.* **D54**, 1264 (1996); D. Blair et al., *Phys. Rev. Lett.* **74**, 1908 (1995); M. Cerdonio et al., *Class. Quantum Grav.* **14**, 1491 (1997).
43. J.C.N. de Araujo, O.D. Miranda, C.S. Castro, B.W. Paleo and O.D. Aguiar *Class. Quantum Grav.* **21**, 521 (2004); J.C.N. de Araujo, O.D. Miranda and O.D. Aguiar *Class. Quantum Grav.* **22** S471 (2005).

C.1(I): GRAVITATIONAL-WAVE DETECTION (I) DETECTOR PERFORMANCE, OPERATION, AND COMMISSIONING

EUGENIO COCCIA

INFN Gran Sasso and University of Rome "Tor Vergata", Rome (Italy)
e-mail: coccia@lngs.infn.it

The Workshop C1(i) regarded the performances, operation and commissioning of Earth based Gravitational-Wave Detectors. Steady progress has been made over the last years in preparing these instruments to search for gravitational waves from the universe. Detectors of two kinds are presently taking data and also being developed towards ultimate efficiency, namely wide-band, km-baseline, interferometers and cryogenic resonant bars. Major advances in the physics and technology of the related areas have been achieved and further advances are needed and indeed expected. Similar progress is continuing in methods for disentangling weak and rare signals from noise. The first generation of long baseline interferometric detectors is now operating or coming into operation. The detectors TAMA in Japan, GEO in Germany and the LIGO Observatory in the USA, have performed the first scientific runs, while the VIRGO detector in Italy is in the commissioning phase. The resonant-mass detectors (ALLEGRO in the USA, AURIGA and NAUTILUS in Italy and EXPLORER at CERN), are all in data taking and new spherical detectors are in preparation (MiniGrail in the Netherland and Shenberg in Brazil). The status and the plans of the VIRGO detector have been presented by Raffaele Flaminio. The two arms of the interferometer have been locked and the start of the first science run is foreseen in 2005. Progresses for the the dual recycled gravitational-wave Detector GEO 600 has been presented by Grote Hartmut while Takahashi Ryutaro reported on status of TAMA 300. The interferometer sector was completed by the contributions by Loupias Magali on the implementation of an automatic mirror alignment system in Virgo and by Ben-Aryeh Yacob on the use of squeezed states in Michelson interferometers. The sector of resonant-mass detectors was represented by all the groups active in the field. Alessio Rocchi presented the status of EXPLORER and NAUTILUS, which are in continuous data taking since years, with high duty cycle. Also ALLEGRO is taking data since February: its status was reported by Warren Johnson. The AURIGA detector, recently upgraded with a broadband readout and a new suspension system, has also started a new science run: the detector performances were reported by Antonello Ortolan. Arlette DeWaard and Odylio Aguiar presented the status of two new detectors: spherical resonators sensitive in the 3 kHz frequency range. In particular MiniGrail has been cooled to very low temperature, and various tests were performed using different readout systems. The features of the most promising MiniGrail readout (a two-stage SQUID system) were reported by Luciano Gottardi. The Workshop experienced a good number of participants and many discussions with the speakers. It can't be less, as the moment of this activity is certainly very interesting. The large interferometers, complemented by the present bars, constitute a network of gravitational wave antennas which may well give a first detection in the next few years.

C1(II): GRAVITATIONAL-WAVE DETECTION (II) CURRENT GW DETECTOR RESULTS

NOBUYUKI KANDA

Department of Physics
Osaka City University, Sugimono 3-3-138
Sumiyoshi-ku, Osaka 558-8585, Japan
E-mail: kanda@sci.osaka-cu.ac.jp

The workshop session C1ii was focused on the results of recent operating detectors. 10 speakers presented the latest results of each experiments: ALLEGRO, GEO, LIGO, TAMA and VIRGO experiments. There were reports about searches for gravitational waves in analysis of observation data. The results are of no detection of gravitational waves, but observational upper-limits of gravitational waves are improved.

1. Introduction : Recent Status of Gravitational Wave Experiments

Gravitational wave detection is of interest not only for the proof of general relativity, but also for physics with strong gravitational fields and for astrophysics and cosmology. Since the gravitational interaction is weak and hard to shield, the direct detection of gravitational waves has not yet been accomplished. However, following on the recent development of high sensitivity detectors, we expect that detection will be achieved in the near future.

Some of the large-scale interferometric gravitational wave detectors have begun observational operation in the last five years, and some resonant mass gravitational wave detectors are continuing operation and development. In this session C1ii, representatives of four laser interferometric experiments and one resonant bar experiment presented a total of 10 talks. The authors reported briefly on the status of the data analysis of these experiments, and picked up on related issues.

One of the most interesting developments in this year is the completion of full optical configuration of the VIRGO detector [1]. Regis Barille presented the latest status of VIRGO. VIRGO had a test run (C3) during

April 23-27th 2004, and a run C4 in June 24-29th 2004. In each run, the newer frequency stabilization worked well. The interferometer was operated with two main cavity common modes locked to the reference cavity in run C4. VIRGO plans to have the C5 run in September 2004, in recycled and recombined mode with complete automatic alignment. The talk also included the improvement of the new laser injection system. The start of the observational operation of VIRGO will mean that the international network of laser interferometric detectors will be lined up waiting for the coincidence observations.

Another European laser interferometric gravitational wave detector is GEO [2]. The talk of Bangalore S. Sathyaprakash was on the status of GEO data analysis, and an introduction to their data evaluations. GEO had scientific observational runs with LIGO during August 23 - September 9 2003 that (S1), and during December 30th 2003 - January 13th 2004 (S3). GEO detector achieved the strain sensitivity (= strain equivalent noise spectrum): $h \sim 3 \times 10^{-21}[1/\sqrt{\text{Hz}}]$ around 1kHz and duty cycle of the operation came up to 98.5% in S3.

Peter Saulson gave a brief summary of LIGO [3] operation and data analysis status. LIGO achieved great success in its sensitivity development. The scientific run S1 in August-September 2002 was held with GEO, up to 96 hours of observation of LIGO detectors with three-folded operations. The search results of the S1 run are published as Refs. [6], [7], [8], [9]. The S2 run was held during 59 days of February - April 2003. Simultaneous observation of LIGO detectors derived 318 hours of three-folded operations. In the S2 run duration, TAMA and ALLEGRO detectors also had observational runs. During 70 days of December 2003 - January 2004, LIGO had S3 run with best strain sensitivity: $h \sim 5 \times 10^{-23}[1/\sqrt{\text{Hz}}]$ at 100-200 Hz. In the S3 run, GEO and TAMA had simultaneous partial runs. Following P.Saulson, four speakers from LIGO scientific collaboration gave descriptions of the search issues; Duncan A Brown (inspiral), Erik Katsavounidis and Lee S. Finn (bursts), Alberto Vecchio (continuous), and John Whelan and Tania C. Regimbau (stochastic). Secs.2.1, 2.3, 2.4 and 2.5 explain more detail of each presentation along with other groups' results.

Daisuke Tatsumi presented the data analysis topics and operation of TAMA [4]. TAMA had first and second observations in August 1999 which were called DT(Data Taking)1 and DT2. Until Summer of 2004, TAMA had nine periods of observation. In DT6, August 1st - September 20th 2001, they had 1038hours of actual data with a duty cycle of 86%. DT8 in February - April 2003 (a coincidence run with LIGO S2) had 1168hours of data, and DT9 in November 2003 - January 2004 was a partial coincidence observation with LIGO S3. TAMA achieved a strain sensitivity of $h \sim 2 \times 10^{-21}[1/\sqrt{\mathrm{Hz}}]$. The TAMA search tasks presented were inspiral gravitational waves from neutron star binaries (Sec.2.1), burst gravitational waves (Sec.2.3), black-hole ringdown (Sec.2.2) and continuous gravitational waves (Sec.2.4).

As the author mentioned already above, following on from the starting up of interferometer observations, there have been some trials of coincidence searches between these detectors. Three LIGO detectors and GEO detectors have commissioned their operations together. LIGO and TAMA achieved a coincidence run in 2002, DT8/S2. LIGO and TAMA also had a partially simultaneous observation in DT7/S1, and in the winter of 2003-2004 DT9/S3. Using the S2-DT8 simultaneous observations, they are analyzing the coincidence in inspiral gravitational waves searches and in burst gravitational waves.

Martin P McHugh displayed the stochastic gravitational waves search in the cooperation between ALLEGRO [5] and LIGO Livingston site. This is an interesting challenge of coincidence detection between the resonant mass GW antenna and laser interferometric GW detector. Sec.2.5 gives more about it.

There were also reports on hardware development for the issues of data analysis. These were on test signal injection and calibration of data quality. Ik Siong Heng reported on the burst gravitational wave signal injection. TAMA reported on on-line calibration and veto analysis. These are described in Sec.3.

2. Searches for Gravitational Waves

The main task of data analysis in gravitational wave detectors is searching for gravitational wave events. None of the talks reported the detection of such events, however detection ranges and observational upper limits were displayed. The following describes progress in data analysis for the currently operating detectors.

2.1. *Gravitational Waves from Binary Inspirals*

Coalescence of compact star binaries; i.e. neutron star pairs, black-hole pairs, or neutron star - black-hole pairs are promising sources of gravitational waves. The gravitational wave radiation will become stronger in amplitude and of higher frequency as each rotating star drops to orbits of lower radii. This is called binary *"inspirals"*. When the star are isolated, each may be treated as point-like and waveforms are well predicted by the Post-Newtoninan approximation. The gravitational wave amplitude and frequency evolution are well characterized with masses of stars. The matched filter method is an optimal filtering with known waveform and with known noise power spectrum. Using an expected waveform template $h(t)$, the correlation between the gravitational wave and the observed data $s(t)$ can be given as;

$$c(\tau) = \int \frac{\widetilde{h}^*(f) \cdot \widetilde{s}(f)}{S_h(f)} e^{-i2\pi f\tau} df, \tag{1}$$

where $\widetilde{h}(f)$ and $\widetilde{s}(f)$ are Fourier transformation of $h(t)$ and $s(t)$, and $S_h(f)$ is a noise power spectrum of the detector. If there is a gravitational wave in the data at t_0 in time domain, $c(t_0)$ will have a peak. The signal to noise ratio (SNR) can be given as $c(t_0)$. In searches for inspiral gravitational wave events, we try to find the best matching for the bank of templates with the variations of source parameters.

LIGO, GEO, and TAMA displayed their searches or performances.

In LIGO S1, they searched for the inspiral binaries in 236 hours of data. Detectable Range in the optimal direction for 1.4-1.4$M_\odot$ binaries is 176 kpc for LIGO Livingston detector (LLO) and 46 kpc for Hanford detector (LHO) with SNR > 8. They obtain the observational upper limit of binary inspiral gravitational waves as < 170 [event/yr] with C.L. 90% in "Milky Way Equivalent Galaxy (MWEG)" [6]. In S2 data analysis, the range reached 1.8 Mpc (LLO), 0.9 Mpc (LHO 4km), and 0.6 Mpc (LHO 2km). The preliminary upper limit in S2 is improved to < 50 [event/yr] with C.L. 90%.

GEO's detection range is monitored online as a sensitivity of the detector. The range reaches > 45 kpc. They participated in the analysis of data from the S1 run that led to new upper limits on the rate of binary neutron star coalescences.

TAMA developed their detection range since 1999 [10] [11]. The latest typical detectable range is 72.5 kpc for $1.4 - 1.4M_\odot$ binaries in DT9. The average ranges obtained were 30.7 ± 6.4 kpc in DT8, and 64.3 ± 10.9 kpc in DT9 for optimal direction with SNR > 10. TAMA's observational upper limit of binary inspiral is ~ 29 [event/yr] with C.L. 90% in our galaxy, for 1-3 $M_\odot$ binaries.

2.2. *Black-Hole Ringdown Gravitational Waves*

Search for the black-hole ringdown gravitational waves is reported by TAMA group. Perturbation theory around the black-hole predicts gravitational wave emission from the quasi-normal modes of the black-hole. The waveform is predicted as damped sinusoidal which is characterized by the mass and Kerr parameter of the black-hole. Since the waveform is well predicted, it is possible to search the data using matched filtering as same as inspiral. Their search focuses on the mass region 4 - 100 $M_{\odot}$, and on events in our galaxy. In this analysis, spurious noises generate many fake events. They introduce a veto method, and suppress the fake 10 times better than in preliminary analyses.

2.3. *Burst Gravitational Wave*

The "burst" gravitational waves are expected as short transient gravitational waves from some candidate sources: supernovae, stellar-core collapse, black-holes, cusps of cosmic string, etc. Since most of the scenarios for burst sources do not predict the exact and generic gravitational waveform, searches in each experiment are a kind of excess filters of power, slope of signal, or time-frequency spectrogram etc. In other cases, the search done is a check on an anomaly of external signals triggered by other observations ; e.g. GRB, neutrino detectors, etc.

LIGO S1 burst search [7] obtained the burst rate upper limit as < 1.6 [event/day] for sensitivity $h_{RSS} \sim 4 \times 10^{-19} [1/\sqrt{\text{Hz}}$. S2 analysis is expected to improve the sensitivity to $h_{RSS} \sim 4 \times 10^{-20} [1/\sqrt{\text{Hz}}]$, at least 10 times better than the S1 result. They also tried to search for evidence in the huge gamma-ray burst GRB030329 during S2 run, using cross correlation between two LHO detectors around the GRB trigger time. No event exceeded the analysis threshold. The upper limit which was associated with this GRB was $h_{RSS} \sim 6 \times 10^{-21} [1/\sqrt{\text{Hz}}]$. LIGO are proceeding the burst coincidence analysis with GEO(S3 run), HETE, TAMA and AURIGA.

TAMA searched for the burst gravitational waves from stellar-core collapses using an excess power filter. The filter evaluates the signal power in a time-frequency window, and tries to find the excess. By adopting the veto with auxiliary monitored signals of the detector instruments, they succeeded suppressing noise. The DT9 analysis improved sensitivity by a factor of 3 - 6 and the fake rate by a factor of 30 - 100 compared with DT8. The preliminary upper limits (fake rate) is $\sim 10^{-3}$[Hz] for $h_{RSS} > \times 10^{-19} [1/\sqrt{\text{Hz}}]$. On the other hand, they implemented a wavelet-based filter. Since the noise - which has a broad frequency spreda - will be localized in wavelet space, it makes it possible to subtract it from wavelet coefficients by setting an adequate threshold. The subtraction of noise and extraction of clearer burst waveforms was demonstrated.

2.4. *Continuous Gravitational Wave Radiations*

Pulsars might emit continuous gravitational waves according to its rotation, if the star has a local distortion of form, precession, accretion gas, or has a R-mode radiation. Since the frequency of gravitational waves is expected to be twice the pulsar's self-rotating period, it is possible to search for gravitational waves from known pulsars.

LIGO S1 analysis gave an upper limit [8] on PSR J1939+2134 as $h < 1.4 \times 10^{-22}$, C.L.95%. In S2 analysis, they tried to search for gravitational waves from 28 known pulsars above 50 Hz. The preliminary results of best upper-limit on amplitude is $h = 1.7 \times 10^{-24}$ for PSR J1910-5959D, and on ellipticity is $\epsilon = 4.5 \times 10^{-6}$ for PSR J2124-3358.

TAMA was tried to search for gravitational waves from SN1987A remnant pulsar, where a rather large amplitude is expected with a large spin-down rate of $2 - 3 \times 10^{-10}$[Hz/s]. Their upper-limit [12] with 1200 hours data of DT4 and DT6 is $h \sim 5 \times 10^{-23}$, C.L.99%.

2.5. *Stochastic Gravitational Wave background*

Stochastic gravitational wave background can arise either from several processes ; inflation, cosmic strings, first phase transition of Electro-Weak interaction. The spectrum of the gravitational wave background Ω_{gw} can be characterized as a fluctuation of energy density ρ_{gw}.

$$\Omega_{gw}(f) = \frac{1}{\rho_c}\frac{d\rho_{gw}(f)}{dlnf}, \tag{2}$$

where ρ_c is a critical density. Waveforms for this "fluctuation" cannot be predicted and we will use the cross correlation between two (or more) detectors.

LIGO S1 gave an upper limit [9] as $\Omega_{gw}h_{100}^2 < 23 \pm 4.6$ for frequency range 64-265 Hz using LHO(2km) and LLO detectors. Preliminary results of the S2 upper limit improved by three orders of magnitude to $\Omega_{gw}h_{100}^2 < 0.018 + 0.007 - 0.003$ for 50-300 Hz. They expect two more orders improvement with S3 data analysis.

ALLEGRO tried to search the stochastic background during a coincidence run with the LIGO Livingston detector. These detectors are separated by only 40km, allowing searches in the frequency band of up to kHz, which is higher frequency than combination of LIGO interferometers. They expect that 75 hours of S2 coincidence run will yield $\Omega \sim 6$ for frequency band 850-950 Hz.

3. Hardware Development on the Issues of Data Analysis

In order to carry out reliable and robust event searches in the data analysis, we need to develop sensitivity and calibration accuracy for not only noise levels but the response for gravitational wave incidents.

GEO achieved the injection test during S3 run. They inject the "burst like" waveforms on the electrostatic drive actuator of the mirror control. Analyzing the hardware injected signals, they evaluate the algorithm. The signal detection consists of a type of time-frequency clustering called HACR, and excess power filtering. All injected signals were confirmed by both two algorithms, and the evaluation of response was successful.

TAMA also tried similar injection on test mass. Moreover, they evaluate the online veto study by applying the calibration method to all of the interesting signals. They obtain the noise spectra and the amount of noise components from laser intensity noise and ℓ_- control in L_- main signals on-line.

4. Conclusions

Large scale laser interferometric gravitational wave detectors are now operating observational runs. Several issues of gravitational wave event search have been incorporated in the data analysis. There is no report of the discovery or the evidence of the gravitational wave in these analyses up until Summer of 2004. However, on search for four sources; binary inspiral, burst, continuous and stochastic background, experiment groups determine observational upper-limits for them.

Some of these experiments have started coincidence operation and analysis, also with the resonant mass detectors. These trials will establish the concrete commissioning of international gravitational wave networks. We hope to detect gravitational waves in the near future.

Acknowledgments

The author is grateful for the contributions of all speakers who gave excellent presentations at the workshop. The viewgraphs which were provided by them supported our writing of this manuscript. The author would like to thank all participants of the workshop session for joining in lively discussions.

The Program of the Workshop

The speakers and titles of this session C1ii of the workshop are listed below according to the order of talks.

speaker (vice speaker)	*title*
Regis Barille	"New results for the injection system of VIRGO"
Bangalore S.Sathyaprakash	"Status of data analysis in GEO600"
Peter Saulson	"Searches for Gravitational Waves by the LIGO Scientific Collaboration"
Duncan A Brown	"Searching for compact binary inspiral signals in LIGO and GEO data"
Lee S. Finn (Erik Katsavounidis)	"Results from LIGO's S2 Run: Gravitational Wave Bursts"
Alberto Vecchio	"The search for gravitational waves from spinning neutron stars: Results from the LIGO Scientific Collaboration"
Tania C. Regimbau (John Whelan)	"The search for a stochastic background of gravitational waves"
Ik Siong Heng	"Analysis of burst hardware injections in GEO600"
Martin P McHugh	"Stochastic background search using the Livingston LIGO and ALLEGRO resonant detectors"
Daisuke Tatsumi	"Status of TAMA data analysis"

References

1. http://www.virgo.infn.it/
2. http://www.geo600.uni-hannover.de/
3. http://www.ligo.caltech.edu/
4. http://tamago.mtk.nao.ac.jp/
5. http://gravity.phys.lsu.edu/

6. B.Abbott, et. al. (LIGO Scientific Collab.), *Phys. Rev. D* **69**, 122001 (2004)
7. B.Abbott, et. al. (LIGO Scientific Collab.), *Phys. Rev. D* **69**, 102001 (2004)
8. B.Abbott, et. al. (LIGO Scientific Collab.), *Phys. Rev. D* **69**, 082004 (2004)
9. B.Abbott, et. al. (LIGO Scientific Collab.), *Phys. Rev. D* **69**, 122004 (2004)
10. H.Tagoshi, et. al. (the TAMA collab.), *Phys. Rev. D* **63**, 062001 (2000)
11. H.Takahashi, et. al. (the TAMA collab.), *Class. Quantum Grav.* **21**, S697-S702 (2004)
12. K.Soida, et. al. (the TAMA collab.), *Class. Quantum Grav.* **20**, S645 (2003)

C.1(III): GRAVITATIONAL WAVE DETECTION (III): ADVANCED DETECTOR RESEARCH AND DEVELOPMENT

NORNA A. ROBERTSON
Edward L. Ginzton Laboratory
Stanford University
450 Via Palou
Stanford, CA 94305, USA
e-mail: nornar@stanford.edu

The session on advanced detector research and development was principally focused on ideas and experiments investigating the reduction of thermal noise in suspensions and test masses for interferometric gravitational wave detectors. Recurring themes were the potential use of silicon, cryogenic operation, and methods of fabricating suspensions. With respect to the optical design of future detectors, we learnt about the potential advantages of using non-Gaussian beams, the possibility of optimizing signal-to-noise ratio by tuning the signal recycling cavity, and recent results on squeezing light at frequencies of interest for gravitational wave detection. We look forward to seeing many of these innovative ideas incorporated in future detectors.

The session started with a presentation from **Warren Johnson** (Louisiana State University) entitled **BLITS: Big, Low-temperature Interferometers of Transparent Silicon**. The speaker referred to an Aspen workshop in February 2004 at which this concept was discussed. The motivation was to consider how one could achieve sensitivity beyond that of Advanced LIGO in the future. By using bigger test masses one can decrease quantum "back-action" noise at low frequencies, which noise would be increased as a consequence of using higher laser power to reduce photon shot noise at high frequencies. Cryogenic operation will decrease another important noise source - thermoelastic noise. The third element, transparency, is motivated by considering the continuing use of conventional interferometer topology which uses transmissive optics. The speaker noted that several of the GR17 talks to follow reveal significant research efforts towards some aspects of this concept. High purity single crystal silicon of diameter up to 22 inches and mass 300 kg is already available commercially, and larger composite masses should be possible to fabricate. High quality surface finish can be obtained. The speaker referred to viewgraphs from a talk by S. Rowan at a previous meeting in which she discussed various attractive material properties of silicon. These included the two zeros in its coefficient of thermal expansion at $\sim$ 120 K and $\sim$ 20 K, at which temperatures thermal deformations and thermo-elastic noise should tend to zero. She had concluded that silicon is well worth studying further for use at cryogenic temperatures. Regarding using silicon as a transmissive optic, the speaker acknowledged input from S. Whitcomb. Silicon is nominally transparent above the band edge, ($\lambda > 1.2$ microns). There exist high power laser systems at $\sim$ 1.5 microns under development for telecom use, and there are reasonably efficient photodiodes at this wavelength. Measurements of the transmissivity of silicon at 1.5 micron of less than 5 ppm per cm at room temperature have been reported. However, one needs to investigate the practicality of obtaining such behaviour in large crystals of the requisite purity at the desired operating temperature and laser power density. At present a commercially available laser at 1.55 microns is an erbium fiber laser with output power of 100 Watts, and the power available from such lasers is likely to increase. However it remains to be seen if this type of laser can have the high amplitude and frequency stability required. Finally the speaker noted that InGaAs photodiodes with up to 70% quantum efficiency at this wavelength are available, but also remarked that one needs to investigate whether such diodes can handle the required optical power.

Results from the LIGO Bench program, used to estimate performance of an interferometer limited by fundamental noise sources, show that a "dream" BLITS, using 700 kg test masses and 400 W of input power and assuming low mirror coating losses are achieved, could increase the observable range for neutron star-neutron star inspirals by a factor of $\sim$ 5 beyond that of Advanced LIGO. Many problems are still to be overcome. However the speaker ended on an optimistic note, based on his experience with cryogenic bars, that some suspension problems become easier at low temperature.

Paola Puppo, (INFN Rome) presented a talk entitled **Development of low vibration cryogenic systems for interferometric gravitational waves antennas**, describing the work done collaboratively with L. Brocco, A. Cinelli, S. Frasca, E. Majorana, C. Palomba, M. Perciballi, P. Rapagnani and F. Ricci from INFN Rome and Università "La Sapienza". She presented the preliminary results on a new active system designed to damp the vibrations coming from a pulse tube refrigerator which can be used to cool down a suspended mirror (see Figure 1). This system should improve the sensitivity of interferometric detectors by reducing one of the dominant noise sources: the mirror thermal noise. They have cooled a two inch diameter by one cm thick mirror made of CaF_2. The mirror is suspended by two copper wires, each of diameter 0.4 mm, wrapped around its lateral surface, which also act as thermal links with the cryocooler cold head. Preliminary results are shown in Figure 2 where the temperatures of the two thermal screens attached to the first and second cold head stages of the cryocooler and of the center of the CaF_2 mirror surface are plotted. The measurements are in agreement with the finite element simulation of the system's thermal behavior.

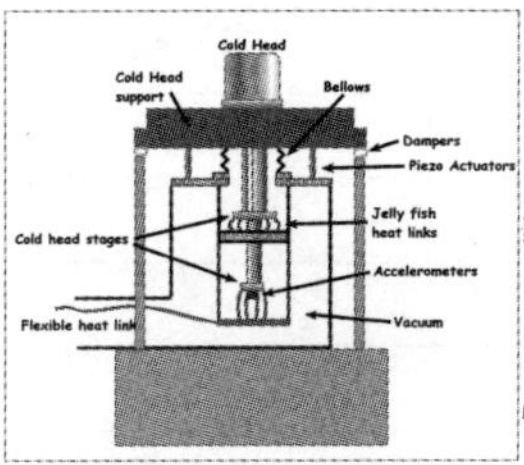

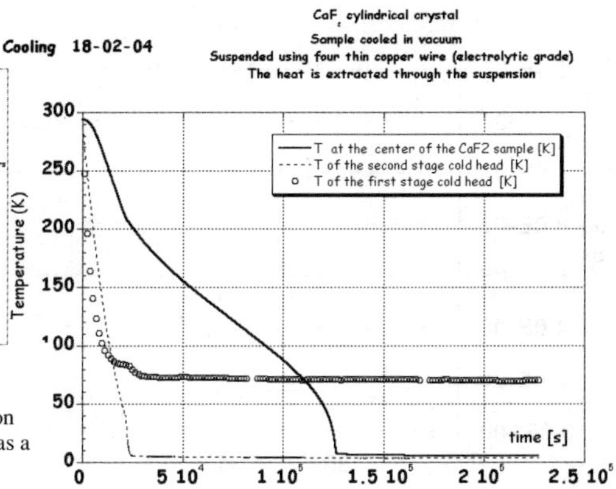

Figure 1 (above). Diagram of apparatus.
Figure 2 (on right). First results on cooling of the suspended mirror as a function of time.

From these data and with the help of the simulation they inferred at which temperatures thermal equilibrium is reached. Moreover, in the steady-state condition, they observed a difference between the temperature values on the center of the mirror surface and the copper wire contact points. This effect could be explained by the presence of a thermal boundary resistance between the CaF_2 mirror and its suspending wires as well as by an unexpected thermal input. Further investigations are in progress.

Continuing the theme of thermal noise reduction, **Sheila Rowan** (University of Glasgow) presented **Developments in test mass and suspension materials for future detectors** describing work done by her and colleagues from Glasgow and Stanford Universities, S. Reid, G. Cagnoli, D.R.M. Crooks, E. Elliffe, M.M. Fejer, A. Heptonstall, J. Hough, P. Murray, O. Solgaard, R. Route, P. Sneddon, and S. Zappe. The overall goal of their programme is to develop low dissipation suspensions suitable for GEO-HF, a proposed upgrade to GEO 600 - the German/UK detector, and possible third generation detectors. Samples of silicon have been fabricated with varying properties and dimensions to study the intrinsic and thermoelastic dissipation in silicon as a function of temperature.

Two sets of experimental results at room temperature were shown. Firstly results were shown of losses in silicon "diving board" samples (5.7 cm by 1 cm by 95 micron) measured by exciting modes electrostatically and monitoring their rate of decay using a shadow sensor. The measured losses over a frequency range $\sim$ 30 to 1800 Hz were consistent with mostly thermoelastic loss - with 'excess' loss being most likely due to surface losses. Work is ongoing to allow cryogenic measurements, with a cryostat commissioned and cooled to 78K.

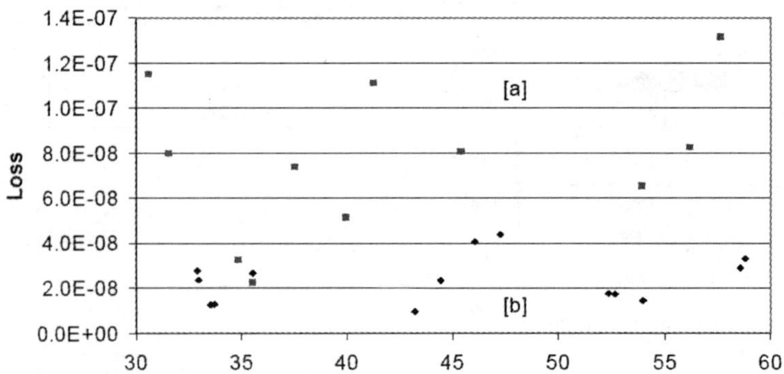

Figure 3. Loss measurements in bulk cylindrical silicon samples each 4 inches diameter by 4 inches thick: [a] (squares) - [100] cut, undoped, [b] (diamonds) - [111] cut, boron doped.

The second set of measurements presented were for losses in two samples of bulk single crystal silicon, see Figure 3. The boron doped [111] cut sample typically showed lower dissipation than the [100] cut, nominally undoped sample, and the lowest loss obtained so far was $(9.6\pm0.3)\times10^{-9}$, comparable with the lowest loss factors measured to date at room temperature. The reason for the difference seen between the samples is under investigation. Further work will include investigation of thermoelastic dissipation as a function of temperature, studies of internal dissipation of doped and undoped bulk silicon as a function of temperature, fabrication and geometry issues for suspension elements and test masses, and techniques for reducing coating dissipation.

In the next talk, **Michele Punturo** (INFN Perugia) presented further work on silicon in **New suspension designs for future GW interferometric detectors**, discussing the work of himself and Virgo colleagues M. Alshourbagy, P.Amico, L. Bosi, L. Gammaitoni, M. Lorenzini, G. Losurdo, F. Marchesoni, F. Martelli, M. Mazzoni, F. Piergiovanni, L. Predolin, R. Stanga, A. Toncelli, M. Tonelli, F.Travasso, F. Vetrano and H.Vocca. He noted that new designs and materials are necessary to realize the suspension system of the optics in second and third generation interferometric gravitational wave detectors, and silicon fibres and ribbons appear to be good candidates. Their group has developed the technology, based on a micro-pull down technique, to produce mono-crystalline silicon fibres longer that 20 cm. The total loss angle measured on these fibres shows a frequency behaviour in agreement with the expected model, see Figure 4, where the thermoelastic peak is located at high frequency, thanks to the high thermal conductivity of the silicon. This will permit the use of monolithic silicon fibres in room temperature suspensions.

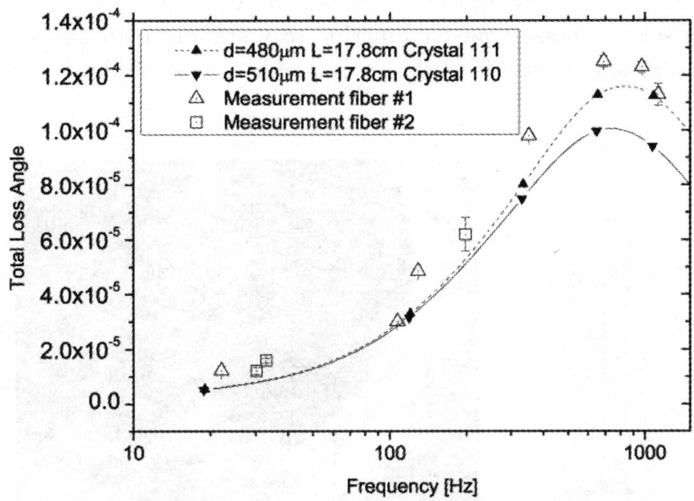

Figure 4. Loss angle for single crystal silicon fibres as a function of frequency.

Further, the temperature dependence of the expansion coefficient of crystalline silicon is such that the expansion coefficient is null at about 120 and 20 K. At these two "zero-expansion" temperatures the thermoelastic loss will vanish. They propose to use a local cooling design to cool down the flexural point of the suspension, to reach at least the highest "zero-expansion" temperature. The speaker observed that third generation detectors will be based on cryogenic suspensions, and silicon would be a good material for the suspensions and the optics in such detectors. Its high thermal conductivity will permit extraction of the heat deposited in the mirrors by the high power laser light circulating in the optical cavities. A preliminary scheme for the last stage of suspension using a monolithic silicon design was proposed.

Angela Di Virgilio, INFN Pisa, presented a talk on progress developing and using the **Low Frequency Facility** (LFF), involving collaborators in the Virgo project from Firenze-Urbino, Napoli, Pisa and Roma. The LFF is designed to study the noise of a 1 cm Fabry Perot (FP) cavity, suspended from a Virgo-like super attenuator (SA) suspension, see figures 5 and 6. The cavity is formed between the VM (Virgo mirror) and the AM (auxiliary mirror). With the use of the SA suspension, the noise of a test mass attached to the SA should be dominated by the suspension thermal noise by 10 Hz, and thus this experiment is aimed at measuring off-resonance thermal noise in the 10 Hz region. In addition the LFF can measure the seismic noise transfer function of a standard SA suspension. The control philosophy used is as close as possible to that used in the Virgo SA.

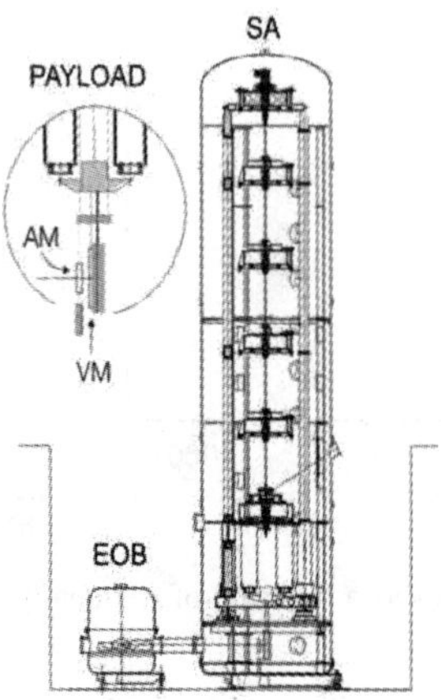

Figure 5 (on left). Schematic diagram of the FP cavity suspended from the super-attenuator, interrogated by light from the optics in the electronics and optics bench (EOB) tank.
Figure 6 (above). Picture of the FP cavity in its tank.

The LFF assembly was completed in the summer of 2001, with first lock of the cavity achieved in summer 2002. Results taken in 2003 showed that there were problems with non-stationary noise in the 10-100 Hz region and that DAC and coil driver noise limited the performance at 10 Hz. Various modifications have been made since then to improve the system, including incorporation of coil drivers which can be operated in two modes, high noise-high power and low noise-low power. Several runs of approximately 10 minutes duration have now been taken running in the low noise state and it has been verified that there is no seismic noise contamination above a few hertz. The general conclusions so far are that the LFF cavity can be locked for several hours with a finesse higher than 3000 and optical gain higher than 10^{10} V/m, and that low noise runs can be performed. Preliminary power spectra have been taken, and work is in progress to calibrate and interpret the data.

Caroline Cantley and **David Crooks** (University of Glasgow) gave a talk on the **Development of a CO_2 laser machine for pulling and welding of silica fibres and ribbons**, describing work done with colleagues G. Cagnoli, R. A. Jones, J. Bogenstahl and P. Holt. The aim of this work is to push silica suspension technology, initially developed for GEO 600, to its limits for room temperature application. To produce higher specification silica fibres and ribbons than those achieved for GEO 600, the Glasgow group are developing a CO_2 laser pulling and welding machine. The conceptual design is currently being developed, see Figure 7, with the research and development partly funded by the EGO organization.

Pulling and welding of silica fibres using CO_2 laser radiation has already been demonstrated in the Glasgow lab. Fibres were pulled using a manually operated jig employing the feed & pull technique. This work will shortly be extended to ribbons. Initial CO2 manually-pulled fibres have been profiled, see figure 8 and the tolerances are already of a similar order to automated flame drawn fibres as used in GEO 600.

The diameter self-regulation effect, in which the heat absorbed by the fibre is balanced by that radiated from the surface at a particular diameter, has been demonstrated and investigated theoretically to see if it could be exploited in this application. However, it has been found from the literature that the absorption depth in silica is $\sim$ 34 microns at 25 °C and decreases with temperature. For their application, typical ribbon dimensions (1 mm x 0.1 mm) are too large to take advantage of this effect.

Characterisation of the laser drawn fibres, including material Q measurements, continues. The group is also investigating automated methods of profiling including edge detection, refraction and absorption techniques.

Juri Agresti, (Caltech and Pisa University) contributed a presentation

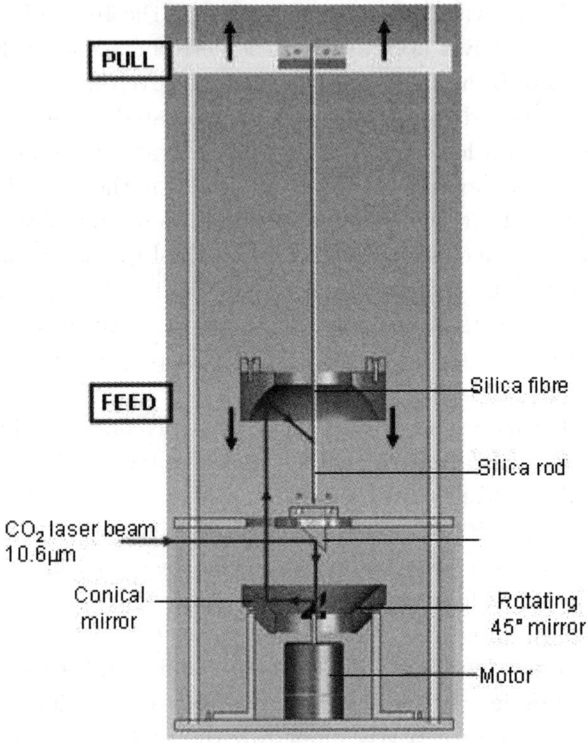

Figure 7. Conceptual design of an automated CO_2 laser pulling machine for fabricating silica fibres.

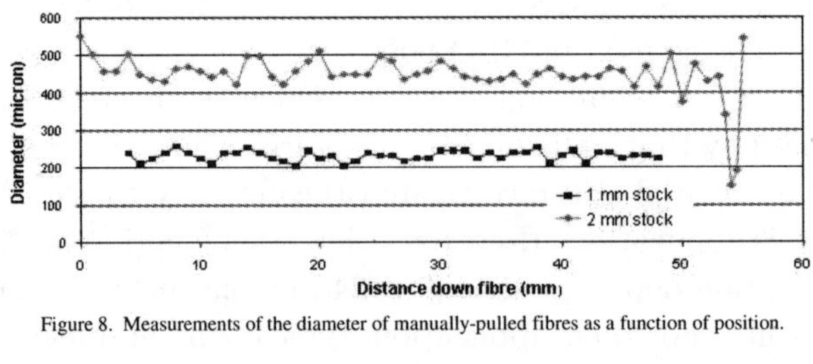

Figure 8. Measurements of the diameter of manually-pulled fibres as a function of position.

entitled **Flat-Top Beam Profile Cavity Prototype**, discussing work in collaboration with colleagues at Pisa, the LIGO Laboratory at Caltech and LMA, Lyon, E. D' Ambrosio, R. DeSalvo, J.M. Mackowski, M. Mantovani, A. Remillieux, B. Simoni and P. Willems. A non-Gaussian, flat-topped, laser beam profile supported by non spherical mirrors known as "Mexican Hat" mirrors, has recently been proposed as a way to lower mirror thermal noise and thus improve the sensitivity of future interferometric gravitational wave detectors including Advanced LIGO. Both substrate and coating thermal noise decrease as beam-spot radius increases. Brownian thermal noise is suppressed with the inverse of the square root of the radius, and the thermoelastic noise is suppressed with the inverse of the square root of the cube of the radius. Coating thermal noise is suppressed as the inverse of the radius. The main motivation of their project is to demonstrate the feasibility of this new concept, designing and building a Fabry-Perot cavity which can support a flat-top beam profile and which represents a scaled down version of Advanced LIGO cavities. The prototype, shown in figures 9 and 10, is a 3.65m long, rigid structure, suspended under vacuum, housing a 7.3m long folded cavity, with finesse of 100. The supporting rods are made of Invar, a material with low thermal expansion coefficient. To reduce ground vibration effects the cavity is suspended by a system of four maraging wires for horizontal isolation and GAS (geometric anti-spring) blades for vertical isolation.

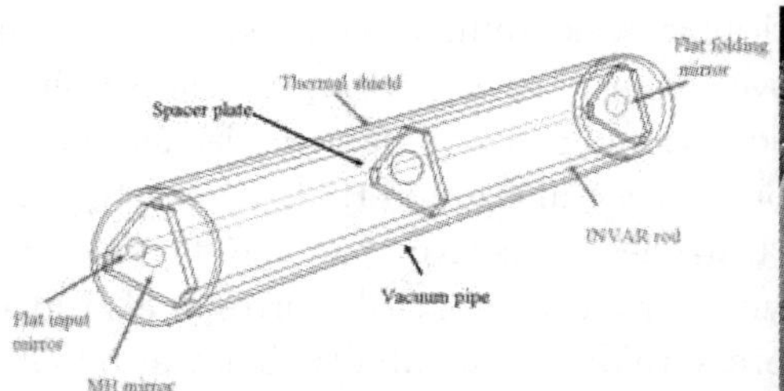

Figure 9 (above). Diagram of cavity prototype apparatus
Figure 10 (on right). Photograph of apparatus.

The Mexican Hat mirror is being produced at LMA-Lyon. The MH profile is obtained by carefully depositing silica from a molecular beam over a flat surface until the desired profile is obtained (precision of less than 10 nm peak-to valley but maximum allowable slope $\sim$ 500 nm/mm).

A test made with spherical optics shows that the device is working as designed. Due to the good seismic isolation and thermal stability they were able to manually align the cavity and lock it with an extremely simple scheme (modulating the input current of the pump laser using the photodiode current signal from the transmitted light). When the Mexican Hat mirror is available, they will characterize the cavity mode structure and study misalignment effects.

The next development may be to switch to a nearly concentric Mexican Hat cavity which would provide the same beam power distribution over the mirror surface, and hence same thermal noise suppression. The advantages of such a configuration was covered in the poster presented by **Juri Agresti** entitled **Equivalence relation between optical cavities and application to Advanced LIGO** on work done in collaboration with Erika D'Ambrosio at LIGO Laboratory, Caltech. They generalized the known equivalence between flat and concentric cavities to the case of non-spherical mirrors supporting a flat-top beam. What this equivalence tells us is that:

(1) The intensity distributions on the mirrors for the modes of the two equivalent resonators are the same,
(2) There is a unique mapping between the eigenvalues of the nearly concentric and nearly flat cavity for all orders,
(3) The two cavities have the same diffraction loss per bounce.

It is known that for spherical mirrors nearly concentric cavities are more stable than nearly flat cavities for misalignments coupled to radiation pressure. They used their proven equivalence to compare the sensitivity to misalignment of nearly concentric and nearly flat cavities supporting the same flat-top beam profile on mirrors, and found that the nearly concentric configuration is about 250 times less sensitive than the flat one, with respect to the mirror's anti-symmetric tilt. That tilt corresponds to the case in which there is radiation pressure driven instability. They also found that the nearly concentric Mexican Hat cavity is more stable than the nearly concentric spherical cavity. These results are in agreement with work done by S. Vyatchanin and P. Savov.

Hans-Peter Nollert, (University of Tübingen) presented a talk entitled **Tuning GEO600 to the signal of the late stage of a collapsing binary** on work carried out with collaborators from the Universities of Hannover and Tübingen, K. Danzmann, H. Ruder and H. Lück. The goal of this work is to optimize parameters of an advanced interferometric gravitational wave detector (specifically GEO600) for the best possible measurement of a specific signal or signals. As a first step they have considered adjusting the signal recycling (SR) cavity during the measurement for an optimum detection of a binary inspiral. The example chosen was the last stage of the inspiral of an equal-mass neutron star binary, $M_{NS} = 1.4 M_{\odot}$.

The dilemma is that most of the power of the inspiral signal is at low frequencies since the binary system spends much more time at lower frequencies than at higher ones. On the other hand, higher-frequency parts of the signal may be more interesting for astrophysical reasons: during the last part of the inspiral (say, the last second) one may study the validity of a higher-order post-Newtonian treatment of the inspiral, and the signal of the actual merger following the inspiral may contain a host of information about the equation of state of a neutron star, relativistic hydrodynamics, and black hole formation.

The solution (hopefully) is to run the detector at the lowest possible SR resonance frequency, examine the output online for signs of an inspiral signal, and then switch the SR cavity to a pre-defined "interesting" higher frequency. The potential gain depends on the reflectivity of the SR mirror. As currently realized for GEO600, the reflectivity is 0.98. One can improve the SNR in the frequency range 600 - 800 Hz by a factor of 1.376 and in the frequency range 1400 - 1600 Hz by a factor of 4.040. If the reflectivity is increased to 0.995, the possible improvement of the SNR in the frequency range 600 - 800 Hz is by a factor of 5.186 and in the frequency range 1400 - 1600 Hz by a factor of 34.807. The lower frequency range corresponds to 100 ms before merger, and the higher frequency range to the end of inspiral and merger.

There remains a question to answer. Is the potential gain sufficient to justify the effort required to implement this scheme? Moreover, how fast can one change the resonance frequency of the SR cavity without it falling out of lock? Further work will be aimed at non-stationary detector SR simulations, measuring the detector response to changes in the SR resonance frequency and implementing and testing the complete procedure.

David McClelland (Australian National University) presented a poster summarizing recent results on **Squeezing in the audio gravitational wave detection band**, describing work done with collaborators Kirk McKenzie, Nicolai Grosse, Malcolm B. Gray, Warwick P. Bowen and Ping Koy Lam from ANU and Stanley E. Whitcomb from the LIGO Laboratory, Caltech. Low frequency squeezing can improve the sensitivity of audio frequency measuring devices such as gravitational wave detectors. At ANU they have demonstrated the generation of broad-band continuous-wave optical squeezing down to 280 Hz using an optical parametric oscillator (OPO) operated below threshold. It was found that low frequency noise sources, such as seed noise, pump noise and detuning fluctuations, present in squeezed states generated by optical parametric amplifiers, have negligible effect on squeezing produced by a below threshold OPO. The experimental setup of the OPO/A is shown in Figure 11. The phase of squeezed vacuum was controlled using a noise locking technique. The OPO squeezing spectrum from 100 Hz-100 kHz is shown in Figure 12. Trace (a) shows the quantum noise limit of the homodyne detection system. The measurement of the squeezed light is shown in trace (b). Trace (c) shows the electronics noise of the detection system. The roll-up in noise power in traces (a) and (b) is attributable to the imperfect homodyne cancellation at low frequencies. They see that broad-band squeezing is obtained from 280 Hz to 100 kHz, with the exception of a locking signal peak at 20 kHz. Squeezing could not be measured at 150 Hz and 250 Hz due to power supply harmonics in the electronic noise. A letter on this work has been accepted for publication in Phys. Rev. Lett.

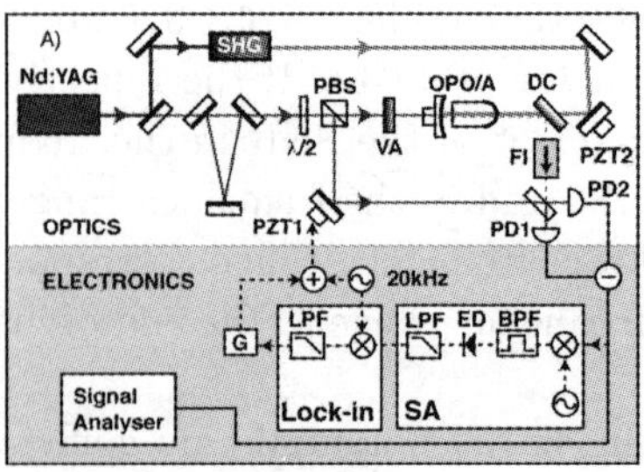

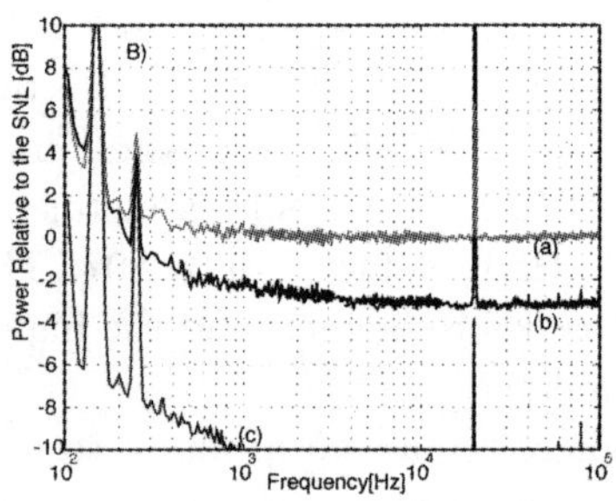

Figure 11 (on left). Schematic of the experiment. The OPA seed power was varied using a variable attenuator (VA). The OPO cavity was isolated from backscatter off the photodetectors using a Faraday isolator (FI).

Figure 12 (on right). Measured noise spectra for (a) the quantum noise limit, (b) the squeezed light and (c) the electronic noise of the homodyne detection system.

Such squeezed states could already be of use for reduction of shot noise in first generation GW detectors. The direction of future research will be to implement an OPO cavity lock, increase the level of squeezing and probe frequencies lower than 100 Hz for use in second generation GW detectors.

Acknowledgements. The author gratefully acknowledges the assistance of all of the workshop contributors for supplying material for this summary, and for commenting on the text.

C.1(IV) GRAVITATIONAL WAVE DETECTION (IV): SPACE-BASED INTERFEROMETERS

P.L. BENDER

JILA, National Institute of Standards and Technology
and
University of Colorado Boulder, CO 80309-0440

There has been a great deal of progress in the last three years in preparing for a gravitational wave (GW) mission in space. The Laser Interferometer Space Antenna (LISA) has continued its development as a joint mission of the European Space Agency (ESA) and NASA. LISA has been an approved ESA mission for a number of years, with the flight schedule to be determined jointly with NASA. In addition, the LISA Pathfinder mission was approved by ESA in November 2003 as a technology validation mission for LISA, with a planned launch date of 2008.

On the US side, LISA has been included in the NASA strategic plan for some time. However, its official approval came in 2003, as part of NASA's Beyond Einstein Program. The targeted launch date is 2012 or 2013. The division of responsibilities recently has been agreed on (August, 2004), with NASA expected to provide the spacecraft and some components of the payload, while ESA and several European National Agencies will take a major part of the responsibility for the payload and provide the main propulsion modules. NASA also is planning to provide a second LISA technology test package, to fly on the LISA Pathfinder mission. This test package is being developed under NASA's New Millennium Technology Validation Program, as the Space Technology 7 (ST7) flight package.

In view of the limited time available for the Workshop, it was decided to concentrate the oral presentations on the experimental side of the LISA and LISA Pathfinder missions. However, two talks on sources, objectives, and data analysis were included as oral presentations, as well as brief introductions to three poster papers in these areas.

The first talk was an overview of the LISA mission, given by Karsten Danzmann of the Albert Einstein Institute, Hannover. The plan for the mission is still basically the same as that resulting from the Industrial

Study at Phase A level, conducted for ESA in 2000. However, the expected launch vehicle is now the Delta IV, which permits a considerably larger launch mass, and thus gives a substantial mass margin for the mission.

The basic plan is to make laser phase measurements between three spacecraft in an equilateral triangle configuration, located 20 degrees behind the Earth, in orbit around the Sun. Each spacecraft would be in an exactly one year period orbit, with roughly 0.010 eccentricity and 0.017 rad inclination. The other orbit elements can be chosen so that the arm lengths stay constant to about 1% over a number of years. The lengths of the three arms of the antenna are about 5 million km. The interplanetary propulsion modules will place the spacecraft in the desired orbits, probably using chemical propulsion. The orbits will be checked using a few days of tracking from the Earth, and then the propulsion modules will separate from the spacecraft.

During a few weeks of commissioning, the directions of the laser beams will be locked onto the proper directions, and adjustments of the optical systems will be optimized. Then the science measurements will begin and last for a nominal lifetime of 5 years. Laser phase measurements with an accuracy of 20 picometers/$\sqrt{\text{Hz}}$ will be made using about 1 W of laser power per transmitter and probably 400 mm diameter transmit/receive telescopes. With one year of observations, this will permit detection of gravitational wave signals with 10^{-23} amplitude with a S/N of 5 near the optimum sensitivity frequency of 5 millihertz.

In the second overview talk, Bernard Schutz from the Albert Einstein Institute, Potsdam, described the types of sources LISA is likely to see and the overall objectives of the mission. The very large number of compact binary stars in our galaxy will form a continuous gravitational wave background below roughly 3 millihertz, but the number of binaries in a one-cycle-per-year bandwidth goes down rapidly at higher frequencies. Thus thousands of close white dwarf binaries will be resolvable at higher frequencies. The study of the distribution of these sources throughout the galaxy will be one of LISA's objectives.

However, the major scientific achievements of LISA are expected to come from detailed observations of three types of gravitational wave sources involving massive black holes throughout the universe. One of particular interest for Fundamental Physics is what are called "extreme mass ratio inspirals." This involves a compact star or stellar mass black hole orbiting around a roughly 10^5 to 10^7 M_{sun} black hole in a galactic nucleus. For $10 M_{sun}$ and $10^6 M_{sun}$ black holes, about 10^5 cycles of GW radiation are expected during the last year before coalescence. The orbits will be eccentric, and the relative velocity near periapsis will be roughly half the speed of light. Thus the GW radiation will be strongly beamed, and the period for precession of periapsis will be comparable with the radial motion period. In addition, the Lense-Thirring effect near a Kerr black hole will be large. Thus sensitive tests of the "no-hair theorem" of general relativity and searches for other deviations from the orbital predictions of the theory will be possible.

Another important source is massive black hole binaries that have been formed when two galaxies merged. If each galaxy already had a massive black hole at its center, the black holes may well have coalesced, subject to some caveats that were discussed. The resulting GW signals would give considerable information on the process of hierarchical formation of the present galaxies.

A third possible source is binaries containing two intermediate mass black holes with masses of roughly $10^2 M_{sun}$ to $10^4 M_{sun}$. The first stars to form probably had higher masses than present ones, and a considerable number per galaxy may have evolved rapidly to form roughly $10^2 M_{sun}$ black holes. If a few of these fall into the center of a galaxy where a roughly $10^4 M_{sun}$ black hole is growing and coalesce with it, the resulting GW signals could be seen even at substantial redshifts. Such signals may provide important information on the formation and growth of intermediate mass black holes.

The next two talks concerned the LISA Pathfinder mission and the test packages to fly on it. Rita Dolesi from the University of Trento described the overall Pathfinder mission and the LISA Technology Package (LTP) being developed by a consortium of European organizations to fly on it. The LTP consists basically of two Gravitational Reference Sensors (GRSs), containing freely moving test masses, and an interferometer system for measuring changes in their separation. Each GRS consists of a vacuum enclosure containing an electrode housing around the test mass, which is a rectangular block of low-susceptibility Au-Pt alloy, 46 mm on a side. The electrodes permit capacitive measurement of the relative displacement of the housing, and thus the spacecraft, with respect to the test mass.

The spacecraft motion is controlled by micronewton thrusters that keep its position constant with respect to one of the test masses to high accuracy. This minimizes changes in the forces on the test masses due to separation-dependent interactions with the housing or the rest of the spacecraft. The overall system is thus called a Disturbance Reduction System (DRS). If the design goal of reducing the disturbing accelerations to $3 \times 10^{-14} m/s^2/\sqrt{\text{Hz}}$ at frequencies above 1 millihertz is met, then the interferometer system will confirm very low relative accelerations of the two test masses in the LTP.

The second test package, provided by the US, is called the ST-7 instrument. It was described by Sasha Buchman of Stanford University. It is similar to the LTP in many ways but involves some planned differences. A displacement sensing system based on "current auto-balance" is used instead of ac voltage-difference sensing between capacitors on opposite sides. The gaps between different test mass faces and the housing are all 2 mm, compared with up to 4 mm for the LTP. The clamping mechanisms for holding the test masses during launch are quite different, and both test-mass-housing combinations are contained in a single vacuum enclosure. However, the overall performance goals are the same as for the LTP. Tests for both systems will be performed with enhanced environmental disturbances (electrical, magnetic, thermal, etc.) to permit extrapolation to the roughly factor of ten better performance planned for LISA.

The fifth talk was by Neil Cornish of Montana State University, on data analysis for LISA. Since the signals from all the GW sources will affect the antenna simultaneously, the different sources have to be separated by their time signatures. The Doppler shifts due to annual motion of the antenna around the Sun provide information on the direction to the source, as do the effects of changes in the antenna orientation during the year. Most resolvable galactic binaries will provide moderate strength signals, with only small frequency changes during a year, and most massive black hole binary signals will be strong. However, clever search procedures will be needed for extreme mass ratio inspirals because of the complexity of the signals.

The subject then returned to experimental issues. William Weber from the University of Trento discussed some of the sources of spurious accelerations of the test masses and torsion pendulum measurements at Trento to quantify them. Of particular interest are electrical forces due to work function differences between the electrodes and the faces of the test mass, plus fluctuations in the charge on the test mass, and time delays in the response to voltage changes. Torsion balance measurements showed that work function differences can be measured initially and then cancelled out with a stability of 2 millivolts over roughly a week. And recent results gave tight limits on the loss tangent that characterizes the response to voltage changes.

Next, Gerhard Heinzel from the Albert Einstein Institute, Hannover, described the interferometry system that has been developed for the LTP. It includes four separate heterodyne Mach-Zehnder interferometers. The main one measures changes in the test-mass separation, with near zero path-length difference to minimize the effect of laser frequency noise. The second one measures the distance of one test mass from a reference location on the optical bench. Both of these will also measure the tip-and tilt-angle of the respective test masses by differential wavefront sensing. The third interferometer provides the stable phase reference for the first two interferometers. And, the fourth interferometer has a large path-length difference to provide a measurement of the laser frequency noise.

Normal incidence reflections are not used in the LTP to avoid possible polarization-dependent effects. Because of the heterodyne approach, accurate measurements of the test mass separation can be made even with substantial changes in the separation.

An engineering model of the interferometer has been built and environmentally tested. Functional and performance tests were all successful. The noise level of the longitudinal (test mass separation and position) channels is nearly at the interferometer-error-budget allowance (10 pm/$\sqrt{\text{Hz}}$ at 3 mHz) and is comfortably below the LPF mission goal. The angular measurements performed even better than expected and have a noise level of only 10 nrad/$\sqrt{\text{Hz}}$ at 3 mHz.

Then Christian Killow from the University of Glasgow described the design, construction, and testing of a prototype optical bench designed to demonstrate precision metrology at a level similar to that required for the proposed LISA and LPF missions. The optical bench consists of a Zerodur baseplate with silica components hydroxide catalysis bonded to it, forming a quasi-monolithic interferometer. The hydroxide-catalysis-bonding technique was originally developed at Stanford and has been extended at Glasgow to bond dissimilar materials and also components requiring adjustment prior to bonding, both of which are vital for the optical bench construction. The techniques and lessons learned during the prototype optical bench construction were adapted and used in the LPF engineering model bench construction.

The completed prototype optical bench was tested and its performance at the LPF displacement goal was verified. In order to achieve this, stabilization of the differential path length of the optical fibers coupling the laser light to the heterodyne interferometer was required. This was a previously unforeseen necessity. Such path length stabilization was investigated at Glasgow and is now included in the LPF design.

The next talk was by Francoise Liorzou from the Office National d'Etudes et de Recherches Aerospatiales (ONERA) in Paris. She described the three-axis gravitational gradiometer that has been developed for the ESA Global Oceans and Climate Experiment (GOCE) mission, which is to be launched in 2006. It consists of 3 pairs of 3-axis inertial sensors mounted on a high-stability frame, with 500 mm separations. The inertial sensors have been derived from the ONERA electrostatic accelerometers flown on the Gravity Recovery and Climate Experiment (GRACE) mission of NASA and the German DRL. The precision has been improved to $10^{-12} m/s^2/\sqrt{\text{Hz}}$.

The last experimental talk was by Jens Gundlach from the University of Washington. He reported on the experimental program for LISA that has been started there. A highly sensitive torsion pendulum has been set up, and test materials to represent the test mass and the capacitative electrodes around it can be brought up very close to each other to magnify the interaction forces. In this way, spurious forces within a factor of 10 of the goal for the LISA Pathfinder mission can be detected. Further improvements in the pendulum design are being made. In this way, the dependence of the interaction force on parameters, such as electric and magnetic field strengths, can be verified, and limits can be set on unexpected forces. Torsion pendulum measurements at the University of Washington, the University of Trento, and possibly elsewhere, are expected to be an important part of the verification of the expected performance of the LISA mission.

The first poster paper was by Jonathan Gair from Caltech and concerned the detection of extreme mass ratio inspirals by LISA. The expected weakness of most such signals and the complexity of the signals make their detection computationally challenging, as mentioned earlier. However, a joint US-European effort has led to a suggested hierarchical approach that seems capable of detecting such signals if the amplitude S/N is about 35. Roughly three week data segments are analyzed coherently, and then the successive segment results are combined incoherently. Quite optimistic estimates of the event rates also have been made.

A second poster paper was presented by Nicolas Yunes from the Pennsylvania State University and Clifford Will from Washington University. The limits that could be set on the parameter ω_{BD} from the Brans-Dicke scalar-tensor theory from different types of LISA sources were discussed. Also, the limits obtainable on the graviton wavelength were presented. The dependences on the LISA arm length and noise amplitude were included in the analysis.

The last poster paper was by Naoki Seto from Caltech. The possibility was discussed of detecting quite small primordial black holes in the halo and central parts of our galaxy, if they make up most of the galactic dark matter. The effect to be looked for would be a Newtonian distortion of a gravitational wave antenna in space when a black hole passed close by. The effect would be too small to be observed by LISA. However, detection might be possible with a future gravitational wave antenna having considerably reduced instrumental noise at low frequencies.

C2: LABORATORY AND OBSERVATIONAL TESTS OF GRAVITATIONAL THEORIES

JENS H. GUNDLACH

University of Washington
CENPA, Box354290,
Seattle, WA98195, USA

Experimental and observational tests of gravity are important in the search for physics beyond General Relativity and the Standard Model and to understand new cosmological findings. Summaries of the talks given in session C2 at GR17 in Dublin in July of 2004 are presented.

1. Introduction

The subject of session C2 at the GR17 meeting in Dublin was experimental and observational tests of gravitational theories that are *not* directly related to gravitational waves. Because of the breadth of the topic and the many recent speculations and theoretical predictions a large number of abstracts were submitted. Only a small subset of the abstracts could be selected to be given as oral presentations. In the allotted time of four hours, 14 talks were presented. The talks were selected to give priority to reports which contained experimental results or which reported about new experiments that have made substantial progress. The talks were, without exception, of highest quality. This made the session particularly fruitful and exciting.

Experimental testing of gravity has emerged as one of the most rapidly growing subfields of modern physics. The underlying motivation to the increasing interest is that break-downs of general relativity will have profound consequences. It is believed that a breakdown of General Relativity, no matter how small, will provide a glimpse of the unification of the Standard Model with General Relativity. Small deviations from General Relativity will also have important implications that could answer some of the most important open questions in the understanding of our Universe. The chance that a deviation from General Relativity will be discovered is considered to be small, but if it is found the pay-off in terms of deep understanding

of fundamental physics would be enormous. It goes without saying that all the experiments in the field are designed to look for very subtle effects and they are therefore considered to be experimentally very challenging, while many are conceptually simple. During the session a wide variety of precision measurement techniques were discussed, ranging from laboratory experiments involving micrometer length scales to tests using ultra precise measurements in the solar system.
The following is a summaries of each of the presentations. Several of the summaries presented here were supplied by the speakers themselves and are printed below with some editorial changes. The summaries are presented in the chronological order as they were given during the session.

2. Summary of the Talks

Short-Range Tests of Newtons Inverse-Square Law

Blayne Heckel, EötWash-group

Blayne Heckel of the University of Washington opened the session with a presentation of the Eötwash group's latest results on testing the $1/r^2$-law at sub-millimeter distances. In the past few years several compelling arguments for violations of Newton's law were proposed. For example, the Eötwash group's tests were motivated by a string theory scenario in which some of the invisible dimensions of the theory attain macroscopic (mm-size) compactification radii. These dimensions would still not be visible using standard model interactions, but the $1/r^2$-law of gravity would be altered. Other motivational arguments include the observed vacuum energy density λ, which corresponds to a length scale of 0.1mm. The smallness of the observed λ, compared to a theoretically expected value, could be explained with a modification of gravity below 0.1mm.
The Eötwash group is using a torsion pendulum, which consists of a horizontal disk that has two rings of 21 holes drilled through it. Concentrically below the pendulum disk is a disk with the same hole pattern in it. This "attractor" disk is rotated slowly so that its gravitational field causes a small variation of the torque on the pendulum. The gravitational torque however is minimized by placing another rotating disk underneath the attractor. This disk also has a similar hole pattern, but the holes are larger and placed exactly out of phase with the upper attractor plate. The lower attractor disk is designed to cancel most of the $1/r^2$-gravity for a mm-separations between the attractor and the pendulum. A thin metal membrane is stretched between the attractor assembly and the pendulum to avoid non-gravitational

coupling. The magnitude of the gravitational coupling, which is reflected in the torsional amplitude of the pendulum angle, was measured as a function of the vertical separation between the pendulum over the attractor and was compared to a predictions using Newtonian gravity. Within uncertainties the results are consistent with Newtonian gravity. New limits on violations in the form of a Yukawa interaction were established that rule out coupling strength of 1.0 relative to Newtonian gravity down to about 100μm. The measurements, for the first time, demonstrate the existence of gravity at a length scale as short as 0.1mm. The results are summarized in a recent paper. [1]

Limits on New Inverse Power Law Forces

Ephraim Fischbach

In this talk by Ephraim Fischbach of Purdue University gave an overview of limits on new power-law forces. Presently the search for new interactions has focused primarily on setting constraints on Yukawa potentials, which have the general form,

$$V(r) = \alpha \frac{Gm_1m_2}{r} e^{-r/\lambda}, \tag{A.1}$$

between point particles of mass m_1 and m_2 separated by a distance r. Here α represents the strength of the force relative to gravity and λ is its range. In his talk Ephraim Fischbach summarized his work with Dennis Krause and reminded us that many theories predict the existence of new forces with potentials having the general form

$$V_n(r) = \alpha_n \left(\frac{Gm_1m_2}{r}\right)\left(\frac{r_0}{r}\right)^{n-1}, \tag{A.2}$$

where α_n is a dimensionless constant and the length scale $r_0 \equiv 10^{-15}$m is chosen by convention. For example, the exchange of two massless quanta (photons, scalars, pseudoscalars, neutrinos, etc.) leads naturally to such inverse power law (IPL) potentials. In addition, recent theories which postulate the existence of small compact extra dimensions would modify Newtonian gravity to a form given by Eq. (A.2) when the interparticle separation becomes smaller than the size of the new dimensions.

Currently the best constraints on new IPL forces come from short-ranged gravity experiments and atomic physics experiments [2]. Surprisingly, recent sub-millimeter gravity and Casimir experiments have not improved upon these limits. This highlights differences between IPL and Yukawa forces: (1) For $n \leq 4$, there is a very weak separation dependence of IPL forces between macroscopic bodies. (2) IPL forces are scale invariant, so r_0 cannot be determined by any experiment (unlike λ). (3) The power n acts like λ in that limits on α_n for higher powers of n come from shorter distance experiments. Since the limits on IPL forces are generally much weaker than limits on Yukawa forces, Fischbach concludes that new experiments dedicated to the search for IPL interactions are needed.

New Limits in Hypothetical Forces in the Casimir Regime Using the Isoelectronic Technique

Riccardo S. Decca

Dr. Decca reported on ongoing measurements of a collaboration involving Indiana University, Purdue University, Lucent Technologies and Wabash College, to improve constraints imposed on short-range Yukawa-like hypothetical forces. He described their microelectromechanical-based system and showed the results obtained using a Au-coated probe and two Au-coated films, made out of Au and Ge. The thickness of the common Au-coating was adjusted so that, in the absence of unknown forces, no differential signal would be detected as the probe oscillates between the Au and Ge. Their experimental approach exploits the high quality factor of a microelectromechanical torsional oscillator (MTO), while avoiding, at the same time, the need to directly excite the system at its resonant frequency, f_r. The required hypothetical signal at resonance was obtained by vertically dithering the MTO at a frequency f_z, while moving the Au-coated sphere over the Au-Ge interface at a frequency f_x. Since the hypothetical force is dependent upon both, the separation and the difference in density of the films in a multiplicative way, the approach yielded signal at $f_r = f_z + f_x$. The measured differential signal between Au and Ge was directly converted into a limit on the standard parameters α and λ, which characterize Yukawa deviations from Newtonian gravity, without having to resort to detailed theoretical descriptions of the Casimir interaction. It was shown that hypothetical forces have to be smaller than $\alpha \approx 10^{12}$ at 200 nm. This represents an improvement of a factor of 10 in this range. The authors also gave a short description on the main limitations of their current design and how to improve it to further constrain hypothetical forces in the submicron range.

University of Washington Equivalence Principle Tests

Blayne Heckel, EötWash-group

This talk was again delivered by Blayne Heckel for the Eötwash group at the University of Washington. Stephan Schlamminger, who had been scheduled to give this presentation was prevented from attending the meeting due to US-immigration problems.
Violations of the equivalence principle are contained within practically every grand unification attempt. It is convenient and plausible that such a violation has a Yukawa potential with a range λ and a coupling strength α and a "charge" that is used to parameterize the composition difference between different materials. The Eötwash group has built a torsion balance instrument to search for EP-violations with ranges from $\lambda = 1$ m to ∞. Their torsion balance is installed on a slowly rotating turntable. The pendulum is highly symmetric and carries four Be-test masses and four Ti-test masses arranged in a composition dipole. For Yukawa forces for which the earth is the source mass ($\lambda = 1$ m to 10000 km) the differential acceleration is modulated by the rotation rate of the turntable. An additional 24h-modulation is provided for accelerations towards the sun and a sidereal modulation for differential accelerations towards the galactic center or fixed reference frames. The challenging aspects of this test is the reduction of systematic uncertainties. For example: The rotation rate of the air-bearing turntable was controlled to within a few nanoradian by using a digital feedback loop. Ordinary Newtonian gravitational coupling of pendulum imperfections to the ambient gravitational gradients had to be avoided. A special gradiometer torsion pendulum was constructed to measure the gravitational gradients and 800 kg of lead were machined to eliminate these gradients. Residual moments of the pendulum were reduced by adjusting small trim screws on the pendulum. The gravitational coupling was routinely measured to subtract a small residual coupling. Their apparatus was continously leveled by using controlled thermal expansion of the support feet of the turntable in a computerized feedback to level sensors on the rotating platform. Preliminary results of a 14 day long data set searching for differential accelerations resulted in $\delta a/a_{sun} = 5 \pm 8 \times 10^{-13}$ towards the sun and $\delta a/a_{MW} = 2.6 \pm 2.8 \times 10^{-5}$ towards the Milkyway's center. The latter results are a test the equivalence principle of ordinary matter in the field of galactic dark matter. The group is now taking an extended data set and expects an improvement by a factor of ten over their last comprehensive published results [3].

A Laboratory Free-Fall Test of the Equivalence Principle

Robert D. Reasenberg

Robert Reasenberg reported about a free-fall test of the equivalence principle in preparation of the Harvard-Smithsonian Center for Astrophysics. The work is carried out in collaboration with James D. Phillips. In this Galilean test two pairs of test mass assemblies (TMA) are in free-fall in a co-moving vacuum chamber. Every 1.3 s, the TMA are tossed upward and are in free fall for about 0.8 s. Each TMA is about 1/3 payload mass and carries a corner cube retroreflector. The relative vertical motion of the TMA of each pair is monitored by an SAO Tracking Frequency laser Gauge (TFG) working between the retroreflectors. The derived relative acceleration from a single toss of a pair of TMA will have a precision of $3 \times 10^{-12} g$. The project goal of $\delta g/g = 5 \times 10^{-14}$ can be reached in a single nights run, but repetition with altered configurations will be used to mitigate systematic error. In addition to test-substance interchange, the reconfigurations include top-bottom swaps and frequent left-right swaps of TMA with sub-micron (TFG) measurement of the absolute vertical separation. Because the accuracy goal can be reached quickly, the SAO group plans to study several pairs of materials. In an extension of the laboratory experiment, they are investigating a version that would run on the International Space Station, with an accuracy goal of $\delta g/g = 5 \times 10^{-16}$. Recently, the SAO group operated the instrument. They demonstrated a mass going upward in free fall, but found too much vibration. The vibration causes the TMA to be launched with excessive rotational and translational velocity. To significantly reduce the vibration, they will replace the current set of rails and ball-bearing wheels with air bearings to guide the vertical motion of the vacuum chamber, as had been planned for a later stage of the project. Their recently improved TFG has proved able to follow the TMA spacing prior to launch, even in the presence of unacceptable, micron-scale high-frequency vibration. The TFG's 1s precision, now about 10 pm, needs to be improved to 1 pm. [4, 5].

Differential Experiments to test the Equivalence Principle on Earth, in Space, and in Parabolic Flight

Anna Nobili

At the University of Pisa a group led by Anna Nobili is preparing Galileo Galilei (GG), a test of the equivalence principle in low Earth orbit with a rapidly spinning differential accelerometer [6]. The differential accelerometer consists of two weakly coupled concentric test cylinders made of different materials. The signal is a small displacement of the cylinders sensed in the rotating frame using capacitance measurements. The high rotation rate (in the few Hz-region) reduces many problems that other equivalence principle measurements operated near DC encounter. In a new INFN laboratory in Pisa the group has built and tested a full-scale prototype that can be operated on the ground called GGG. Because of the weight of the rotating parts in the ground test, a sophisticated cardanic suspension is necessary, making the mechanical coupling between the two cylinders much stiffer than for the zero-g version. This yields a proportionally smaller differential acceleration sensitivity and complicates the dynamics of the rotation. A whirl motion arises which displaces the cylinders relative to each other. The group has made great progress in reducing this motion to the sub-micron level. A low frequency stability of 10^{-10} m is expected in the next years, which would correspond to a sensitivity in testing the equivalence principle in the field of the Sun to one part in 10^9. The flight version of the experiment will be intrinsically much more sensitive (absence of motors noise, ground vibrations, floor tilt, etc.), and will also have the advantage over the laboratory experiment of using the earth itself as the source mass rather then the sun. The acceleration towards the earth is 1400 times larger than towards the sun. The goal for the GG is to verify the equivalence principle to one part in 10^{-17}.

Towards the end of the talk Anna Nobili presented an idea of conducting an entirely different free fall experiment on board ESA's zero-g aircraft. Approximately 20 seconds of near-zero-g can be achieved during parabolic dives.

General Relativity Accuracy Test (TEPEE/GReAT): New Designs for the Differential Accelerometer

Valerio Iafolla

This talk was presented by V. Iafolla of the Institute of Interplanetary Space Physics (IFSI; Rome, Italy) representing a collaboration with the Harvard-Smithsonian Center for Astrophysics and Boston University. The group aims to verify the Weak Equivalence Principle (WEP), or universality of free fall, with a precision of several parts in 10^{15}. Their project is called "The General Relativity Accuracy Test" (TEPEE/GReAT). It consists of differential accelerometer dropped from a high altitude balloon. To perform such measurements, the detector must have a very high sensitivity to differential accelerations acting on a pair of test masses of different materials in free fall. At the same time, the detector must be able to reject common-mode external accelerations and gravity gradients. The differential accelerometer will be cooled down to the liquid helium temperature and spun at 1 Hz around the horizontal axis during its free fall inside an evacuated capsule. The capsule released from a stratospheric balloon, will shield the experiment from the effects of residual air at altitudes of 35-40 km. A violation signal (if present) will be recorded at the rotation frequency while gravity gradients act at twice the rotation frequency. In order to reach the accuracy goal of the experiment over an integration time of 25 s, the accelerometer must attain a sensitivity close to $10^{-14}g/\sqrt{Hz}$. The free-fall time is determined by the time that the detector takes to span the co-falling capsule from top to bottom. The new designs of the differential accelerometer strive to minimize the effects of gravity gradients and to reduce the higher-order mass moments acting on the test masses. Various designs of the test masses and arrangements for the mass suspensions have been considered with the goal of satisfying geometrical requirements and detector construction constraints. Furthermore, the design must provide a high-level of common mode rejection of the residual, external accelerations acting on the detector. The proposed experiment represents a good compromise between the on-ground experiments already performed and the future space experiments that could reach an even higher accuracy with greater cost and complexity.

Numerical Simulation of the Test Mass Dynamics for MICROSCOPE - A Space Mission for Testing the Weak Equivalence Principle

Hanns Selig

H. Selig of ZARM - Center of Applied Space Technology and Microgravity in Bremen Germany reported on the current state of his work concerning numerical simulations for the MICROSCOPE mission. This work was carried together with his colleagues H. Dittus and C. Lämmerzahl. MICROSCOPE is a French space mission (CNES, ONERA) that will be carried out in 2007/2008. It aims to be a high precision test of the Weak Equivalence Principle with an accuracy of $\eta = 10^{-15}$.
For the determination of η the differential acceleration of test masses made of different materials will be measured capacitively. Although disturbing forces acting on the satellite will be minimized by use of a drag free control

system, residual accelerations may superimpose a possible η - signal. For the analysis of the resulting signals it is necessary to know the effects of all mission parameters and error sources.
For this reason numerical simulations of the test mass dynamics are performed, where all mission parameters, disturbance and error sources can be taken into account. One aim is to optimize the combination of mission parameters in order to separate the η - signal frequency from the major disturbance frequency bands, the other aim is to get powerful tools for the mission data analysis that will be carried out at ZARM and at several French institutes.
First results of their work show the influence of different mission parameters on the simulated mission data and the corresponding differential acceleration spectra. Even models with only a few parameters show complex results. The basic principles, several examples and an outlook to the future tasks were presented in this talk.

New constraints on Scalar-Pseudoscalar Couplings

Christian Trenkel

In this talk Christian Trenkel reported on recent progress in the search for scalar-pseudoscalar couplings that is being conducted at the University of Birmingham, UK. Their experiment is operated at cryogenic temperatures and is designed to be sensitive to spin-coupling forces with ranges down to a few mm.
He first described the superconducting torsion balance that is used as a force sensor, the details of which have been published elsewhere. [9]
He then described the spin source that generates the new spin-coupling force if it exists. It is based on the principle of magnetic shielding inside a high permeability tube, and uses the magnetic field from the spins to minimize the total field at the location of the unpolarized test mass.
Possible systematic background effects were described in detail, and measurements of residual field levels and magnetic moments were presented. At the current level of sensitivity, direct magnetic interactions with the test masses were shown to be insignificant.
The first series of experiments had been completed. The dominant systematic effect was found to be due to residual magnetic fields from the spin source leaking into elements of the torsion balance itself. This resulted in a non-zero, residual torque. Phase information could be used to rule out a new spin-coupling force.

Converting the measured torque into upper limits on mixed coupling constants, it appeared that existing constraints had been improved by about one order of magnitude at a range of 1mm.
Work was being carried out to improve the magnetic shielding, and a publication of the results was said to be imminent.

Testing Planck-Scale Lorentz Violations with High Energy Astrophysics

Stefano Liberati

Dr. Liberati of the ISAS in Trieste Italy started his talk by discussing the possibility that quantum gravity effects may lead to a violation of the Lorentz invariance in the form of modified dispersion relations for elementary particles. After arguing that such violations could be induced by several models of quantum gravity he showed how such Lorentz violating (LV) dispersion relations are amenable to observational tests even if LV terms are suppressed by powers of the Planck scale. He described several phenomena where these LV effects are amplified and chose a specific LV framework to impose the constraints on. In particular Dr. Liberati considered the QED sector of an effective field theory with dimension five LV operators. This is characterized by dispersion relations for photons and electrons with LV terms of order $E/M_{\rm Pl}$ ($M_{\rm Pl}$ being the Planck scale). The actual strength of these terms is determined by dimensionless coefficients which *a priori* are expected to be $O(1)$ (given that the Planck suppression is already factored out). Dr. Liberati showed that ξ and η, the LV coefficients respectively of photons and electrons, can be severely constrained using available astrophysical observations. In particular he considered the case of the Crab nebula where the simultaneous detection of Inverse Compton emission, extending up to 50 TeV gamma rays, and of synchrotron radiation, reaching up to 0.1 GeV, allows to limit the parameter space of ξ and η to a tiny region around the origin ($\xi = \eta = 0$, i.e. Lorentz invariance). The boundaries of such a region correspond to values of η and ξ much less than one and hence indicate that the framework is severely constrained if not completely ruled out. Perspectives to improve such constraints, and impose similar ones on higher order violations, were discussed in the final remarks. The work presented is discussed in references [10, 11].

The next four talks were on precision tests using space as a laboratory as well as astronomical observations.

Status of Gravity Probe B

Sasha Buchmann

Sasha Buchmann of Stanford University represented the Gravity Probe B space probe collaboration to report on the status of the mission. Gravity Probe B had a picture-book launch and was precisely injected into the desired 640km polar orbit on April 20th, 2004. This Relativity Mission, which has a 40 years history, was designed to measure two effects predicted by General Relativity: the geodetic effect and the Lense-Thirring frame dragging. The core of GP-B houses four redundant spherical gyroscopes. For General Relativity to hold, these gyroscopes must precess by 6.6 arcsec/year for the geodetic effect and perpendicular to that by 0.42 arcsec/year to account for frame dragging. The design precision of GP-B is 0.0003 arcsec/year, while ground testing predicts an even higher accuracy. At the time when the talk was presented GP-B was in the checkout and calibration phase which was originally scheduled to last only 60 days. This phase lasted somewhat longer than expected, but no severe abnormalities were reported. A lock to the guide star (in Pegasus) had been established and verified by briefly pointing the star tracker on a neighboring star. The gyroscopes were in the process to being spun up and performed according to specifications. Of particular interest to the audience were the electrical charging measurements of the freely floating gyros. Charging rates are generally small and were well controlled using UV light.
GP-B has now entered the science mode and continues to be functioning as expected.

Satellite Laser Ranging -based Measurement of Relativistic Effects in the Orbits of LAGEOS and LAGEOS II Satellites

Roberto Peron

Roberto Peron of Università di Lecce Italy presented his work on tracking the LAGEOS satellites. Satellite Laser Ranging (SLR) is one of the most powerful techniques to track satellites. The tracking is done my measuring the travel time of a laser pulse shot by an Earth station and reflected back by corner cube retroreflectors (CCR) mounted on the satellite. The intrinsic precision of this technique allows highly accurate orbit determina-

tion, especially using the kind of geodetic satellites, which are designed to resemble a test mass. For these satellites (in particular the two LAGEOS satellites) the accuracy is below 1 centimeter in single–shot, and reaches sub–millimeter levels for normal points.
This high level of accuracy allows for a series of tests in geophysics, geodynamics and relativistic physics. In particular, study of the LAGEOS and LAGEOS II tracking data can be used to test the Lense–Thirring effect in the Earth field. This frame–dragging effect is caused by a peculiar distortion of space around a rotating mass. Frame–dragging perturbs the orbit of a satellite in such a way that both longitude of the node and argument of pericenter precess. The magnitude of these effects (calculated in the weak–field, low–velocity approximation) is such that it can be experimentally detected: it is of the order of some tens of milliseconds of arc.
The data analysis requires satellite orbit to be modeled as accurately as possible, to extract the tiny relativistic signal. In order to perform this task GEODYN II software (NASA GSFC) together with the most recent models for the satellite dynamics were used. These include descriptions of the static and time–variable gravitational geopotential, PPN relativistic corrections (without including Lense–Thirring) and the most important non–gravitational perturbation sources (like solar radiation pressure). The residual time series for the node reveal a clear secular trend due to frame–dragging.
Non–gravitational perturbations play a crucial role in assessing the uncertainties. Some of the perturbations are not easy to model, and could degrade the quality of the fit. Work is under way to better understand thermal thrust effects and a possible variation of the reflection coefficient of the LAGEOS satellites.
As a useful by–product, the analysis gives a validation of the models used, in particular those of the geopotential. The group used several of these models including recent ones (CHAMP– and GRACE–derived). Not all fits have the same accuracy, and it was found that some models are more suited than others in modeling the dynamics of the LAGEOS satellites.

Review of the Measurements of the PPN parameters γ Carried out with the Cassini Space Mission

Bruno Bertotti

Bruno Bertotti of the Unversità di Pavia presented a review of the astoundingly precise measurements of the PPN parameter γ carried out with the

Cassini space mission, and already reported 2003 [12]. In the past, several experiments, using either the differential deflection of two radio sources near the Sun, or the delay Δt induced in radio propagation, have confirmed Einstein's prediction γ to ≈ 0.001; Cassini has used a third method, the frequency shift, proportional to $d(\Delta t)/dt$ for radio waves sent from a ground station and transponded back from the Cassini spacecraft as they passed near the Sun. The (1σ) result $\gamma = 1 + (2.1 \pm 2.3) \times 10^{-5}$ is consistent with Einstein, with an accuracy ≈ 50 times better than published measurements. This accuracy approaches a sensitivity at which, theoretically, deviations from General Relativity due to the remnant scalar field of early cosmology are expected. Since the experiment has been carried out near solar conjunction, when the signal is strongest, the elimination of the plasma contribution of the solar corona was crucial; this has been successfully accomplished with a five-fold microwave link, which has also used the new K_a microwave band (32-34 GHz). The new radio station DSS 25 of NASA's Deep Space Network at the Goldstone complex (California), with its advanced instrumentation, in particular a water vapor radiometer for the tropospheric contribution, has been used. On board, a very stable transponder provided the appropriate lock between the incoming and the outgoing beams. The spacecraft, on its cruise towards Saturn, was in very stable dynamical state; the main non gravitational forces, due to the radiation emitted by the three thermonuclear generators and the solar radiation pressure have been independently measured, and also modeled in the final fit.

Radio Tests of the Light Deflection Produced by a Moving Body

Edward Fomalont

Ed Fomalont of the National Radio Astronomy Observatory in Charlottesville N.C. reported a light deflection test of signals from quasars in the field of Jupiter. In the past the PPN parameter gamma had been measured from a variety of experiments associated with the gravitational field of the sun. Together with his colleague Sergei Kopeikin they used the close passage of Jupiter on September 8, 2002 to within 3.7' of the line-of-sight of a the quasar J0842, to measure the deflection of light (radio waves) by a significantly moving object and, thus, to measure the (v/c) deflection term. On five days around the Jupiter encounter, the Very Long Baseline Array (VLBA) with the Effelsberg telescope measured, at 8 GHz, the relative separation of J0842 with J0839 about one degree east, and with J0854, about four degrees west. By using two comparison quasars on opposite sides of J0842, the angular dependence of the tropospheric delay above the telescopes could be removed. The position obtained for J0842 on September 8, with respect to four other observing days between September 4 and 13, was consistent with the Shapiro-radial deflection of 1.1 mas plus the aberrational deflection of 0.05± 0.01 mas.

The measurement of the aberrational deflection term agrees with that predicted by GR to 20% accuracy. The question,"what property of gravity is measured by this experiment", has led to two years of discussion in the physics community. Dr. Fomalont argued that the aberrational deflection is related to the propagation speed of gravity. Others believe that this deflection is not at all related to the speed of gravity which only should occur in $(v/c)^2$ terms.

Using the same observing technique as that for the Jupiter experiment, the deflection of the sun can now be measured with a VLBA experiment to an accuracy of 5×10^{-5}, similar to the Cassini experiment of 2002. Since the VLBA can easily detect the Cassini spacecraft, there was some discussion about a combined VLBA/Cassini experiment when the spacecraft goes behind the sun in July 2005. The two somewhat different techniques could reinforce each other to produce an even more accurate value of gamma.

References

1. C.D. Hoyle, D.J. Kapner, B.R. Heckel, E.G. Adelberger, J.H. Gundlach, U. Schmidt, and H.E. Swanson, Phys. Rev. **D 70**, 042004 (2004).
2. D. E. Krause and E. Fischbach, in *Particle Physics and the Universe,* Proceedings of the 9th Adriatic meeting, ed. by J. Trampetic and J. Wess (Springer, 2004), to be published.
3. Y. Su, B.R. Heckel, E.G. Adelberger, J.H. Gundlach, M. Harris, G.H. Smith and H.E. Swanson, Physical Review **D 50**, 3614 (1994).
4. R. D. Reasenberg and J. D. Phillips .Testing the Equivalence Principle on a Trampoline,. in Proceedings of the International Symposium on Testing the Equivalence Principle in Space, N.A. Lockerbie, Guest Ed., Class. Quantum Grav., **18**(13), 2435-2445 (2001).
5. "An Equivalence Principle Test During a 1 Second Suborbital Flight", in the proceedings of the meting on Testing the Equivalence Principle in Space and on Ground, Pescara, 20-23 September, 2004, in preparation.
6. G.L. Comandi, A.M. Nobili., D. Bramanti, R. Toncelli E. Polacco, M.L. Chiofaloa, Phys. Lett. **A 318**, 213-222 (2003).
7. V. Iafolla, S. Nozzoli, E.C. Lorenzini, V. Milyukov, Rev. of Sci. Instr. **69**, 4146- 4151 (1998).

8. E.C. Lorenzini, I. Shapiro, F. Fuligni, V. Iafolla, M.L.Cosmo, M.D.Grossi, P.N.Cheimets and J.B.Zielinski. Il Nuovo Cimento **09** B N. 11 Nov. (1994).
9. Giles D. Hammond, Antonio Pulido-Paton, Clive C. Speake, and Christian Trenkel, Rev. Sci. Instr. **75**(4), 955 (2004).
10. T. Jacobson, S. Liberati and D. Mattingly, Nature **424**, 1019 (2003), [arXiv:astro-ph/0212190].
11. T. A. Jacobson, S. Liberati, D. Mattingly and F. W. Stecker, Phys. Rev. Lett. **93**, 021101 (2004), [arXiv:astro-ph/0309681].
12. B. Bertotti, L. Iess and P. Tortora, Nature, **425**, 374 (2003).

D.1: QUANTUM GENERAL RELATIVITY

THOMAS THIEMANN
Albert Einstein Institut, MPI f. Gravitationsphysik, Potsdam, Germany
and
Perimeter Institute f. Theoretical Physics, Waterloo, ON, Canada
e-mail: thiemann@aei.mpg.de, tthiemann@perimeterinstitute.ca

There were altogether four oral sessions on quantum gravity at the GR17 conference in Dublin: The first two focussed on issues related to Loop Quantum Gravity (LQG), the third on aspects concerned with other approaches to quantum gravity and the fourth concentrated on black hole physics. Altogether there were 31 talks and 15 posters whose contents we briefly summarize below:

I. *Oral Contributions*

A *LQG I: Canonical Approach*

Ai. Lee Smolin described his views on the low energy limit of LQG, in particular the perturbation theory around the Kodama state, the conjecture that the limit is governed by DSR (doubly special relativity) and the construction of a positive definite and hermitian hamiltonian operator.

Aii. Luca Bombelli described Riemannian statistical geometry methods in order to study the low energy limit of LQG and more general quantum gravity theories. These are reminiscent of random graph constructions familiar from statistical physics.

Aiii. Jurek Lewandowski reported on an error in a previous LQG calculation of the black hole entropy within the framework of isolated horizons and was able to fix it together with Molgado based on a work by K. Meissner. Technically this changes the value of the Immirzi parameter, conceptually it means that all the possible spin configurations contribute and not only the spin

1/2 one.

Aiv. Christian Fleischhack reported on the status of the uniqueness of the kinematical representation of LQG discovered by Sahlmann. Under the assumption of spatial diffeomorphism invariance and certain additional technical assumptions which one also uses in the proof of the Stone von Neumann theorem for the representation theory of the Weyl algebra of quantum mechanics, a uniqueness result can be established.

Av. Alberto Molgado applied the framework of refined algebraic quantization (RAQ) for solving quantum constraints to a simple model which is SL(R;p,q) invariant which extends previous work. This is technically nontrivial due to the non-amenability of the gauge group.

Avi. Bianca Dittrich reviewed the Master Constraint Programme for LQG and discussed several non trivial examples. The MCP replaces a set of first class constraints by their weighted square sum (integral) which allows to solve systems with structure functions by RAQ methods.

Avii. Martin Bojowald discussed aspects of the volume operator in several symmetry reduced models such as homogeneous (and isotropic) or spherically symmetric ones, focussing on level splitting when different symmetries are compared. This can be compared to Zeeman and Stark effects applied to the hydrogen atom.

Aviii. Loop quantum cosmology (LQC) due to Bojowald is not the cosmological sector of LQG but rather a truncation to one to three degrees of freedom of this QFT, a toy model. Oliver Winkler explained that the recent finding in LQC, the boundedness of the inverse scale factor, is due to the use of a non standard representation of the canonical commutation relations which results in the space of square integrable functions over the Bohr compactification of the real line.

B *LQG I: Path Integral Approach (Spin Foams)*

Bi. Jorge Pullin described the consistent descretization approach to quantum gravity which consists in discretizing the constraints classically and fixing lapse and shift through imposing the vanishing of the commutators of the inconsistent algebra of the discretized constraints. The resulting theory is free of the constraints which is of conceptual advantage.

Bii. Annalisa Marzuoli described how elements of the spin network basis for LQG can be used for quantum computing.

Biii. Robert Oeckl presented his proposal for a general boundary formulation of quantum mechanics. In this formalism the TQFT axioms are generalized to all boundaries, not only cobordisms. "Dynamics" is no longer restricted to mean a time-evolution, but becomes the association of amplitudes to generalized states. This approach is motivated by an analysis of the measurement process in a context where space and time are supposedly quantum mechanical entities. It is aimed at solving interpretational problems arising from the special role of time in standard quantum mechanics. He described how the problem of time is avoided and how a local prescription of measurement processes is achieved, without recourse to asymptotic states or boundary conditions.

Biv. Luisa Doplicher applied the general boundary proposal to path integral quantization. Her framework can be applied in particular to LQG and could be used to describe graviton scattering.

Bv. Jonathan Engle carried out a non-perturbative quantization of the sector of general relativity admitting isolated horizons with fixed multipoles. In order to do this, one constructs multipole operators and defines the micro-canonical ensemble for the entropy calculation. The counting of states yields the usual Bekenstein Hawking entropy for the same value of the Barbero-Immirzi parameter as in the non rotating case.

Bvi. Chris van den Broeck successfully verified that current proposals for semiclassical states in linearized gravity due to Varadarajan correspond to a Coulomb phase to the theory, i.e. one recovers Newton's potential in a Wilson loop expectation value.

Bvii. Emanuele Berti reported on a calculation that tried to establish whether the correspondence between the quasinormal modes of a ringing Schwarzschild black hole and the black hole area quantum can be extended to the Reissner-Nordstrom and Kerr family respectively. The numerical and analytical results seem to exclude this possibility.

Bviii. Alfio Bonanno studied Euclidean quantum gravity with a of a class of non-perturbative renormalization group flow equations which employ a regulating smearing function in the proper-time integration. Analytical and numerical results for a non-trivial ultraviolet fixed point are discussed for arbitrary dimensions, proper time regulator and gauge fixing parameter. These and related calculations in a cut off theory space seem to indicate the existence of a non Gaussian fixed point.

C *Alternative Quantum Gravity Theories*

Ci. John Klauder gave a brief introduction of his Affine Quantum Gravity programme, which is based on affine rather than canonical commutation relations. Tools and procedures useful for this general analysis include: (i) affine field coherent states; (ii) reproducing kernel Hilbert spaces; (iii) continuous-time regularized functional integrals; (iv) projection-operator treatment of quantum constraints; and (v) unconventional counterterms to account for potential nonrenormalizability.

Cii. Steven Carlip described a possible mechanism for creating a homogeneous early universe from sums over topologies within the Hartle-Hawking "no boundary" proposal for the wave function of the universe. This is based on constructive interference among the amplitudes for different topologies which in the presence of a negative cosmological constant, can lead to sharp peaks in the wave function, which are centered precisely on maximally homogeneous spatial geometries. While such results are unlikely to replace inflation as an explanation for the homogeneity of the Universe, they may fill in an important gap, the open problem of how to obtain an initial homogeneous patch required to start inflation.

Ciii. Seth Major reported on possible quantum gravity phenomenology resulting from a discrete spacetime structure. Effects could be existent in the form of violations of Lorentz invariance for which high-precision experiments and astrophysical observations are already constraining the recently proposed ad hoc models. Two attempts to constrain Lorentz invariance violations, in the preferred frame in the doubly special relativity contexts were presented in that context.

Civ. Ruth Williams discussed the general form of Lagrangians for Scalar fields on triangulated manifolds, focusing on the two-dimensional case. Their structure is restricted by their behaviour under conformal transformations in the associated Regge calculus.

Cv. Ingemar Bentsson showed that using ideas borrowed from statistical geometry and thermodynamic fluctuation theory, one finds a natural metric on the thermodynamic state space of black holes. The resulting geometry (for the Kerr, RN, and BTZ cases) is surprisingly simple. The hope is that the investigation should shed light on the underlying statistical mechanical system (if any), and perhaps reveal special properties of black holes Gibbs surfaces.

Cvi. Victor Valdaya reported on important features of his jet-gauge and jet-diffeomorphism groups unified approach to gravity and internal interactions. In this scheme new fields, the gauge parameters themselves, enter the theory playing a non-trivial role at the quantum level. In fact, they contribute to the mass of some of the intermediate vector bosons without damaging gauge symmetry. In addition, a natural action of the (jet-)diffeomorphism group on the internal symmetry appears which accounts for some non-trivial mixture of interactions.

Cvii. Fernando Barbero reported on work together with

Mena Marugan, and Villase on the asymptotic behavior of regulated field commutators for linearly polarized, cylindrically symmetric gravitational waves. Important physical effects are brought about by the introduction of a physical cut-off in the study of the microcausality. The result is relevant for the study of the full S matrix of the model.

Cviii. David Craig described joint work with James Hartle on the decoherence functional for Bianchi IX cosmologies. The consistent history proposal is a generalization of quantum mechanics for situations without an outside observer. The construction is in accordance with the formal application of the Copenhagen interpretation (wave function of the universe).

D *Quantum Black Hole Physics*

Di. Petr Hajicek showed that Dirac observables for asymptotically flat general relativity admit a nontrivial and physically interesting Hamiltonian dynamics, which is determined by the asymptotic symmetries. A new distinction between gauge transformations and symmetries in general relativity is used.

Dii. Tejinder Singh reported on joint work together with C. Vaz and Louis Witten where they construct an exact quantum gravitational state describing the collapse of an inhomogeneous spherical dust cloud, using a lattice regularization of the Wheeler-DeWitt equation. In the semiclassical approximation around a black hole, this state describes Hawking radiation. The leading quantum gravitational correction to Hawking radiation renders the spectrum non-thermal.

Diii. Robert Mann demonstrated how to extend the formalism of Hamiltonian string bit models of quantum gravity in two spacetime dimensions to include couplings to particles, which are represented by marked bits. The single-particle closed and open universe models respectively behave like empty open and closed universes. A system of two distinguishable particles in a closed universe behaves like an empty closed universe.

Div. Vyacheslav Rychkov gave a status report on the possibility of black hole production in elementary particle collisions within the string theory inspired framework of Large Extra Dimensions scenarios. Previous studies used classical gravity (Aichelburg-Sexl shock waves) to estimate its cross section. It is however conceivable that higher curvature corrections and fluctuations of the gravitational field drastically change this picture.

Dv. Joseph Medved analyzed in very general terms what the spectrum of quasinormal modes versus can tell us about theories of Quantum Gravity.

Dvi. Michael Fil'chenkov pointed out a possible cosmological origin of the rotation of astronomical objects, since the specific angular momentum, corresponding to to Birch's upper limit on the angular velocity of the Universe, much exceeds the specific angular momenta of spiral galaxies. The problem of rotation of the astronomical objects is expected to be resolved in the framework of quantum cosmology. The proposal is to use Raychaudhuri's equation for expansion, rotation, shear and acceleration of a homogeneous medium and to quantize it to give Wheeler-DeWitt's equation for a set of two-component cosmological models with slow rotation.

Dvii. Charles Wang extended a nonlinear quantization of Friedmann and to a field-theoretical description. This is based on parametrizing the action integral for the functional Schrodinger equation in non-perturbative quantum field theory. The new theory provides a natural extension from conventional to nonlinear quantum formulation. This generalization enables a unitary description of the quantum evolution for a broad class of constrained Hamiltonian systems with a nonlinear kinematic structure. The new theory was applied to the quantization of spacetimes with one spacelike Killing field and orthogonal hypersurfaces.

II. *Posters*

Pi. Mamdouh Wanas reported on progress of his absolute parallelism geometry (APG).

Pii. Magd Kahil described quantum properties of APG.

Piii. Dimiter Chakalov described his ideas on what human brain dynamics has to do with quantum gravity.

Piv. Ali Mostafazadeh solved the Hilbert Space Problem (representation and Dirac observables) in certain Minisuperspace Quantum Cosmologies.

Pv. David Wiltshire presented new results on the stability of gravastars which are alternatives to black holes.

Pvi. Jakob Lamey discussed the quantum physical relevance of the Einstein tensor.

Pvii. Takashi Tamaki proved that the highly damped quasinormal mode of the Gibbons-Maeda solution completely gives the same area spectrum in that of Schwarzschild black hole in LQG.

Pviii. Martin Damien reported on the status of the relationship between 2D CFT's and the Bekenstein Hawking entropy formula.

Pix. Carlos Barcelo analyzed the stability of the Einstein static universe in thermal bath filled with radiation.

Px. Cesar Moral presented a noncommutative model of quantum cosmology with matter.

Pxi. Eric Minassian reported on singularity avoidance of collapsing spherically symmetric quantum dust shells.

Pxii. Edward Anderson presented a new 3-space approach (TSA) to geometrodynamics which is different from the ADM formalism.

Pxiii. Petr Velan presented an idea for how to combine canonical methods with doubly special relativity (DSR).

Pvix. Michael Fil'chenkov reported on progress concerning the problem of matter creation in the early universe by considering interactions of scalar fields with the spinor and vector fields of matter.

D2: STRING AND M THEORY

GARY T. HOROWITZ
Department of Physics, UCSB, Santa Barbara, CA 93106
e-mail: gary@physics.ucsb.edu

1. Introduction

As expected at an international relativity meeting, the string and M theory workshop focused mostly on gravitational aspects of the subject. There were two main topics of discussion. The first was gravity in more than four spacetime dimensions. This, of course, is motivated by the fact that string theory predicts higher dimensional space. The talks included discussion of new types of black holes which only exist in higher dimensions (e.g. black rings) and new types of instabilities. For example, Gregory and Laflamme have shown that a uniform black string (the product of four dimensional Schwarzschild and a circle) is unstable if the Schwarzschild radius is less than the size of the circle. Some of the talks discussed generalizations of this result to higher dimensions and others discussed nonuniform black strings that are related to this instability. One explanation for why we have never seen these extra spatial dimensions is the so-called brane-world scenario. This is the idea that we live on a three dimensional "brane" in the higher dimensional space, and all the elementary particles we see (except gravity) are confined to move only on the brane. Some speakers discussed properties of black holes in these brane-world scenarios.

The second main topic of discussion was gravity in asymptotically anti de Sitter (AdS) spacetimes. This is largely motivated by the remarkable AdS/CFT correspondence which roughly states that string theory on asymptotically AdS spacetimes is completely equivalent to a conformal field theory (CFT). There were talks about newly discovered supersymmetric black holes in AdS, black hole hair, and appropriate boundary conditions for linear fields in AdS. There were also several talks focused on trying to use the AdS/CFT correspondence to better understand gravitational singularities. These include singularities inside black holes, possible naked singularities, and cosmological singularities.

Finally, a few talks explored other aspects of string theory, including a possible de Sitter/CFT correspondence and the role of closed timelike curves. The string and M theory workshop contained a total of 17 speakers. The following is a brief description of each talk.

2. Gravity in more than four dimensions

Henriette Elvang (UCSB): "Non-uniqueness of black holes in string theory" Elvang discussed her recently discovered supersymmetric solutions describing rotating black rings in 4+1-dimensional asymptotically flat spacetimes [1]. A spatial slice of the event horizon of a black ring has topology of a two-sphere times a circle. The black rings have five conserved charges, three net gauge charges and two unequal angular momenta. The solutions, however, contain seven independent parameters and thus they provide the first example of non-uniqueness of supersymmetric black holes: classically there are infinitely many black rings with the same conserved charges.

The black rings can be lifted to be solutions of ten-dimensional supergravity, and the three net charges can then be interpreted as D1-brane, D5-brane, and momentum P in string theory. The D1-D5-P system has played an important role in the success of string theory accounting for the microstates responsible for the entropy of extremal and near-extremal black holes. String theory uses the conserved charges of a black hole to identify an appropriate brane-system for which the number of states can be counted. With non-uniqueness for the black rings, more information must be supplied about the brane system, for instance how the system carries angular momentum. If the D1-D5 branes themselves carry the angular momentum, then the system is known as a supertube; this takes the system to a phase different from the one for which previous calculations of entropy has been made. It is an exciting challenge for string theory to account for the entropy of the supersymmetric black rings.

Evgeny Sorkin (Racah Institute of Physics): "Critical dimensions in black-brane phase transitions" Sorkin first considered black strings wrapped over the compact circle of a D-dimensional spacetime which is asymptotically $\mathbf{R}^{D-1} \times S^1$. He constructed static perturbative non-uniform string solutions around the instability point of a uniform string [2]. In such a perturbation theory: (i) He computed numerically the mass of a critical string for a large range of dimensions, D, and found that it follows essentially an exponential law γ^D, where $\gamma \approx 0.68$ is a constant. Considering the large D limit he showed analytically that $\gamma = \sqrt{e/2\pi} \approx 0.66$ in agreement with the numerical results. (ii) He determined that there is a critical dimension, $D_* = 13$, such that for $D \leq D_*$ the phase transition between the uniform and the non-uniform strings is of first order, while for $D > D_*$, it is, surprisingly, of higher order.

Sorkin then gave an expression for the critical mass of a black brane in a background with an arbitrary compactification manifold. This expression is completely explicit for $\mathbf{T}^n$, the n dimensional torus of an arbitrary shape. An indication was given that by employing a higher dimensional torus, rather than a single compact dimension, the total critical dimension for the phase transition can be as low as $D_* \leq 11$.

Takashi Tamaki (Kyoto University): "Radionic non-uniform black strings"

Non-uniform black strings in the two-brane system were investigated using the effective action approach. It was shown [3] that the radion acts as a non-trivial hair of the black strings. From the brane point of view, the black string appeared as the deformed dilatonic black hole which becomes a dilatonic black hole in the single brane limit and reduces to the Reissner-Nordstrom black hole in the close limit of two-branes. The stability of solutions was demonstrated using catastrophe theory. From the bulk point of view, the black strings were proved to be non-uniform. Nevertheless, the zeroth law of black hole thermodynamics still holds.

Cristiano Germani (DAMTP, University of Cambridge): "On the quest for a holographic interpretation of the black hole evaporation process" It has been suggested that classical black holes in the brane-world reproduce the semiclassical behavior of four-dimensional black holes, hence giving an interpretation of the Hawking evaporation process in holographic terms. Germani studied the gravitational collapse of a spherical star in the brane-world in order to clarify the underlying dynamics that leads to this interpretation. He found [4] that effective brane-world equations naturally lead to the collapsing star emitting a flux of energy that approaches the Hawking behavior as the (apparent) horizon is being formed. The trace anomaly of four-dimensional quantum field theory was also interpreted in this context as the brane-world contribution to the trace of the energy-momentum tensor of matter.

Toby Wiseman (Harvard): "Black hole-black string phase transitions in gauge theory" General relativity compactified on a torus admits both black hole and black string solutions (where the horizon wraps the torus cycles). Recent constructions of such solutions indicate a first order phase transition occurs between them. Wiseman described how to use a gauge-gravity duality to relate such a gravity system to a large N gauge theory [5]. At strong coupling, this gravity behavior predicts a corresponding gauge theory phase transition. He described various methods to then study the phenomenon at weak coupling, and found good evidence for such a transition there. This suggested this 'Gregory-Laflamme' behavior is generic in the gauge theory, and persists for all coupling.

3. Gravity in asymptotically anti de Sitter spacetimes

Eric Woolgar (University of Alberta): "AdS/CFT and uniqueness of the AdS soliton spacetime" The generalized positive energy conjecture of Horowitz and Myers arose from positing an AdS/CFT correspondence for spacetimes without supersymmetry. The conjecture posits that the AdS soliton spacetime, a negative mass static solution of the Einstein equations with negative cosmological constant and toroidal boundary at infinity, is the lowest energy spacetime obeying these conditions. By applying Galloway's null splitting theorem, Woolgar showed how to go some distance toward resolving this conjecture by proving a uniqueness theorem (in general dimension) for the AdS soliton [6]. In fact, a certain type of nonuniqueness is allowed. Page has exploited this fact to show that the AdS/CFT correspondence predicts zero temperature phase transitions in the CFT on the torus at infinity.

Harvey Reall (KITP): "Anti de Sitter black holes" String theory calculations of black hole entropy are all applications of the AdS/CFT correspondence for the special case of a two dimensional CFT. In a sense, the only black holes whose entropies have been calculated are BTZ-like solutions. It is clearly desirable to use the correspondence to calculate the entropies of higher dimensional AdS black holes. So far this has not been possible because the corresponding CFTs are strongly coupled. In his talk, Reall explained how it might be possible to overcome this problem for supersymmetric AdS black holes [7]. The first examples of five dimensional, supersymmetric, asymptotically AdS black hole solutions were described and the calculation of their entropies discussed.

Kengo Maeda (Kobe City College of Technology): "Non-uniqueness in gauged supergravity theories" Maeda showed that $\mathcal{N} = 8$ gauged supergravity in four or five dimensions admits a more general class of asymptotic AdS black hole solutions with a nonzero scalar field whose mass lies in the range $-d^2/4 \leq m^2 < -d^2/4 + 1$, where $d = 3$, 4 [8].

The boundary condition on the scalar field is relaxed to admit both the slower and faster fall-off modes. Naively, the mass diverges, but a finite modified mass can be defined which is conserved and associated with the asymptotic global time translations. This means that a black hole uniqueness theorem does not hold for the supergravity theory: for given total energy there exists a static black hole with scalar hair, in addition to the Schwarzschild-AdS solution. A physical interpretation of the hairy black holes was given in terms of a dual CFT. Since the supergravity theory is a consistent truncation of ten dimensional IIB theory or eleven dimensional M-theory, the hairy black holes can be lifted to the higher dimensions as "non-uniform black branes". Hence these results provide a new way to study the Gregory-Laflamme instability, and cosmic censorship violation in this context.

Akihiro Ishibashi (DAMTP, University of Cambridge): "Dynamics in anti-de Sitter spacetime" AdS spacetime fails to be globally hyperbolic, so a classical field satisfying a hyperbolic wave equation on AdS spacetime need not have a well defined dynamics. Nevertheless, AdS spacetime is static, so the possible rules of dynamics for a field satisfying a linear wave equation have been shown to correspond to choices of positive, self-adjoint extensions of a certain differential operator, A. Ishibashi reduced the analysis of electromagnetic, and gravitational perturbations in AdS spacetime to scalar wave equations. He then applied his general results to analyze the possible dynamics of scalar, electromagnetic, and gravitational perturbations in AdS spacetime [9]. In AdS spacetime, the freedom (if any) in choosing self-adjoint extensions of A corresponds to the freedom (if any) in choosing suitable boundary conditions at infinity, so his analysis determined all of the possible boundary conditions that can be imposed at infinity. In particular, he showed that other boundary conditions besides the Dirichlet and Neumann conditions may be possible, depending on the value of the effective mass for scalar field perturbations, and depending on the number of spacetime dimensions and type of mode for electromagnetic and gravitational perturbations.

Veronika Hubeny (Stanford University): "Holographic description of singularities" In the context of the AdS/CFT correspondence, Hubeny discussed the information about bulk geometry that can be extracted from boundary CFT correlators [10]. In particular, she argued that the CFT correlators contain distinct, albeit subtle, signals of the black hole singularity. This was contrasted with a collapsing cosmology embedded in AdS/CFT.

Mukund Rangamani (UC Berkeley): "Cosmic censorship in AdS/CFT" Rangamani discussed recent attempts to violate the cosmic censorship conjecture in the AdS/CFT context. He re-examined the proposal of Hertog, Horowitz, and Maeda (HHM) who argued that for certain initial data there is insufficient energy available to make a black hole whose horizon is big enough to cloak the singularity that forms. He investigated this proposal in the models HHM discuss and was unable to find initial data that provably satisfy this criterion, despite development of an improved lower bound on the size of the singular region [11]. For certain initial data, the energies of certain configurations were not far above the lower bound on the requisite black hole mass, and so it is possible that in the exact time development naked singularities do form. He went on to argue that the finite radius cut-off AdS_5 situation discussed by HHM displays instabilities when the full 10D theory is considered. He proposed an AdS_3 example that may well be free of this instability.

Thomas Hertog (UCSB): "Towards a big crunch dual" Hertog showed there exist smooth asymptotically anti-de Sitter initial data which evolve to a Big Crunch singularity in a low energy supergravity limit of string theory [12]. A preliminary study of the dual CFT evolution suggests the big crunch singularity is an endpoint of evolution even in the full string theory. He also related these results to his earlier work on cosmic censorship violation in $N = 8$ supergravity.

4. Other topics

Robert Mann (University of Waterloo): "Maximal mass and entropy conjecture violations in Taub-NUT de Sitter spacetimes" Motivated by the AdS/CFT correspondence, some people have suggested that there should also be a dS/CFT correspondence. This has given rise to two conjectures about asymptotically de Sitter spacetimes. The entropic N-bound conjecture asserts that the largest entropy that any such spacetime can have is that of pure de Sitter spacetime. The maximal mass conjecture states that any asymptotically dS spacetime with mass greater than dS has a cosmological singularity. Mann demonstrated that spacetimes with NUT-charge provide counterexamples to both of these conjectures [13]. This suggests that there are significant limitations to both of these conjectures. Their calculation of conserved mass and entropy is based on an extension of the path integral formulation to asymptotically de Sitter spacetimes.

Carlos Herdeiro (Oporto University): "Recent developments

concerning closed timelike curves in string theory" Closed timelike curves are present in a number of solutions of classical general relativity and are, in general, regarded as pathological. It was suggested by Hawking in the early 90s that the 'Chronology Protection Mechanism' should involve quantum effects. It becomes therefore interesting to explore the nature of closed timelike curves in a framework of quantum gravity like string theory. The recent discovery of supersymmetric Godel type universes in string theory which are T-dual to pp-waves and the existence of supersymmetric rotating black holes (which have closed timelike curves) possessing a CFT description motivated several suggestions on how string theory might cope with closed timelike curves [14]. In his talk, Herdeiro reviewed some of these recent suggestions.

Belkis Cabrera Palmer (Syracuse University): "Counting supertubes" Palmer discussed her recent work in which the quantum states of the supertube are counted by directly quantizing the linearized Born-Infeld action near the round tube [15]. The result is an entropy $S = 2\pi\sqrt{2(Q_{D0}Q_{F1} - J)}$, in accord with conjectures in the literature. As a result, supertubes may be the generic D0-F1 bound state. This approach also shows directly that supertubes are marginal bound states with a discrete spectrum. She also discussed the relation to recent suggestions of Mathur et al. involving three-charge black holes.

Sergey Cherkis (Institute for Advanced Study): "Self-dual gravitational instantons without isometries" Self-dual gravitational instantons are four-dimensional Riemannian manifolds with self-dual Riemann curvature. These are solutions of the vacuum Einstein equation and preserve half of the supersymmetry. The only compact self-dual gravitational instantons are a flat four-torus and K3 manifold. Noncompact gravitational instantons are distinguished by their asymptotics and topology. Until recently the only self-dual gravitational instantons known were multi-Taub-NUT and Atiyah-Hitchin spaces (as well as the limits of these). Via a chain of string theory dualities, Cherkis obtained the twistor spaces of certain gravitational instantons which generically have no isometries [16]. From the twistor data he computed explicit metrics on these spaces.

Lee Smolin (Perimeter Institute): "String theories with deformed energy momentum relations, and a possible non-tachyonic bosonic string" Smolin discussed a prescription for introducing deformed dispersion relations in the bosonic string action [17]. In a subset of such theories it remains true that the embedding coordinates propagate linearly on the worldsheet. While both the string modes and the center of mass propagate with deformed dispersion relations, the speed of light remains energy independent. Smolin considered the canonical quantization of these strings, and found that it is possible to choose theories so that ghost modes still decouple, as usual. However the Virasoro algebra now exhibits an energy dependent central charge. He also found that there are examples where the tachyon is eliminated from the spectrum of the free bosonic string.

Acknowledgments

I want to thank the organizers of GR17 for putting together such a lively and productive conference. I also thank all the participants of the D2 workshop for their excellent talks. This work was supported in part by NSF Grant PHY-0244764.

References

1. H. Elvang, R. Emparan, D. Mateos and H. S. Reall, "A supersymmetric black ring," Phys.Rev.Lett. **93** (2004) 211302 [arXiv:hep-th/0407065]; "Supersymmetric black rings and three-charge supertubes," Phys.Rev. **D71** (2005) 024033 [arXiv:hep-th/0408120].
2. E. Sorkin, "A critical dimension in the black-string phase transition," Phys. Rev. Lett. **93** (2004) 031601 [arXiv:hep-th/0402216]; B. Kol and E. Sorkin, "On black-brane instability in an arbitrary dimension," Class.Quant.Grav. **21** (2004) 4793-4804 [arXiv:gr-qc/0407058].
3. T. Tamaki, S. Kanno and J. Soda, "Radionic non-uniform black strings," Phys. Rev. D **69** (2004) 024010 [arXiv:hep-th/0307278].
4. R. Casadio and C. Germani, "On the quest for a holographic interpretation of the black hole evaporation process," arXiv:hep-th/0403258; "Gravitational collapse and black hole evolution: Do holographic black holes eventually 'anti-evaporate'?," arXiv:hep-th/0407191.
5. O. Aharony, J. Marsano, S. Minwalla and T. Wiseman, "Black hole - black string phase transitions in thermal 1+1 dimensional supersymmetric Yang-Mills theory on a circle," Class.Quant.Grav. **21** (2004) 5169-5192 [arXiv:hep-th/0406210].
6. G. J. Galloway, S. Surya and E. Woolgar, "Non-existence of black holes in certain Lambda < 0 spacetimes," Class. Quant. Grav. **20** (2003) 1635 [arXiv:gr-qc/0212079].
7. J. B. Gutowski and H. S. Reall, "Supersymmetric AdS(5) black holes," JHEP **0402** (2004) 006 [arXiv:hep-th/0401042]; "General supersymmetric AdS(5) black holes," JHEP **0404** (2004) 048 [arXiv:hep-th/0401129].
8. T. Hertog and K. Maeda, "Black holes with scalar hair and asymptotics in N = 8 supergravity," JHEP **0407** (2004) 051 [arXiv:hep-th/0404261]; "Stability

and thermodynamics of AdS black holes with scalar hair," Phys.Rev. **D71** (2005) 024001 [arXiv:hep-th/0409314].

9. A. Ishibashi and R. M. Wald, "Dynamics in non-globally hyperbolic static spacetimes. III: anti-de Sitter spacetime," Class. Quant. Grav. **21** (2004) 2981 [arXiv:hep-th/0402184].
10. L. Fidkowski, V. Hubeny, M. Kleban and S. Shenker, "The black hole singularity in AdS/CFT," JHEP **0402** (2004) 014 [arXiv:hep-th/0306170].
11. V. E. Hubeny, X. Liu, M. Rangamani and S. Shenker, "Comments on cosmic censorship in AdS/CFT," JHEP **0412** (2004) 067 [arXiv:hep-th/0403198].
12. T. Hertog and G. T. Horowitz, "Towards a big crunch dual," JHEP **0407** (2004) 073 [arXiv:hep-th/0406134].
13. R. Clarkson, A. M. Ghezelbash and R. B. Mann, "Entropic N-bound and maximal mass conjectures violation in four dimensional Taub-Bolt(NUT)-dS spacetimes," Nucl. Phys. B **674** (2003) 329 [arXiv:hep-th/0307059].
14. D. Brace, C. A. R. Herdeiro and S. Hirano, "Classical and quantum strings in compactified pp-waves and Goedel type universes," Phys. Rev. D **69** (2004) 066010 [arXiv:hep-th/0307265].
15. B. C. Palmer and D. Marolf, "Counting supertubes," JHEP **0406** (2004) 028 [arXiv:hep-th/0403025].
16. S. A. Cherkis and N. J. Hitchin, "Gravitational instantons of type D(k)," arXiv:hep-th/0310084.
17. J. Magueijo and L. Smolin, "String theories with deformed energy momentum relations, and a possible non-tachyonic bosonic string," Phys.Rev. **D71** (2005) 026010 [arXiv:hep-th/0401087].

D3: QUANTUM FIELD THEORY ON CURVED SPACETIME

KLAUS FREDENHAGEN
Universität Hamburg

As long as the back reaction of quantum fields on the spacetime metric can be ignored, quantum field theory on a curved background is an excellent approximation to a theory combining quantum effects with gravitational interactions. Famous predictions have been obtained in this framework, in particular Hawking's prediction of thermal radiation of black holes [16] and Unruh's prediction of a thermal environment experienced by an accelerated observer [29]. In the topical session of GR17 on this theory three main subjects were discussed, namely the construction of interacting fields by renormalized perturbation theory, the construction of analogous models for gravitational interactions and the rôle of fluctuations of the energy momentum tensor.

The present status of renormalized perturbation theory was reviewed in a talk of Robert Wald (Chicago). Older attempts starting from path integrals or euclidean field theory suffer from conceptual problems which are due to two major obstacles: one is the nonexistence of a distinguished state which can replace the vacuum of Minkowski space; such a state seems to exist only in very special situations with high symmetry. Therefore there is no distinguished Feynman propagator, hence a formal (real time) path integral is not well defined even in the noninteracting case. In addition the concept of a particle looses its definitive meaning, as is nicely illustrated by the description of the Unruh effect in terms of so-called Rindler particles. As a consequence the S-matrix can no longer be the central object of the theory, and the question arises which quantities of the theory can be brought into contact with real measurements.

The other obstacle is more on the technical side: It has become customary in quantum field theory to perform calculations in the euclidean domain and to go back to real time at the very end by the Wick rotation. In the curved situation this amounts to work first on a Riemannian space and then try to pass to a Lorentzian spacetime. But in the general case, for smooth spacetimes, this is meaningless, and for analytic spacetimes it is only possible in very special situations as e.g. for static spacetimes.

These obstacles have been circumvented during the last years by essentially four ingredients: one is the algebraic formulation of quantum field theory, as proposed long ago by Haag and Kastler [14, 15]. In this framework, the basic objects are the algebraic relations between the fields, not their action on a Hilbert space. As was first emphasized by Kay [22], the *non-uniqueness of quantization* as observed by Fulling [11] for quantum field theories on curved spacetime does not occur in the algebraic approach, due to the fact that the commutator function, defined as the difference between the retarded and the advanced propagator, is unique (on globally hyperbolic spacetimes).

The second ingredient is the use of techniques from microlocal analysis [21]. This was long ago proposed by Duistermaat and Hörmander [7] and was first used by Radzikowski [23, 24] in his analysis of the Hadamard condition on two point functions. By using these techniques it became possible to enrich the structure of algebras of fields such that manipulations became possible which previously required the representation on a Hilbert space. In particular composite fields could be defined [2, 3, 6, 17].

The third ingredient is a formulation of perturbation theory which works entirely in real space and is local. Such a method was developed by Epstein and Glaser [8] on the basis of previous work of Stückelberg [28] and Bogoliubov [1]. This method exploits the fact that the time ordered products of n fields are determined by the time ordered products of less than n fields provided not all arguments coincide. Renormalization then amounts mathematically to an extension problem for operator valued distributions, whose solution on a curved spacetime was found by Brunetti and Fredenhagen [3]. The method, however, leaves a freedom of a choice of renormalization conditions which could, due to a lack of symmetries, vary from point to point.

The latter difficulty was solved by the fourth ingredient, namely an incorporation of the concept of general covariance into quantum field theory. On a first sight, the path integral seems to deliver a generally covariant framework. This is, however, invalidated by the nonuniqueness of the Feynman propagator. On the algebraic side, an obstacle is the fact that diffeomorphisms alter the causal relations; since the latter enter the commutation relations, diffeomorphisms cannot be algebraic automorphisms. The solution comes from a new principle of locality which appears as an idea e.g. already in papers by Wald on the energy momentum tensor [33,34] (whose renormalization should involve the metric only locally). Its precize formulation is due to Brunetti, Fredenhagen and Verch [4](cf. also [17,30]). It characterizes quantum field theory as a functor from the category of globally hyperbolic manifolds (with causality preserving isometric embeddings as morphisms) to the category of operator algebras (with injective homomorphisms as morphisms). General covariance is then incorporated in the covariance of this functor. As a matter of fact, this concept implies the Haag-Kastler framework by restriction to subregions of Minkowski space.

On the basis of this structure, covariant quantum fields can, loosely spoken, be described as a family of fields φ_M on all spacetimes M in the class considered which transform covariantly under the embeddings of one spacetime into the other. On a more formal level, they can be characterized as natural transformations from, e.g., the functor of test function spaces over spacetimes to the functor of quantum field theory. This covariance requirement on the fields amounts to the possibility of identifying fields living on different spacetimes. In this new framework it is then possible to use renormalization conditions which are independent not only of the chosen point of space time but even of the spacetime itself. This analysis was performed in a series of famous papers by Hollands and Wald [17–20].

The new concept of a generally covariant field opens also the way for a physical interpretation of the theory which can no longer be based on the concept of particle scattering. Namely one may start from some standard spacetime (e.g. Minkowski space) and look at a family of states with well known interpretation (for instance thermal equilibrium states). A given state on a given spacetime may then be compared with states in this family by measuring the expectation values of the same covariant field. Such an interpretation was already proposed in a new approach to nonequilibrium thermodynamics by Buchholz, Ojima and Roos [5] and could provide a basis for a thermal interpretation of states on a nonstationary spacetime.

Whereas Robert Wald's talk gave a general overview on the subject, the following talk by Stefan Hollands (Chicago) was concentrated on a more specific problem, namely the question of which renormalization condition implies the conservation of the energy momentum tensor of the interacting theory. The idea is that a *principle of perturbative agreement* should be invoked. This principle means that the theory should not depend on the way the Lagrangian is split into free and interacting part. The implementation of this plausible requirement is not completely straightforward and requires the introduction of retarded fields, which depend not only on the spacetime point but in addition on a symmetric tensor which changes the metric in the future of some Cauchy surface. It was shown in the talk that the condition can be met at least for scalar fields in more than two dimensions. Gravitational anomalies are visible already at the very beginning in this approach [20].

Other talks concentrated on the questions of fluctuations of the energy momentum tensor. Due to the Einstein equations, these fluctuations limit the validity of the semiclassical approximation. In a talk by Larry Ford (Tufts University) the effect of fluctuations on the Raychaudhuri equation was discussed. The Raychaudhuri equation characterizes the contraction of a geodesic congruence and leads under positivity conditions on the energy momentum tensor to the singularity theorems. Ford treated this equation as a Langevin equation with a fluctuating Ricci tensor. It was argued that these fluctuations constitute a gap in the proof of singularity theorems, different from the violations of the positivity conditions by the expectation value of the energy momentum tensor. Furthermore, it was shown that fluctuations might produce fluctuations in the apparent luminosity of a distant source. Interesting effects are also expected in thermal states.

Another aspect of fluctuations was illuminated in a talk by Thomas Roman (Connecticut State University). The by now classical example of the effect of an accelerated mirror was analyzed with respect to correlations of the energy momentum tensor. The acceleration was chosen as to produce an initial burst of negative energy followed by a period of constant acceleration and ending with a compensating burst of positive energy. It was found that nontrivial correlations occur between spacetime regions before and after the pulses, whereas the expectation values within these regions vanish.

The third main topic was the search for analogues of gravitational systems in condensed matter physics. One of the motivations is that the predicted effects of the theory of quantum fields on curved spacetimes are usually too small for being measured with our present technology. For instance, the Hawking temperature of stellar black holes is much below the temperature of the cosmic microwave background. Another motivation is the wish to find an analogous structure in situations which are more familiar to us and might help us in our understanding of the somewhat counterintuitive predictions of the theory. Last not least, if found, such analogous systems might provide interesting new applications.

Matt Visser (Wellington, New Zealand) compared in his talk the equatorial slice of the Kerr geometry with the acoustic geometry encountered by sound propagating in a vortex [31]. There are striking similarities, but also differences, and it was discussed how the analogy could be refined. Typically, the analogues are based on the behaviour of sound in a moving fluid, and the corresponding quantum field is massless. But Silke Weinfurtner (Wellington) presented in her talk a model based on Bose-Einstein condensation which leads to a massive Klein Gordon field on a curved back ground [31]. This may open new possibilities for the actual construction of black hole analogues. It is remarkable, that also in this model, similar to the massless case, a Lorentz invariant field equation arises. The third talk in this series was given by Bill Unruh (Vancouver). He desribed the present situation for the realizations of analogs of black holes and pointed out that there are several candaidates for which the theory clearly predicts a behaviour as black holes. In particular, Hawking radiation is to be expected [27]. Unfortunately, the experimental restrictions are too strong, so that it seems to be unlikely that the predicted effects can be seen in the immediate future.

Besides these main topics there were talks concentrating on other questions. Brendan Foster (University of Maryland) discussed a model in which together with the expansion of the universe the number of degrees of freedom increases [10]. The idea is that spacetime can be approximately described as a lattice which grows in time. Daniel Terno (Perimeter Institute) analyzed geometric entropy which arises in quantum field theory if horizons are present, from the point of view of quantum information [26]. Shinichi Horata (Sokendai, Graduate University for Avanced Studies, Japan) discussed anisotropies of the primordial cosmic background curvature.

In addition to the talks, there were a lot of interesting presentations in the poster session. To mention a few, there was a poster by Fewster (York) and Roman presenting the result that the energy momentum tensor for a minimally coupled massless scalar quantum field is unbounded from below if averaged along a null direction [9], in contrast to timelike averaging, which generically leads to finite lower bounds. Other posters dealt with the Hawking effect. The end point of black hole evaporation was studied in a poster by Grumiller (Leipzig). Arguments were given that at the end there is a first order phase transition to a specific remnant geometry [12,13]. An ambitious attempt towards the nonequilibrium situation in black hole physics was made in a poster by Saida (Nagoya, Japan). There two black holes at different temperatures were coupled to the radiation field [25]. The entropy of the radiation field was determined, and the generalized second law was obtained; i.e. the total entropy of black holes and radiation never decreases. Arguments were also given which indicated that gravothermal catastrophes are prevented, to the effect that black hole evaporation has to stop and to stabilize with the surrounding environment.

References

1. Bogoliubov, N.N., and Shirkov, D.V., *"Introduction to the Theory of Quantized Fields"*, New York (1959)
2. Brunetti, R., Fredenhagen, K., and Köhler, M., The microlocal spectrum condition and Wick polynomials of free fields on curved spacetimes, *Commun. Math. Phys.* **180**, 312 (1996)
3. Brunetti R. and Fredenhagen, K., Microlocal analysis and interacting quantum field theories: Renormalization on physical backgrounds, *Commun. Math. Phys.* **208**, 623-661 (2000).
4. Brunetti, R., Fredenhagen, K. and Verch, R., The Generally Covaraint Locality Principle: A new paradigm for Local Quantum Physics, *Commun. Math. Phys.***237**, 31-64 (2003).
5. Buchholz, D., Ojima, I., and Roos, H., Thermodynamics properties of non-equilibrium states of quantum field theory, *Annals of Physics* **297**, 219-242 (2002).
6. Dütsch, M. and Fredenhagen, K., Algebraic quantum field theory, perturbation theory, and the loop expansion, *Commun. Math. Phys.* **219** 5 (2001).
7. Duistermaat, J.J. and Hörmander, L., Fourier Integral Operators II, *Acta Math.* **128** 183 (1979).
8. Epstein, H. and Glaser, V., The role of locality in perturbation theory, *Ann. Inst. H. Poincaré A* **19** (1973) 211.
9. Fewster, C.J. and Roman, T.A., Null energy conditions in quantum field theory, *Phys. Rev. D* **67** (2003) 044003.
10. Foster, B.Z. and Jacobson, T., Quantum field theory on a growing lattice, *JHEP* **0408** (2004) 024.
11. Fulling, S.A. Nonuniqueness Of Canonical Field Quantization In Riemannian Space-Time, *Phys. Rev. D* **7** (1973) 2850.
12. Grumiller, D., Kummer, W. and Vassilevich, D.V., Dilaton gravity in two dimensions, *Phys. Rept.* **369** (2002) 327.
13. Grumiller, D., Long time black hole evaporation with bounded Hawking flux, *JCAP* **0405** (2004) 005.
14. Haag, R. and Kastler, D., An algebraic approach to quantum field theory. *J. Math. Phys.*, **5**, 848 (1964).
15. Haag, R., *Local quantum physics: Fields, particles and algebras*, Springer-Verlag, 1996 2nd edition.
16. Hawking, S., Particle creation by black holes. *Comm. Math.Phys.*, **43**, 199 (1975).
17. Hollands S. and R. M. Wald, Local Wick polynomials and time ordered products of quantum fields in curved space, *Commun. Math. Phys.*, **223**, 289-326 (2001).
18. Hollands S. and R. M. Wald, Existence of local covariant time-ordered-products of quantum fields in curved spacetime, *Commun. Math. Phys.*, **231**, 309-345 (2002).
19. Hollands S. and R. M. Wald, On the renormalization group in curved spacetime, *Commun. Math. Phys.*, **237**, 123-160 (2003).

20. Hollands S. and R. M. Wald, Conservation of the stress tensor in perturbative interacting quantum field theory in curved spacetimes, *Rev. Math. Phys.* **17** (2005) 227-312.
21. Hörmander, L., *The Analysis of Linear Partial Differential Operators (I–V)*, Springer, Berlin (1983)
22. Kay, B.S., Linear Spin 0 Quantum Fields In External Gravitational And Scalar Fields. I. A One Particle Structure For The Stationary Case, *Commun. Math. Phys.* **62** (1978) 55.
23. Radzikowski, M.J., The Hadamard condition and Kay's conjecture in (axiomatic) quantum field theory on curved space-time. *Dissertation, Princeton 1992*, unpublished.
24. Radzikowski M. J., Micro-local approach to the Hadamard condition in QFT on curved space-time, *Commun. Math. Phys.* **179**, 529-553 (1996).
25. Saida, H. Steady State Thermodynamic Functions for Radiation Field, arXiv:cond-mat/0310685.
26. Terno, D.R., Entropy, holography and the second law, *Phys. Rev. Lett.***93** (2004) 051303.
27. Unruh, W.G. and Schutzhold, R., On the universality of the Hawking effect, *Phys. Rev.* **D71** (2005) 024028.
28. Stueckelberg, E.C.G. and Petermann, A., La normalization des constants dans la theorie des quanta, *Helv. Phys. Acta* **26** 499 (1953) and earlier references therein.
29. Unruh, W.G., Notes On Black Hole Evaporation, *Phys. Rev. D* **14** 870 (1976).
30. Verch, R. A spin-statistics theorem for quantum fields on curved spacetime manifolds in a generally covariant framework, *Commun. Math. Phys.* **223** (2001) 261.
31. Visser, M. and Weinfurtner, S.E.C., Vortex geometry for the equatorial slice of the Kerr black hole, *Class. Quant. Grav.* **22** (2005) 2493-2510.
32. Visser and S. Weinfurtner, Massive phonon modes from a BEC-based analog model, arXiv:cond-mat/0409639.
33. Wald, R.M., The back reaction effect in particle creation in curved spacetime, *Commun. Math. Phys.* **54**, 1-19 (1977).
34. Wald, R.M., Axiomatic renormalization of the stress tensor of a conformally invariant field in conformally flat spacetimes. *Ann. Phys. (NY)*, **110**, 472–486 (1978).